BRIEF TABLE OF CONTENTS

CONSERVATION ECOLOGY

CONSERVATION
ECOLOGY

7 A L

RSITY

WCB   **Wm. C. Brown Publishers**
Dubuque, Iowa • Melbourne, Australia • Oxford, England

Book Team

Editor *Kevin Kane*
Developmental Editor *Megan Johnson*
Production Editor *Marlys Nekola*
Designer *Christopher E. Reese*
Art Editor *Mary E. Swift*
Photo Editor *Carol Judge/Diane Saeugling*
Permissions Editor *Vicki Krug*
Art Processor *Mary E. Swift*

Wm. C. Brown Publishers
A Division of Wm. C. Brown Communications, Inc.

Vice President and General Manager *George Bergquist*
National Sales Manager *Vincent R. Di Blasi*
Assistant Vice President, Editor-in-Chief *Edward G. Jaffe*
Marketing Manager *Paul Ducham*
Advertising Manager *Amy Schmitz*
Managing Editor, Production *Colleen A. Yonda*
Manager of Visuals and Design *Faye M. Schilling*

Design Manager *Jac Tilton*
Art Manager *Janice Roerig*
Photo Manager *Shirley Charley*
Production Editorial Manager *Ann Fuerste*
Publishing Services Manager *Karen J. Slaght*
Permissions/Records Manager *Connie Allendorf*

Wm. C. Brown Communications, Inc.

Chairman Emeritus *Wm. C. Brown*
Chairman and Chief Executive Officer *Mark C. Falb*
President and Chief Operating Officer *G. Franklin Lewis*
Corporate Vice President, Operations *Beverly Kolz*
Corporate Vice President, President of WCB Manufacturing *Roger Meyer*

Cover © David Milburn, Lightworks, Inc.; © Mark Gibson;
© Wolfgang Kaehler, USA, Wyoming, Grand Teton National Park,
Teton Range; © Allan and Sandy Carey

The credits section for this book begins on page 342 and is considered
an extension of the copyright page.

Printed in the United States of America by Wm. C. Brown Communications, Inc.,
2460 Kerper Boulevard, Dubuque, IA 52001

10 9 8 7 6 5 4 3 2 1

TO DARLA

for helping in field studies throughout the world, and for patience, encouragement and love during the preparation of this textbook.

TABLE OF CONTENTS

A PERSONAL LIBRARY IS A LIFELONG SOURCE OF ENRICHMENT AND DISTINCTION. CONSIDER THIS BOOK AN INVESTMENT IN YOUR FUTURE AND ADD IT TO YOUR PERSONAL LIBRARY.

Planet Earth is entering a critical period for the survival of its natural ecosystems and their plant and animal members. The continuing growth of human populations, through consumption of resources and discharge of the resulting wastes, is modifying ecosystem processes on a global scale. Global warming, acid deposition, the ozone holes, tropical deforestation, desertification, marine pollution, and other world-wide changes are combining to threaten a massive loss of biodiversity. Yet our ultimate well-being depends on the health of the biospheric processes that are experiencing these massive changes—the myriad interactions among plants, animals, microorganisms, and their abiotic surroundings that provide us with food, fiber, medicine, pure water, and healthful air. The crisis of survival of natural ecosystems is a crisis of survival of the human species.

This global challenge requires our personal involvement. As responsible citizens, we must become aware of the nature of this crisis, and of the choices that must be made in meeting it. An understanding of basic ecology has become a requirement of every person.

Each of us is now faced with decisions, on an almost daily basis, about environmental issues relating to the global ecosystem. Should we buy only tuna that were caught by dolphin-safe techniques, eat hamburgers that were not produced on tropical pastures created by deforestation, or use paper products instead of plastic and styrofoam items that might pollute the oceans? Should we support preservation of old-growth forests in the Pacific Northwest, oppose offshore oil drilling near coral reef areas of the Florida Keys, support the establishment of new park and wilderness areas in the Mojave Desert of California, or oppose the construction of water storage dams in Colorado that might affect rivers used by migratory cranes in Nebraska? Should we support measures to reduce power plant emissions that contribute to acid rain, to ban the use of chlorofluorocarbons and other chemicals that damage the stratospheric ozone layer, and to reduce the use of fossil fuels that contribute to the greenhouse effect and global warming? Should we require shrimp fishermen to use devices that prevent sea turtles from entering their nets, developers to avoid areas where endangered species occur, or hunters to use steel instead of lead shot for hunting waterfowl? These, and countless issues like them, demand the collective decision of society.

The goal of this text is to introduce the reader to the nature of the biosphere, to the threats to its integrity that are arising, and to the ecologically sound responses to these threats. The biosphere is not only the remarkable product of billions of years of evolution, it is our biosurvival system.

The text is based on the upper-division course "Conservation of Wildlife" that I have taught at San Diego State University since the early 1960s. Its content has evolved considerably over the years, although remaining concentrated on the global ecosystem and its health. This same course was the stimulus for *Readings in Conservation Ecology*, first published in 1969. The students in this course (in some cases now the offspring of earlier students!) thus deserve credit for helping to maintain their instructor's interest in conservation ecology, especially over a long period when conservation was placed on the back burner by society at large. My sincere appreciation goes to all the students in this course.

I have also prepared an instructor's manual to accompany this text. It contains learning objectives, questions for discussion and testing, suggested audio-visual supplements, and other hints for enriching a course in conservation ecology.

I also thank the many colleagues, including several former students, who have read and commented on chapters or groups of chapters: Edith B. Allen, Ellen T. Bauder, A. T. Bergerud, Lawrence J. Blus, Mark S. Boyce, Beth Braker, Richard W. Braithwaite, Bayard Brattstrom, Richard Brewer, John Celecia, Charles F. Cooper, Deborah M. Dexter, Thomas A. Ebert, Thomas C. Emmel, David A. Farris, Christopher G. Gakahu, Renatte K. Hageman, Rick A. Hopkins, Don Hunsaker II, Jodee Hunt, Stuart H. Hurlbert, Joseph R. Jehl Jr., David W. Johnston, Steve Kingswood, Barbara E. Kus, Michael Kutilek, Stephen Leatherwood, Theodore Lopushinsky, Leroy R. McClenaghan, Dale R. McCullough, Pierre Mineau, Dieter Muller-Dombois, David A. Munro, Harry M. Ohlendorf, Dwain Parrack, William F. Perrin, Louis F. Pitelka, Donald B. Porcella, Philip R. Pryde, Keith Ronald, William J. Rowland, Victor B. Scheffer, J. Michael Scott, Roger S. Sharpe, Robert C. Stebbins, Douglas H. Strong, William Toone, Ross A. Virginia, Linda L. Wallace, David Western, Kathy S. Williams, Susan L. Williams, Joy B. Zedler, and Paul H. Zedler.

In addition, I thank the individuals who helped me obtain illustrations for the text. These include Anthony F. Amos, Spencer Beebe, Charles Birkeland, Richard O. Birregaard Jr., Richard W. Braithwaite, James J. Brett, David S. Brookshire, Graeme Chapman, Walter Courtenay, A. T. Cringan, Don Despain, Deborah Dexter, John Francis, Donald W. Kaufman, Joseph R. Jehl Jr., Peter Johnson, Janet Jorgenson, Charles D. Keeling, Stephen Leatherwood, Thomas E. Lovejoy, A. W. Maki, Peter Meserve, Robert Mesta, J. P. Myers, David A. Norton, Bruce Nyden, Storrs Olson, Donald B. Porcella, Rolf O. Peterson, Dan Reed, John D. Schroer, W. Roy Siegfried, Robert L. Smith, Tom Stehn, Brent Stewart, Harrison B. Tordoff, H. W. Vogelmann, LaRue Wells, Dorn Whitmore, Susan Williams, Thomas D. Williams, and David Zoutendyk.

George W. Cox

PART ONE

Basic Concepts

The ecosystem concept is central to conservation ecology. Unless conservation efforts are based on this concept, they can do little more than treat the symptoms of the biodiversity crisis. We shall first examine this concept and show why it is essential to protection of biodiversity and to sustainable management of the biosphere. Second, we shall examine the process of extinction, and consider how it is exaggerated by the disruption of basic ecosystem processes.

CHAPTER 1

The Ecosystem Concept and Conservation Ecology

Conservation ecology is concerned with the origin and preservation of biotic diversity. It seeks to understand how the rich variety of plant and animal life around us arose, how it has been maintained by natural processes, and how we can prevent its destruction by our own actions. Like other branches of science that address matters of societal concern, conservation ecology has emerged because of a basic need: the human population stands on the verge of causing the massive extinction of species throughout the biosphere.

In this chapter we shall first examine conservation ecology in relation to the science of ecology and to the philosophy of environmentalism. Next, we shall discuss the concept of the ecosystem, the concept central to the theory and practice of conservation ecology. We shall define and illustrate this concept, first at the scale of individual ecosystems, and then at the levels of complex landscapes and the biosphere. Finally, we shall consider why the ecosystem approach is essential to the goal of preserving biotic diversity.

Ecology, Conservation Ecology, and Environmentalism

Ecology is the science concerned with relationships of organisms with each other and with their non-living environment. Ecology encompasses many sub-disciplines. Some of these deal with particular groups of organisms; these include plant ecology, animal ecology, and microbial ecology. Others, such as terrestrial ecology, freshwater ecology, and marine ecology, focus on particular environments. Still others, such as organismal ecology, population ecology, community ecology, and ecosystem ecology, are concerned with different levels of organization. Finally, fields such as mathematical ecology, systems ecology, and remote sensing are technique-oriented branches of ecology.

In any of these sub-disciplines, however, studies may be basic or applied. Basic studies are concerned with filling gaps in the inventory of ecological knowledge (descriptive studies) or with understanding how ecological systems function (theoretical studies), quite apart from any consideration of the concern of society about these matters. Applied studies, on the other hand, focus on aspects of ecological knowledge or functions that are of concern to society, and for which society has often defined goals to be achieved. Through the political process, for example, society may decide to manage national parks to minimize the risk of disastrous fires, to reduce the pollution of streams and lakes, to increase the productivity of game of fish populations, or to prevent the extinction of species. To achieve such goals, applied ecologists must investigate ways of obtaining specified results, regardless of whether or not our knowledge of basic ecological principles is increased. The results of these studies must then be put into use through management.

Conservation ecology, in reality, is a field that is partly basic and partly applied. Its foremost goal is to obtain and apply the knowledge necessary to conserve and manage biotic diversity for the benefit of humanity. To achieve this goal, however, conservation ecologists must gain both a sound understanding of the structure and function of natural ecosystems, and an effective ability to manipulate these systems to desired ends.

Thus, the key goal of conservation ecology is to find ways both to gain scientific knowledge and to apply it to the protection and wise use of biotic diversity. The available knowledge must be put to work in programs of protection, sustained use, and restoration of ecological systems. These applied conservation efforts must also go hand-in-hand with continuing studies. The unfortunate truth that conservation efforts must often begin with knowledge is still rudimentary. The critical state of the global environment demands that the best information available be put to work at once.

Because conservation ecology is engaged in activities that interact with the diverse, often conflicting interests of complex human society, it must draw knowledge and use methods not only from the various branches of ecology, but also from the other basic and applied branches of the life sciences, the physical sciences, engineering, economics, sociology, law, history, and political science (Brussard 1991). To achieve a defined goal, such as the protection and recovery of an endangered species, conservation ecologists must work in a real-world context of institutions and laws that govern the day-to-day operation of society. They must understand the biology of the species, the ways in which human activities are likely to affect its environment, and how efforts to preserve the species can be carried out under the constraints of laws, available funds, and public opinion. Like a doctor of medicine, a successful conservation ecologist must be able to integrate the areas of research, diagnosis, and treatment. The emerging crises of the global environment, in fact, reflect to a great extent the lack of effective integration of ecological research, monitoring of environmental change, and corrective actions needed to avoid environmental disaster.

Conservation ecology also interfaces with environmentalism. **Environmentalism** is different from ecology, applied ecology, and even conservation ecology. It is a societal movement, in the words of Scheffer (1989), ". . . toward understanding humankind's natural bases of support while continuously applying what is learned toward perpetuating those bases." In other words, environmentalism is the effort to live in harmony within the global ecosystem. Environmentalism is thus a philosophical conviction that the future of humankind depends on the establishment of a sound relationship between humans and the natural world. This conviction provides the hard-core political support for many efforts of conservation ecology. Many, if not most, conservation ecologists are environmentalists at heart, but to do their job right, they must also be objective scientists and diligent workers in the studies and programs they conduct.

The Ecosystem Concept

An **ecosystem** is a unit of the environment made up of living and non-living components that interact with interchanges of nutrients and energy. The living components of the ecosystem are various groups of organisms that occur together and interact with each other to form the **biotic community.** The coexisting and interacting individuals of each of the species of the community constitute a **population.** The non-living components of the ecosystem make up the physical **habitat** in which these organisms exist, and with which they interact. As Evans (1956) noted long ago, the ecosystem is the basic unit of ecology; appreciation of this fact is central to conservation ecology.

A lake provides an easily appreciated example of an ecosystem (Fig. 1.1). The living components of the lake ecosystem can be classified in various ways, depending on the objective of a particular study. In studies of energy and nutrient relations, five major components are often distinguished: **producers,** the organisms (usually green plants) that store energy in new organic matter by photosynthesis or chemosynthesis; **herbivores,** the organisms that feed on pro-

FIGURE 1.1

■ This lake in Rocky Mountain National Park is a distinct ecosystem unit. The water mass, sediments, and near-surface air constitute the major non-living components of the ecosystem and the plants, animals, and microorganisms the living components.

ducers; **carnivores,** the animals that feed on herbivores; **top carnivores,** the animals that feed on carnivores; and **decomposers,** the organisms that consume dead organic matter produced by all organisms (including themselves). In a lake, producers include forms such as **phytoplankton,** the tiny single-celled or colonial plants that live floating free in the water, filamentous algae that live attached to submerged surfaces, duckweeds that live floating on the surface, and larger flowering plants that are rooted in the bottom. Herbivores include **zooplankton,** the small free-swimming organisms that feed on phytoplankton, along with snails, turtles and other plant-feeders. Carnivores consist of predatory invertebrates and fish, and top carnivores of forms such as fish-eating birds. The major non-living components of the lake ecosystem are the water mass, with its particular physical and chemical conditions, the sediments forming the lake bottom, and the air space above the surface within which organisms functionally connected with the lake system are active.

The concept of the ecosystem can be applied to environmental units of greatly differing size and permanence. A tiny temporary pond is an ecosystem, as is a large body of water such as Lake Michigan or the cold upwelling waters that surround the continent of Antarctica. Terrestrial ecosystems can range in scale from small islands of soil and life on bare granitic outcrops (Fig. 1.2) to vast areas such as Yellowstone National Park and its surrounding areas of national forest, which together form the Greater Yellowstone Ecosystem. The **biosphere,** the zone of the earth's surface occupied or influenced by living organisms, is an ecosystem, as well.

The key feature of an ecosystem is the interaction among these living and non-living components. These interactions, in fact, provide the functional integration that enables specific ecosystems to be distinguished. Within an ecosystem the interactions among various living and non-living components are strong; across the boundaries of different ecosystems they

FIGURE 1.2

■ The concept of the ecosystem applies to small islands of soil, together with plants, animals, and microorganisms, that develop on granitic rock outcrops in many parts of the world.

are weaker. Interactions are of four basic types (Fig. 1.3). Some may be of a purely physical or chemical nature, such as change in water density with change in temperature, or the chemical reactions by which acids are neutralized by buffering minerals in the water. Others, such as predation, competition, and mutualism among organisms, are purely biotic. Still others occur between living and non-living components. The non-living components influence organisms; increased acidity, salinity, or turbidity of lake waters, for example, can kill certain organisms and benefit others. Finally, the organisms themselves strongly influence their physical habitat. In a lake, for example, the amount of photosynthetic production in the surface waters determines how much dead matter falls into the deep water zone. Decomposition of this dead matter, in turn, determines how much oxygen is used up, and how much remains for the use of other organisms of deep waters. In fact, much of the chemistry of the lake waters and sediments is influenced by the activities of the lake organisms. Appreciation of the magnitude of the influences of organisms on their physical environment during the 1920s was really the key factor in the emergence of the ecosystem concept.

These interactions exist not only at the scale of a lake ecosystem, but at the scale of the biosphere. That living organisms affect the physical environment at the global scale is evidenced by the fact that the composition of the atmosphere, the chemistry of the oceans, the structure of soils, and the composition of marine sediments are all determined primarily by life processes. Increasingly, the activities of humans affect these components of the biosphere, which, in their turn, influence the biosphere's biotic processes.

The relationships that exist within an ecosystem are the result of interaction through evolutionary time, as well. Evolution has adapted groups of organisms to utilize different resources available in the physical and chemical environment.

Some producer organisms use light energy for **photosynthesis;** others use energy from chemical reactions to carry on **chemosynthesis,** for example. The remarkable ecosystems of the deep ocean thermal vents depend on chemosynthetic production. The living members of ecosystems have also become adapted to each other through patterns of mutual evolutionary adjustment, or **coevolution.** The degree of this adjustment depends on the length and intensity of their interaction. In some cases the interaction is species specific, as for certain orchids and the specialized insect that pollinates them. In other cases it is quite generalized, as for hummingbirds and the various flowers that they feed at and pollinate. Through such evolution, the biotic structure of the ecosystem tends to adjust so that energy is more effectively captured and transformed, nutrients are retained and recycled more efficiently, and biotic interactions become more beneficial to the participants.

Many factors also act as external controls on ecosystems, influencing conditions within an ecosystem without being influenced significantly by the ecosystem itself. The regional climate, or **macroclimate,** is one of the most obvious external controls. Regional weather systems impose seasonal patterns of temperature, precipitation, solar radiation, wind, and other factors that strongly influence the conditions that prevail within an ecosystem such as a lake. Regional geology and geography also set limits on the development of ecosystems. The type of bedrock in the watershed of a lake, for example, influences the chemistry of the lake waters. The available **regional biota**—the kinds of plants and animals that exist in a particular geographical area—also places limits on the nature of the living community that develops. A newly formed lake, for example, will become inhabited by species that are present in nearby aquatic ecosystems, not those of lakes on distant continents, even though many of those species might be able to flourish. In the case of ecosystems of small and medium scale, humans are also an external influence. Humans add materials such as sediment and sewage to lakes, and remove other materials, such as water and fish. At the scale of regional ecosystems and the biosphere, however, humans are inextricably components of the system. Indeed, this is the fact that makes global environmental change a threat to human welfare.

Despite the importance of their internal interactions, ecosystems are never completely closed systems. In all instances, ecosystems experience inflows and outflows of energy and matter. In the case of a lake, light energy enters the surface waters, and an equal amount of energy eventually leaves the ecosystem in some fashion, much of it as heat radiation from the lake surface. Dissolved and particulate materials enter the lake from the watershed, and quantities of these materials often leave through streams that flow from the lake. In some ecosystems, such as streams, the inflows and outflows are the dominant factors of ecosystem function. Not even the biosphere is a closed system; solar energy enters and heat energy is radiated back to space.

FIGURE 1.3

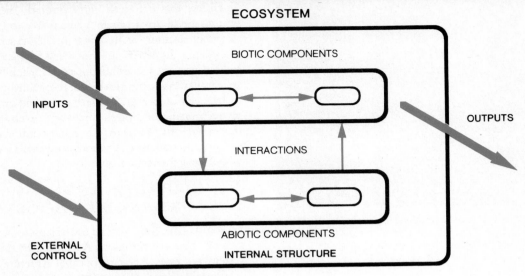

■ Ecosystems possess major interactions within and between their living and non-living components. External forces exert controlling influences on the ecosystem, as well. Ecosystems are not closed systems; both materials and energy may enter and leave.

FIGURE 1.4

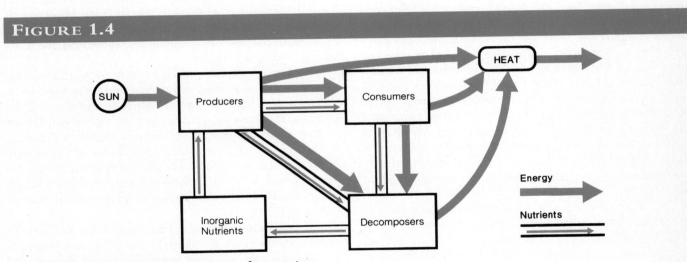

■ Energy moves through an ecosystem as a one-way flow, entering as solar radiation and leaving as heat radiated back into space. Nutrients, on the other hand, are capable of moving in true cycles.

The external controls and internal interactions of an ecosystem combine to produce a certain **ecosystem structure.** The macroclimate interacts with internal components of the ecosystem to create an internal **microclimate.** In a lake, the microclimate constitutes the regime of temperature, water density, light, and current pattern. Regional geology and geography interact with internal physical and biotic factors to create specific conditions of water and sediment chemistry. From the regional biota, certain species are sorted out by physical and biotic challenges to become the members of the lake community. These species develop certain patterns of abundance, biomass, stratification, and seasonality. Where

human influences impact a lake, many structural conditions may be modified, as well.

Internal relationships are not only structural, but involve many forms of **ecosystem dynamics.** Most important among these are the processes of **energy flow** and **nutrient cycling.** Important also, especially for conservation ecology, are the regulatory processes that determine stability and resilience.

Energy and nutrients move through an ecosystem in basically different fashions (Fig. 1.4). Energy flows through the system in a one-way fashion, entering almost entirely as solar radiation that is captured and stored by photosynthesis

FIGURE 1.5

■ The California chaparral shows high resilience, and recovers quickly after fire. After only six months, this stand is recovering by resprouting of many perennials and by the germination of both annuals and perennials.

in the chemical energy of organic molecules, and leaving as heat energy released by the metabolic breakdown of organic molecules. Although the pathway of flow may branch in a complex fashion as chemical energy is passed from organism to organism along food chains, once energy has been converted to heat it cannot be recycled into the chemical energy of organic molecules. Nutrients, on the other hand, can move in a truly cyclical manner, the same atoms of nitrogen or phosphorus passing back and forth from the organisms of a lake to the water mass many times. Nutrient cycles in ecosystems are not perfect, however. Quantities of nutrients enter the active cycle from outside the system, and other quantities are lost from the system, so that an ecosystem may be envisioned as having a certain **nutrient capital.** This nutrient capital may increase or decrease, depending largely on the actions of external controls, increasingly those exerted by human activity.

The dynamics of ecosystems confer particular patterns of stability and resilience. In this sense, ecosystems are somewhat like organisms; they have a certain ability to regulate their internal conditions in the face of external challenges of the environment. **Stability** is the ability of an ecosystem to *resist* change in the face of a disrupting external influence. Many desert ecosystems, for example, are quite stable in the face of highly variable macroclimatic conditions; a few years of unusual heat, cold, drought, or dampness cause little change in structure. Many stream ecosystems, on the other hand, are very unstable; flood and drought conditions may cause enormous change in conditions from year to year.

Resilience, in contrast, is the ability of an ecosystem to *recover from* a change in structure caused by some disturbance (Westman 1986). The chaparral shrublands of California are highly resilient in the face of fire. Following the destruction of above-ground material by a wildfire, many woody plants sprout from specialized root crowns and regrow rapidly (Fig 1.5). Other plants are stimulated to germinate

profusely in the early post-fire environment. Within just a few years, the chaparral community has returned to a state much like that before the fire. Similarly, many stream ecosystems are highly resilient, and are able to recover from the effects of drought or flooding in a short time. Most desert ecosystems, however, are not very resilient. If the above-ground parts of the dominant plants are killed over a large area, it may take centuries before re-establishment of the original vegetation occurs. Similarly, semiarid grasslands, stressed by overgrazing or drought, may be converted to desert scrub in a matter of a few years. An understanding of the degree of stability and resilience of various ecosystems is essential to their wise management by man.

EQUILIBRIUM AND DISTURBANCE IN ECOSYSTEMS

Ecologists have long viewed ecosystems as being equilibrium systems, first and foremost. According to this view, an ecosystem possesses a certain optimal structure at which relatively stable flows of nutrients and energy occur among its members. In a popular sense, the ecosystem possesses a "balance of nature." Disturbance that upsets this equilibrium is followed by a type of healing process, through which the system returns to its optimal state, its balance of nature restored. This viewpoint suggests that the most important relationships are those that involve interactions between ecosystem components at equilibrium; everything else is relatively unimportant noise.

In recent decades, ecologists have found that equilibrium conditions in most ecosystems are rare or short-lived. Most systems are in a state of change in structure and dynamics, in some cases gradual, in other cases rapid. Disturbance is often a dominant influence in ecosystem dynamics, as it is in terrestrial ecosystems prone to fire or aquatic ecosystems prone to flooding. For such ecosystems, in fact, disturbance may be the factor that maintains such features as high biotic diversity and a high yield of materials useful to humans (Botkin 1990).

Furthermore, the idea of an optimal equilibrium structure is itself a myth. It is unlikely that any system returns to its original state following some disturbance; the renewed ecosystem may be similar, but it may often be quite different in structure and function. Thus, no specific balance of nature exists. Many different balances might occur, assuming that the system reaches an equilibrium at all. This is not to say that all is chaos, and that stability and resilience are meaningless concepts, but that the external controls on ecosystems are themselves highly variable, changing, and complex.

In spite of the fact that disturbance is natural, and that many different equilibria are possible, not all equilibria are equally beneficial from a human perspective. Some forms of disturbance, such as the introduction of an exotic species of plant or animal, may disrupt coevolved relationships among native species, leading to their extinction. Other disturbances, such as pollution, may kill the carnivores and top carnivores of the ecosystem, which may be species that are valuable to humans for food or recreation. Still other disturbances, such as acid rain, may flush away nutrients, leaving a less fertile and less productive ecosystem. All of these ecosystems may develop or approach a new equilibrium, but the conditions of that equilibrium may be detrimental to human welfare.

ECOSYSTEMS AND LANDSCAPES

Individual types of ecosystems form a mosaic across the landscape. Since these ecosystem units are open systems, they interact with each other through the flows of materials and the movements of organisms across their borders. The importance of these processes has only recently been given its deserved recognition, with the emergence of the field of landscape ecology (Forman and Godron 1981, Naveh and Lieberman 1984). **Landscape ecology** seeks to understand how the pattern of juxtaposition of ecosystems, both natural ecosystems and those dominated by humans, influences their individual function and that of the landscape as a whole.

In continental areas, for example, terrestrial ecosystems often exist in a mosaic with ponds, lakes, and streams. The terrestrial systems form the watersheds for the aquatic systems, so that how these watersheds are managed influences the quantities of water, nutrients, and pollutants that flow into the aquatic systems. Likewise, how the aquatic systems are managed influences the kinds of species that are able to live in the adjacent land systems. The kinds and abundance of fish in a river, for example, influences the presence and numbers of animals such as ospreys and otters. Even in the ocean, different types of ecosystems are influenced by currents and animal migrations. The influx of warm tropical water into areas of normally cold, upwelling water in the eastern Pacific—an **El Niño**—can catastrophically affect populations of kelps, fish, and seabirds. The seasonal migrations of larger marine mammals and fish likewise affect the activity and abundance of their prey.

Human activities are increasingly important in structuring landscapes, both on land and in aquatic environments. New types of ecosystems created by human activity, such as cultivated cropland, landscaped residential land, and dredged waterways, border or surround natural ecosystems. In eastern North America, for example, fragments of native forest exist as islands in landscapes of diverse agricultural ecosystems. Thus, to employ an ecosystem perspective in conservation ecology, we must consider not only the interactions within ecosystems, but how the different ecosystems, both natural and human-made, interact in a landscape.

THE BIOSPHERE AND GLOBAL CHANGE

The highest ecological level of organization is that of the biosphere. At this level of organization, the only major inflow is that of solar radiation and the only major outflow is that of heat radiation to space. This flux of energy, together with a small flow of heat energy from the earth's core to its surface, is, in a sense, the earth's macroclimate. Over several billion years, these energy sources, interacting with the primeval geochemistry of the land, surface waters, and atmosphere, have produced the conditions of climate, soil structure, water chemistry, and biodiversity.

Human activities now impact the biosphere at large, influencing basic interactions at this global scale (Schlesinger 1991). Burning of fossil fuels releases increasing quantities of CO_2, sulfur and nitrogen oxides, and heavy metals into the atmosphere. The release of industrially produced halogen compounds, such as the chlorofluorocarbons, and the widespread use of nitrogen fertilizers that are partially returned to the atmosphere as nitrogen oxides are influencing the chemical composition of the atmosphere (Lester and Myers 1989). Deforestation and overgrazing of arid lands are directly changing the biota of the continents and modifying the energy exchange characteristics of the land surface (Myers 1988). Crop monoculturing and overgrazing are increasing the erosion of soil by water and wind, increasing the inflow of sediment into aquatic ecosystems and the input of dust into the atmosphere. Reduced river inflows to the oceans are fostering coastal erosion and depriving coastal marine waters of nutrients. Nutrients from disturbed landscapes and from urban sewage discharges are destructively over-enriching other aquatic ecosystems. Natural ecosystems are being replaced by farmland and urban landscapes, and major trophic components of aquatic ecosystems are being exploited with growing intensity.

The rapidly increasing magnitude of these changes in the biosphere poses an imminent threat to the diversity of life on earth. Changes in climate, oceanic circulation patterns, and the water budgets of rivers and lakes will likely occur with greater acceleration than at any time in earth's history (Lester and Myers 1989). At the same time, the ability of terrestrial plants and animals to shift their geographic ranges is highly constrained by developed landscapes (Mintzer 1988), and that of aquatic ecosystems highly affected by patterns of exploitation and pollution by humans. The limits of stability and resilience of all of the earth's ecosystems will certainly be tested throughout the biosphere within the next century.

THE NECESSITY OF THE ECOSYSTEM APPROACH

Well-intentioned actions that do not recognize the integration of the components of an ecosystem, the dynamic interactions that occur among these components, and the influence of adjacent ecosystems on each other, can destroy the object of the conservation effort (Walker 1989). Protection of an area against disturbances, such as fire, that are integral to the dynamics of an ecosystem, or centering management policy on a particular species, with inadequate consideration of its effects on other species, are good examples of such approaches (White and Bratton 1980).

A clear example of the need for an ecosystem view in conservation ecology is provided by many of the wildlife parks

FIGURE 1.6

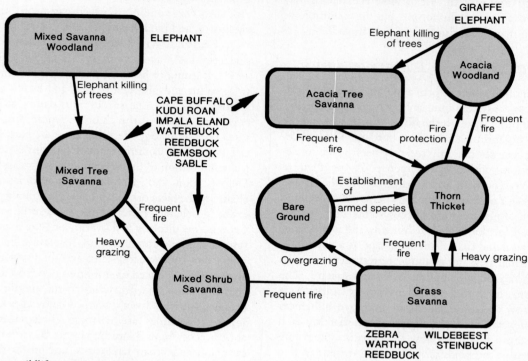

Many East African wildlife areas show a complex interaction between animals, vegetation, and fire that favors the maintenance of a mosaic of plant communities, each with its associated species of large animals.

Communities identified by rectangles are most closely adjusted to soils and climate (mixed savanna woodland), frequent fire (grass savanna), and intense animal effects (acacia tree savanna).

of East Africa (Boughey 1963). In Hwange National Park, Zimbabwe, for example, a mosaic of woodlands, savannas, grasslands, and other communities exists in an area of relatively homogeneous soils and climate (Fig. 1.6). Each of these plant communities is the favored habitat of certain large animal species, so that the park supports a rich fauna. What is responsible for this diversity of plant and animal life? Much of the diversity of large animals obviously results from the vegetational diversity within the park. Analysis of biotic interactions within the park ecosystem, however, shows that the vegetational diversity is largely created by the activities of large animals and by the influence of these activities on the frequency of fire.

The scenario for this relationship is complex. Elephants, which favor the denser woodlands, kill many trees by stripping their bark for food, or simply by pushing them over to gain access to the foliage. When this activity goes on for a long time, or is carried out by many animals, the density of large trees is reduced and the woodland becomes more open. Initially, this reduces the tendency for one tree species to dominate the woodland canopy, increasing tree diversity. Eventually, however, it converts the community into a much more open savanna woodland. Here, increased light reaches ground level and permits the growth of a dense grass and herb layer. Many species of grazing and browsing animals, such as impala, greater kudu and sable antelope, frequent the savanna

woodland (Fig. 1.7). The intensity of feeding by these species determines how much grass and herb material remains into the dry season to become fuel for wildfires. If the populations of these animals are high, fires will be infrequent and low in intensity. If their populations are low, fires are likely to be frequent and intense, killing many of the remaining trees and transforming the vegetation into a shrub savanna, in which the shrubs are trees and smaller woody plants that are able to resprout after fire. Shrub savannas, too, are the preferred habitat of some species, such as Cape buffalo and gemsbok. Elephants also frequent certain areas of shrub savanna, browsing the woody plants so frequently and heavily that they cannot grow into trees, even in the absence of fire (Fig. 1.8). If fire continues to be frequent, however, it may eliminate virtually all woody plants, creating an open grassland, which will become the home of still other grazing animals such as wildebeest, zebra, and gazelles. When these animals are numerous, their feeding may favor the invasion of spiny acacias that are resistant to the feeding activities of grazers. In time, this may result in a thicket or woodland dominated by acacias, the preferred habitat of species such as the giraffe.

Thus, the diversity of large animals depends on the existence of a mosaic of distinct plant communities, but this mosaic, in turn, depends on the activities of the various large animals themselves. Strictly protective management is inadequate for such a situation. A policy of complete fire control,

FIGURE 1.7

■ Impala and many other ungulates occupy East African savanna woodlands that are maintained by the interaction of the vegetation with browsing and grazing mammals and fire.

FIGURE 1.8

■ Heavy elephant browsing on trees such as the mopane, *Colospermum mopane*, can maintain the vegetation as shrub savanna, as in this location in Chobe National Park, Botswana.

for example, might quickly lead to the disappearance of the savanna woodlands, shrub savannas, and grasslands, and of the animals that occupy these communities. Over-protection of elephants might lead to the loss of the denser woodlands, and excessive protection of the animals of open grasslands might lead to the transformation of grasslands into thorn woodlands, ultimately causing the grassland animals to disappear. To be successful, management of areas such as this must recognize and maintain the basic interactions of the ecosystem (McNaughton 1989).

Although the above example comes from East Africa, this same principle applies to terrestrial and aquatic environments in other areas, including North America (Wagner 1977). Overprotection of deer and elk populations in many areas has led to overpopulation, depletion of range food plants, starvation of the protected animals, and long-term degradation of habitat conditions. In other instances, simple protection, such as that afforded the California Condor, was inadequate to offset the problems created by deterioration of the ecosystem in which this species existed. This species, in fact, provides a good example of failure to employ an ecosystem perspective in conservation efforts (See chapter 25). In the chapters that follow, we shall encounter many other examples of management problems created by the lack of an ecosystem perspective.

HUMANS AS MEMBERS OF THE GLOBAL ECOSYSTEM

As we noted earlier, humans act as external controls on small-scale ecosystems, but are an integral component of regional ecosystems and of the biosphere as a whole. As the human population grows, and as its use of natural resources also ex-

pands, our interactions with other living and non-living components of the global ecosystem intensify. The impressive accomplishments of our technology sometimes lead us to imagine that we are becoming more independent of the rest of the global ecosystem. In reality, however, our dependence on the rest of the global ecosystem is increasing. Abnormal seasonal patterns of regional rainfall, temperature, ocean upwelling, and other conditions now cause major disturbances in the world economy. Slow, progressive changes in global climate, sea level, area of forests and deserts, and other features threaten to cause major impacts on human food and fiber production within one or two decades. Conservation must thus be a global effort. In the following chapters, our examination of varied topics gives particular attention to international efforts to implement conservation practices.

The reality of human membership in the global ecosystem means that, to be successful, conservation strategy must address the human as well as the non-human components of the global ecosystem. The grandeur of the tropical forests, the teeming wildlife of the African plains, and the wealth of fish and invertebrate life of tropical reefs cannot be preserved without solving the problems of the human populations that seek to exploit these ecosystems for the bare necessities of life. Consequently, we shall emphasize the need to couple conservation with improvement of economic conditions for impoverished human populations.

Countering the threats of global environmental change and continued growth of human populations is the challenge of the next two decades. If this challenge is not met, the year 2010 will find the biosphere a biologically, culturally, and economically impoverished ecosystem, perhaps analogous to a heavily polluted lake.

Key Principles Relating to the Ecosystem Concept

1. An ecosystem, the living community together with its non-living habitat, is influenced both by external controls and internal interactions of its physical and biotic components. Humans and all other species are members of ecosystems, on which they depend for survival.

2. Ecosystems are affected to differing degrees by disturbance, and show varying degrees of stability and resilience.

3. The ecosystems that occupy a landscape are linked by the flow of materials and movement of organisms across their boundaries.

4. The global nature of environmental change, together with its roots in human activity, indicate that conservation must be a global effort that simultaneously addresses human population growth and economic development.

Literature Cited

Boughey, A. S. 1963. Interaction between animals, vegetation, and fire in Southern Rhodesia. *Ohio J. Sci.* **63:**193–209.

Botkin, D. B. 1990. *Discordant harmonies: A new ecology for the twenty-first century.* Oxford Univ. Press, New York, NY.

Brussard, P. F. 1991. The role of ecology in biological conservation. *Ecol. Applications* **1:**6–12.

Evans, F. C. 1956. Ecosystem as the basic unit in ecology. *Science* **123:**1127–1128.

Forman, R. T. T. and M. Godron. 1986. *Landscape ecology.* John Wiley and Sons, New York.

Lester, R. T. and J. P. Myers. 1989. Global warming, climate disruption, and biological diversity. Pp. 177–221 *in* W. J. Chandler (Ed.), *Audubon wildlife report 1989/1990.* Academic Press, San Diego, CA.

McNaughton, S.J. 1989. Ecosystems and conservation in the twenty-first century. Pp. 109–120 *in* D. Western and M. Pearl (Eds.), *Conservation for the twenty-first century.* Oxford Univ. Press, New York.

Mintzer, I. 1988. Global climate change and its effects on wild lands. Pp. 56–67 *in* V. Martin (Ed.), *For the conservation of earth.* Fulcrum, Inc., Golden, CO.

Myers, N. 1988. Tropical deforestation and climatic change. *Env. Cons.* **15:**293–298.

Naveh, Z. and A. S. Lieberman. 1984. *Landscape ecology. Theory and applications.* Springer-Verlag, New York.

Scheffer, V. B. 1989. Environmentalism: Its articles of faith. *Northwest Env. J.* **5:**99–109.

Schlesinger, W. H. 1991. *Biogeochemistry: An analysis of global change.* Academic Press, San Diego, CA.

Wagner, F. H. 1977. Species vs. ecosystem management: Concepts and practices. *Trans. N. Amer. Wildl. Nat. Res. Conf.* **42:**14–24.

Walker, B. 1989. Diversity and stability in ecosystem conservation. Pp. 121–130 *in* D. Western and M. Pearl (Eds.), *Conservation for the twenty-first century.* Oxford Univ. Press, New York.

Westman, W. W. 1986. Resilience: Concepts and measures. Pp. 5–19 *in* B. Dell, A. J. M. Hopkins, and B. B. Lamont (Eds.), *Resilience in Mediterranean-type ecosystems.* W. Junk, Dordrecht.

White, P.S. and S.B. Bratton. 1980. After preservation: Philosophical and practical problems of change. *Biol. Cons.* **18:**241–255.

TABLE 2.2

Recorded Extinctions of Various Groups of Organisms Since 1600 A.D.

TAXON	CONTINENTAL	ISLAND	OCEANIC	TOTAL	PERCENT OF SPECIES
Mammals	30	51	2	83	2.1
Birds	21	90	2	113	1.3
Reptiles	1	20	0	21	0.3
Amphibians	2	0	0	2	0.01
Fish	22	1	0	23	0.1
Invertebrates	49	48	1	98	0.01
Vascular plants	245	139	0	384	0.2

Sources: W. V. Reid and K. R. Miller, *Keeping Options Alive: The Scientific Basis for Conserving Biodiversity,* 1989, World Resources Institute, Washington, D.C.; Data from IUCN 1988 (fish and invertebrates); Nilsson 1983, supplemented by species listed as extinct in IUCN 1988 (vertebrates other than fish); Threatened Plants Unit, World Conservation Monitoring Centre, 15 March 1989, personal communication (plants); G. Vermeij, personal communication (marine invertebrates); Wilson 1988 (number of species).

TABLE 2.3

Major Causes of Vertebrate Extinctions.

	PERCENT OF EXTINCTIONS				
GROUP	HUMAN EXPLOITATION	INTRODUCED SPECIES	HABITAT DISRUPTION	OTHER	UNKNOWN
Mammals	24	20	19	1	36
Birds	11	22	20	2	37
Reptiles	32	42	5		21
Fish	3	25	29	3	40

Sources: W. V. Reid and K. R. Miller, *Keeping Options Alive: The Scientific Basis for Conserving Biodiversity,* 1989, World Resources Institute, Washington, D.C.; Data from Day 1981, Nilsson 1983, and Uno, et. al. 1983.

General Population Decline to Extinction

Data compiled by Reid and Miller (1989) indicate that human exploitation, the effects of introduced species, and habitat disruption are the major causes of recent historical extinctions of vertebrates (Table 2.3). King (1980), in a more detailed analysis for birds, showed that killing by humans and the effects of introduced predators were the major cause of extinctions since about 1650 A.D. (Table 2.4). In North America, the Steller's sea cow and great auk were driven to extinction by their killing for oil and meat, respectively. Giant tortoises were eliminated from several islands in the Indian Ocean and from several of the Galápagos Islands off Ecuador by exploitation by early whaling crews (Honegger 1981). Killing by humans still threatens many species, a recent example being the poaching of rhinoceroses in East Africa that has led to their near-extinction in several countries (See chapter 10). Introduced rats, cats, mongooses, and other predators have eliminated many birds, mammals, lizards, and snakes from oceanic islands (See chapter 9). The brown tree snake, native to New Guinea and neighboring areas, for example, was introduced accidentally to the island of Guam in the late 1940s or early 1950s, probably in shipments of fruit. An arboreal, nocturnal predator on eggs, young, and adult birds, this snake has nearly eliminated native forest bird species, driving six

TABLE 2.4

Causes of Historical Extinctions of Birds in Island and Continental Regions.

CAUSE	CONTINENTAL REGIONS	OCEANIC ISLANDS
Hunting	61.5%	14.9%
Predation		41.8%
Competition		6.7%
Disease		5.6%
Genetic Swamping		0.7%
Weather		0.4%
Habitat Disturbance	23.1%	19.4%
Unknown	15.4%	10.4%
Number of Species	11	92
Number of Subspecies	2	83
Total number of Species and Subspecies	13	175

Source: Data from W. B. King, *Ecological Basis of Extinction in Birds,* 1980. Proc. XVII Int. Orn. Congr., 1978: 905–911.

species to extinction and reducing the remaining four species to less than 100 individuals on the main island (Savidge 1987). Introduced herbivores have often decimated the native vegetation of islands, causing extinctions of both plants and animals. On San Clemente Island, one of the Channel Islands off southern California, goats introduced by early Spanish seafarers have eliminated a number of endemic plants.

Exotic plants and animals have contributed to extinction in several other ways. Several extinctions of island mammals and birds seem to have resulted from introduced competitors. Introduced black rats have eliminated native rice rats from four of the six islands they originally occupied in the Galápagos Islands, for example (Eckhardt 1972). Exotic plants are damaging competitors for native species in both continental and island areas (See chapters 4, 9). Introduced diseases have also been the cause of extinctions. Avian malaria (in the blood of introduced continental birds) and suitable mosquito vectors have contributed to extinction of several native Hawaiian birds since discovery of the islands by Captain Cook in 1778 A.D., for example (van Riper et al. 1986).

Genetic swamping has contributed to the extinction of 15 or more species of fish in North America (Miller et al. 1989). In New Zealand, genetic swamping has resulted in the near-extinction of the grey duck, through hybridization with the introduced mallard duck (Gillespie 1985). In North America, the native black and Mexican ducks are facing a similar threat, due to the spread of the mallard duck into regions where it was formerly absent or uncommon. The original wild populations of the red wolf, which once occurred in the lower Mississippi Valley and along the Gulf Coast, have also disappeared due to genetic swamping by coyotes (Pimlott and Joslin 1968). The species now survives only in captivity, and on a North Carolina barrier island wildlife refuge to which it has recently been reintroduced.

Habitat disruption has long been an important cause of extinction (Table 2.3), and it is now a factor of increasing importance. Habitat conversion, the extreme of such disruption, effectively eliminates even the populations of mobile animals that are able to escape immediate destruction. In all likelihood, the areas to which they flee are unsuitable or already fully occupied. Extinction of the migratory Bachman's warbler in North America, for example, was probably due to the combined destruction of breeding habitat in southeastern United States swamp forests and wintering habitat in Cuban rain forests (See chapter 11). In the tropics, the clearing of moist forests is likely to bring about the extinction of thousands, if not millions, of species in the coming century (Wilson 1989). Introduced animals and plants, together with the changes that humans have caused in frequency of fire and intensity of grazing have modified the habitats of native species almost everywhere. The changes in habitat conditions can often be subtle, yet significant. The spread of crop and livestock farming over much of North America, for example, has greatly extended the portion of the continent suitable for the brown-headed cowbird, a species that lays its eggs in the nests of other birds. Originally confined to the Great Plains, where

FIGURE 2.5

■ This Kirtland's warbler, an endangered species that breeds only in a small area in the southern peninsula of Michigan, is feeding a young brown-headed cowbird that has displaced its own young.

it foraged in association with herds of bison, the cowbird is now found from coast to coast. Nest parasitism by this species now affects many other birds that are not adjusted to this threat through evolution. The extra pressure of cowbird parasitism may be a major factor in the decline of species such as the Kirtland's warbler in Michigan (Mayfield 1977) and the Bell's vireo in southern California (Fig. 2.5).

Many other habitat changes are occurring. Fragmentation of natural ecosystems is a threat to many species. Agriculture and urbanization are rapidly transforming continuous areas of natural ecosystems into islands of natural habitat in a developed landscape sea. Fragmentation reduces the populations of many species to scattered groups of a few individuals, which possess a high risk of local extinction. In addition, small islands of natural habitat have a high edge-to-interior ratio, compared to larger areas, and thus are subject to stronger influences from the surrounding agricultural and urban systems. Increasingly, as well, natural ecosystems are subject to chemical contamination by oil spills, pesticides, industrial chemicals, and urban waste discharges. These have contributed to the decline of populations of many species toward extinction, as we shall see in later chapters.

For birds (King 1980) and for amphibians and reptiles (Honegger 1981), by far the majority of extinctions have been of island forms, which are especially vulnerable because of their small populations and lack of adaptation to man and introduced predators (Table 2.4). This highlights the special conservation problems of oceanic islands, such as the Galápagos and Hawaiian archipelagos, where some of the world's most remarkable biotas exist (See chapter 9).

On close examination, almost all extinctions are attributable to human activities. Among birds, King (1980) was able to recognize only a single extinction evidently due to natural causes—the disappearance of the Puerto Rican bullfinch on the West Indian island of St. Kitts at about the time that island forests were decimated by a hurricane. This sug-

FIGURE 2.6

FIGURE 2.7

■ The passenger pigeon was originally abundant in the eastern forested region of North America.

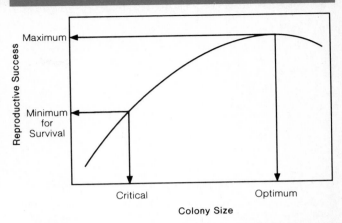

■ Hypothesized relationship of colony size and reproductive success for a colonial bird such as the passenger pigeon (Modified from Halliday 1980).

gests that the extinctions attributed to "natural" causes by Fisher et al. (1969) are better characterized as of "unknown" causes.

Survival Problems and Extinction in Small Populations

Habitat conversion, fragmentation, and change may ultimately create a situation in which a species can no longer survive over a large region, even when areas of seemingly suitable habitat still exist. The extinction of the passenger pigeon (Fig. 2.6), once one of the most abundant birds in North America, may illustrate this process. This species, which nested in colonies and laid only a single egg per nest, depended heavily on the fruits of oaks, beeches, and other trees. The birds moved from place to place in large flocks, and settled to breed where food supplies were abundant. Clearing of the eastern forests for farming in the 1700s and 1800s reduced and fragmented the area of suitable habitat for the species. The birds were also subjected to heavy commercial hunting. Nevertheless, when the species finally became extinct in the early 1900s, much suitable habitat still existed.

The final decline of the passenger pigeon may have been related to its colonial habit (Halliday 1980). Coloniality presumably confers a benefit on the members of the colony—greater survival or reproductive success (Fig. 2.7). Individuals in colonies that are too small may not be able to detect or repel predators, or to locate good feeding areas as efficiently as those in large colonies. Likewise, if colonies are too large, the birds may suffer from competition for food or problems of disease and parasitism. Thus, an optimal colony size should exist, at which the greatest survival and reproduction is realized. In addition, a critically small colony size must also exist, at which survival or reproduction are barely able to permit the population to maintain itself. From this, one can see that

if the overall population of the species declines to the point that colonies fall below this critical size, survival and reproduction become inadequate, and the species declines to extinction.

This model may apply to many species, and not only those that are colonial in their habitats. Once a population is forced below a critical density, the frequency of contacts between individuals may become inadequate for the species to maintain a normal social and reproductive behavior. Factors of this sort may have contributed to the decline of the California condor, for example. Extinction may then follow automatically, in spite of whatever protection may be granted.

Furthermore, in small populations, chance events affecting survival and reproduction of individuals become significant in terms of the overall population. Random, or **stochastic,** variability can involve chance events within the population itself, or random variations in conditions of the habitat. **Demographic stochasticity** refers to random variation in mortality and survivorship due to chance accidents involving individuals. The few breeding females in a small population may fail to mate, produce young, or successfully rear young in a given year for any of several accidental reasons. Or, by chance, all the individuals of one sex might die in a given year. Similarly, **environmental stochasticity,** or random variation in habitat conditions that affect survival and mortality, may cause unusual effects on small populations. Environmental stochasticity can range from the normal variations that occur around average conditions in most years to catastrophic variations, such as those caused by hurricanes or volcanic eruptions. For the same reason that flipping a coin just a few times may produce a run of all heads, random events like these in small populations can be the final cause of extinction.

FIGURE 2.8

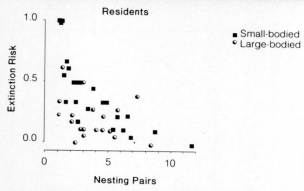

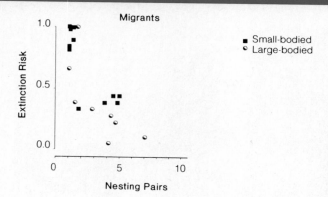

Vulnerability of populations of breeding birds on small islands off the British coast to extinction in relation to population size (Modified from Pimm et al. 1988).

WHAT MAKES SPECIES VULNERABLE TO EXTINCTION?

The features that make populations of particular species highly vulnerable to extinction have been investigated by several workers. Through an analysis of long-term censuses of land birds breeding on islands off the British Isles, Diamond (1984) and Pimm et al. (1988) showed that small population size was the strongest determinant of vulnerability (Fig. 2.8), confirming theoretical predictions by other workers (e.g., Leigh 1981). Overall, large-bodied species were more vulnerable to extinction, not surprisingly, because their populations tended to be smaller. For populations of the same size, however, these data also showed that species with large body size, low reproductive rates, slow growth, and long lives were more vulnerable that those with opposite characteristics only when populations were larger than about seven breeding pairs. Apparently, small-bodied species were less at risk because their faster population growth rates enabled them to recover from a severe reduction in number due to demographic or environmental accidents. On the other hand, when populations were very small (fewer than seven pairs), large-bodied species were less vulnerable to extinction than small-bodied forms. Long life apparently enabled large species to spread their reproductive efforts over a longer period, and thus avoid the catastrophic consequences of a few bad years, which may drive shorter-lived species to extinction. In addition to body size, Pimm et al. (1988) noted that migratory behavior also made species more prone to extinction, apparently because extinction could occur not only because of failures of survival but also because of failure of surviving birds to return to an island.

For Australian mammals, Burbridge and McKenzie (1989) found that non-flying species of intermediate body size accounted for most extinctions. They concluded that these species, with higher individual metabolic requirements than small mammals but less mobility than the large kangaroos,

were least able to adjust to decline in food availability due to human settlement. In addition, these mid-sized species were more attractive to introduced predators than were the smallest mammals.

Several other characteristics predispose species to extinction (Ricklefs and Cox 1972, Terborgh 1974). Species at the ends of long food chains tend to be low in abundance and often large in body size, making them vulnerable for reasons noted above. Species that are highly specialized for particular habitats, breeding sites, or food resources tend to be more at risk than do generalists. The occupation of insular habitats or ranges, where populations are often small and species may become specialized for highly distinctive environmental conditions, increases the risk of extinction when the environment is disturbed or changed. And, of course, species with characteristics of value that attract humans to exploit them are intrinsically vulnerable to extinction.

To emphasize the characteristics that tend to make species vulnerable to extinction, we may summarize them as follows:

1. *Large species with a low reproductive potential*
   Mammals such as the great whales, rhinoceroses, and great apes, and birds such as the California condor and whooping crane are especially vulnerable because their populations are slow to recover, even when they receive protection. Once reduced in numbers, they remain at high risk for a long time.

2. *Species with a high economic value*
   Animals such as whales, sea turtles, elephants, rhinoceroses, and spotted cats are hunted legally and illegally for their meat and oil, tusks and horns, or pelts. The value of such products is sometimes enormous, and the incentive for killing even the last survivors in an area is very strong, especially in underdeveloped countries.

TABLE 2.5

Estimates of the Pre-Extinction Number of Species of Organisms on Earth.

GROUP	IDENTIFIED	CONSERVATIVE ESTIMATE OF TOTAL	LIBERAL ESTIMATE OF TOTAL	PERCENT IDENTIFIED
Mammals, Reptiles, Amphibians	12,410–12,962	13,644	13,800	90–95
Birds	8,715–9,040	9,000	9,224	94–100
Fish	19,056–21,000	22,000	23,000	83–95
Plants	322,399–400,000	443,644	526,024	67–90
Invertebrates	1,027,634–1,300,000	4,400,000	33,000,000	3–27
TOTAL	1,390,115–1,743,002	4,843,644	33,526,024	4–39

Sources: Data from World Resources Institute and International Institute for Environment and Development, *World Resources 1986* (Basic Books, New York, 1986), Table 6.1, p. 86; Edward C. Wolf, *On the Brink of Extinction: Conserving the Diversity of Life* (Worldwatch Institute, Washington, D.C., 1987), Table 1, p. 7.

3. *Species at the ends of long food chains*
Animals such as hawks and owls, cats and canids, and many reptiles depend on food animals that are themselves not very abundant. Disruption that affects basic ecosystem productivity creates the greatest effect at these higher links in the food chain, making it more and more difficult for higher-level predators to meet their needs. In addition, a number of toxic materials can become concentrated along food chains, ultimately to a level toxic to species at their ends (See chapter 13).

4. *Species restricted to local, insular habitats*
Species restricted to small islands or isolated habitats such as small lakes are at high risk because of their small total populations and limited geographical distribution. Only a small-scale disturbance is needed to eliminate such forms. In addition, the isolation of their habitats may have protected them against many of the biotic challenges of non-insular environments, and they may have lost their ability to cope with such challenges.

5. *Species specialized for habitats, breeding sites, or foods*
Species of old-growth forests, such as the spotted owl, are likewise vulnerable because even limited modification of their habitats may change conditions to their disfavor. Animals that breed in colonies are more subject to catastrophic accidents and the activities of predators, including humans. Animals such as the giant panda, a feeding specialist on bamboos, or the black-footed ferret, a specialist predator on prairie dogs, are especially vulnerable because they have few alternatives when something affects their food.

6. *Migratory species*
Migratory birds, mammals, fish, and even insects such as the monarch butterfly, depend on habitat conditions in different geographical areas at particular parts of their seasonal cycle. Thus, they are vulnerable to habitat disruption in any of these regions.

GLOBAL RATES OF EXTINCTION

Global rates of extinction can only be estimated in the crudest fashion. First of all, we still have only an approximate idea of the total number of species that exist on earth (Table 2.5). Although most species of vertebrates and perhaps two-thirds of all plants have been discovered and named, only a small fraction of invertebrate species have been described. Conservatively, perhaps 5 million species exist on earth, less than half of them described. On the other hand, if tropical forest canopies and other poorly explored habitats are richer in species, the world's inventory of species may exceed 30 million, with only a few percent yet described. This difference obviously influences the rate of extinction that now prevails due, for instance, to tropical deforestation.

For terrestrial vertebrates (mammals, birds, and most reptiles), which are perhaps the most carefully watched and guarded species, the rate of extinction is probably 1 to 10 species per year. Aquatic vertebrates (fish, amphibians, and some reptiles) probably experience a rate of 10 to 100 species per year, probably largely the result of disturbance of species-rich lakes, coral reefs, and other aquatic ecosystems in the tropics. For plants, the rate may be about the same, 10 to 100 species per year, largely as a result of tropical deforestation. Aquatic invertebrates may experience an extinction rate of 100 to 1000 species per year. Terrestrial invertebrates doubtlessly have the highest rate of extinction, 1,000 to 10,000 species per year.

KEY CONCEPTS RELATING TO EXTINCTION

1. Humans have long been agents of plant and animal extinction due to hunting, introduction of predators to new areas, and habitat conversion.

2. Increasingly, habitat fragmentation, modification of biotic conditions, and chemical pollution of natural ecosystems contribute to extinctions.

3. The extinction of most species is the consequence of many pressures that combine to reduce survival and reproduction below levels adequate to sustain a population.

4. Species that are most highly specialized in habitat and food requirements are most vulnerable to extinction, especially if they have a low reproductive potential or high economic value.

LITERATURE CITED

Alvarez, L. W., W. Alvarez, F. Asaro, and H. V. Michel. 1980. Extraterrestrial cause for the Cretaceous-Tertiary extinction. *Science* **208**:1095–1108.

Archibald, J. D. and L. J. Bryant. 1990. Differential Cretaceous/Tertiary extinctions of nonmarine vertebrates; evidence from northeastern Montana. *Geol. Soc. Amer. Special Paper* **247**:549–562.

Burbridge, A. A. and N. L. McKenzie. 1989. Patterns in the modern decline of Western Australia's vertebrate fauna: Causes and conservation implications. *Biol. Cons.* **50**:143–198.

Diamond, J. M. 1984. "Normal" extinctions in isolated populations. Pp. 191–245 *in* M. H. Nitecki (Ed.), *Extinctions*. Univ. of Chicago Press, Chicago.

Eckhardt, R. C. 1972. Introduced plants and animals in the Galápagos Islands. *BioScience* **22**:585–590.

Ehrlich, P. and A. Ehrlich. 1981. *Extinction*. Random House, New York.

Fisher, J., N. Simon, and J. Vincent. 1969. *Wildlife in danger*. Viking Press, New York.

Gillespie, G. D. 1985. Hybridization, introgression, and morphometric differentiation between mallard (*Anas platyrhynchos*) and grey duck (*Anas superciliosa*) in Otago, New Zealand. *Auk* **102**:459–469.

Halliday, T. R. 1980. The extinction of the passenger pigeon *Ectopistes migratorius* and its relevance to contemporary conservation. *Biol. Cons.* **17**:157–162.

Honegger, R. H. 1981. List of amphibians and reptiles either known or thought to have become extinct since 1600. *Biol. Cons.* **19**:141–158.

Janzen, D. H. and P. S. Martin. 1982. Neotropical anachronisms: The fruits the gomphotheres ate. *Science* **215**:19–27.

Kaufman, L. and K. Mallory (Eds.). 1986. *The last extinction*. The MIT Press, Cambridge, Massachusetts.

King, W. B. 1980. Ecological basis of extinction in birds. *Proc. XVII Int. Orn Congr.* **1978**:905–911.

Leigh, E. G. 1981. The average lifetime of a population in a varying environment. *J. Theor. Biol.* **90**:213–239

Martin, P. S. 1970. Pleistocene niches for alien animals. *BioScience* **20**:218–221.

Martin, P. S. 1973. The discovery of America. *Science* **179**:969–974.

Martin, P. S. 1984. Prehistoric overkill: The global model. Pp. 354–403 *in* P. S. Martin and R. G. Klein (Eds.), *Quaternary extinctions*. Univ. Ariz. Press, Tucson.

Mayfield, H. 1977. Brown-headed cowbird: Agent of extermination? *Amer. Birds* **31**:107–113.

Miller, R. R., J. D. Williams, and J. E. Williams. 1989. Extinctions of North American fishes during the past century. *Fisheries* **14**:22–38.

Mosimann, J. E. and P. S. Martin. 1975. Simulating overkill by paleoindians. *Amer. Sci.* **63**:304–313.

Myers, N. 1979. *The sinking ark*. Pergamon Press, Oxford, England.

Nitecki, M. H. (Ed.). 1984. *Extinctions*. Univ. of Chicago Press, Chicago, Illinois.

Norton, B. G. (Ed.). 1986. *The preservation of species*. Princeton Univ. Press, Princeton, New Jersey.

Officer, C. B. and C. L. Drake. 1983. The Cretaceous-Tertiary transition. *Science* **219**:1384–1390.

Olson, S. L. and H. F. James. 1984. The role of Polynesians in the extinction of the avifauna of the Hawaiian Islands. Pp. 768–780 *in* P. S. Martin and R. G. Klein (Eds.), *Quaternary extinctions*. Univ. Ariz. Press, Tucson.

Pimlott, D. H. and P. W. Joslin. 1968. The status and distribution of the red wolf. *Trans. N.A. Wildl. Nat. Res. Conf.* **33**:373–389.

Pimm, S. L., H. L. Jones, and J. Diamond. 1988. On the risk of extinction. *Amer. Nat.* **132**:757–785.

Reid, W. V. and K. R. Miller. 1989. *Keeping options alive: The scientific basis for conserving biodiversity*. World Resources Institute, Washington, D.C.

Ricklefs, R. E. and G. W. Cox. 1972. The taxon cycle in the West Indian avifauna. *Amer. Nat.* **106**:195–219.

Savidge, J. A. 1987. Extinction of an island forest avifauna by an introduced snake. *Ecology* **68**:660–668.

Temple, S. A. 1986. The problem of avian extinctions. *Current Ornithology* **3**:453–485.

Terborgh, J. 1974. Preservation of natural diversity: The problem of extinction prone species. *BioScience* **24**:715–722.

van Riper, C. III and S. G. van Riper. 1986. The epizootiology and ecological significance of malaria in Hawaiian land birds. *Ecol. Monogr.* **56**:327–344.

Wilson, E. O. 1989. Threats to biodiversity. *Sci. Amer.* **261**(3):108–116.

Terrestrial Ecosystems

In this section we shall examine principles and problems of conservation ecology in terrestrial ecosystems. For this we shall group terrestrial ecosystems into broad ecological and geographic categories. Major types of terrestrial ecosystems, or *biomes,* are usually characterized by their dominant plants. Examples of biomes are temperate deciduous forest, temperate grassland and arctic tundra. Most chapters in this section will consider several related biomes. Other chapters will consider terrestrial environments defined in a purely geographical fashion, such as coastal and island ecosystems

CHAPTER 3

Temperate Forests, Woodlands, and Shrublands

Temperate forests include the deciduous, coniferous, and broad-leaved evergreen forest biomes. These forests are wide-spread in North America and Eurasia, but also occur in small parts of the southern hemisphere (Fig. 3.1). Temperate woodlands and shrublands include the evergreen communities of regions with Mediterranean climates of hot dry summers and cool, moist winters, such as the chaparral of California, together with other, more localized evergreen or deciduous communities. In North America these include the pinyon-juniper woodlands of the western mountains and deciduous shrublands of the central Rocky Mountains. These biomes cover some of the most heavily populated regions of the developed nations, and thus present the challenge of sound management under conditions of intensive land use.

In these biomes, biotic succession is a dominant ecological process. **Biotic succession** is the change in composition and structure of an ecosystem through time, under constant conditions of climate and geology. Managing biotic succession is therefore a key strategy for conservation ecology in these biomes. Successful management of succession requires consideration of 1) the amount of habitat in different stages of succession, 2) the size and location of areas in different successional stages, and 3) the nature and intensity of human use of different stages of succession. We shall first examine the process itself, and then consider specific issues of conservation management.

BIOTIC SUCCESSION

Biotic succession can be illustrated by the sequence of communities that occurs on a newly exposed rock outcrop (Fig. 3.2). Initially, organisms such as lichens and mosses colonize the bare rock surface, forming a **pioneer community.** These simple organisms trap soil particles carried by wind and water, and contribute to the breakdown of the rock surface by chemical action. Eventually, they accumulate enough soil to permit the growth of annual grasses and broad-leaved herbs. These species, in turn, accelerate breakdown of the rock by penetrating their roots into crevices, and forming a layer of dead plant litter that is an even more effective trap for soil. In time, perennial herbaceous plants invade. As the soil system develops, still larger plants such as shrubs and trees appear, and many of the pioneer and early successional species drop out. Eventually, after a century or two, a community of species that can reproduce and replace themselves indefinitely—a **climax community**—may develop. This process is termed **primary succession,** since it takes place on a site that was not previously occupied by a community. Succession that follows a disturbance, such as a forest fire or the clearing of land for farming, is termed **secondary succession.** Secondary succession usually proceeds faster than primary succession in terrestrial ecosystems, since well-developed soil is often already present.

Succession does not always lead to a stable climax community, however. In the later stages of succession, change may simply become very slow, and before a true climax is achieved, some disturbance may strike the community. A fire may occur and initiate secondary succession anew. Disease, insect outbreaks, or an invasion of a new species may perturb the community, triggering a successional response. In some instances, what seem to be climax species can bring about the degradation of their habitat, causing the system to undergo succession anew. In addition, a long-term shift in climate may favor a slightly different group of climax species. Ecologists now recognize that primary succession does not often lead to a highly predictable, stable climax, and that secondary succession rarely restores the exact system that experienced disturbance.

If biotic succession is a process that occurs under constant climatic and geologic conditions, what are its causes? The most important processes are themselves biotic. Ecologists now recognize four biotic mechanisms of succession (Fig. 3.3), which can act alone or in combinations (Connell and Slatyer 1977, Pickett et al. 1987). The first of these is **differential dispersal.** Plant and animal species differ enormously in their dispersal capabilities. Some, such as winged insects and plants with wind-blown seeds, are able to colonize new or vacant sites very quickly. Others with slower dispersal

FIGURE 3.1

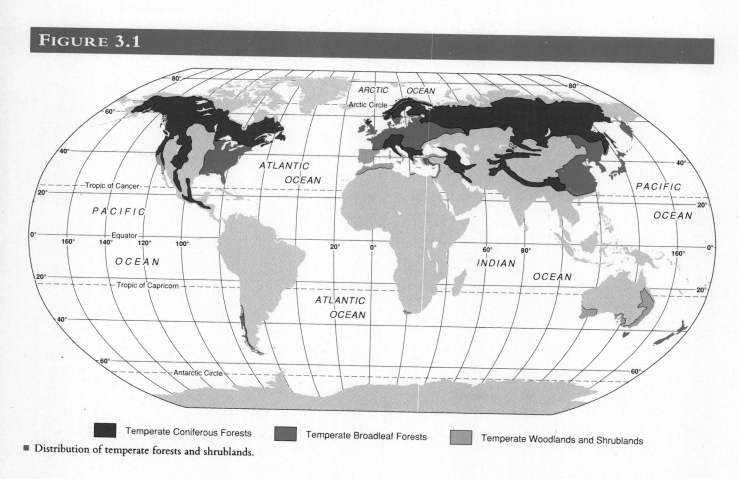

Temperate Coniferous Forests   Temperate Broadleaf Forests   Temperate Woodlands and Shrublands

■ Distribution of temperate forests and shrublands.

FIGURE 3.2

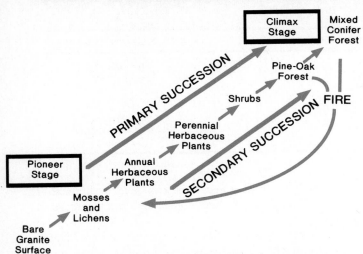

■ A schematic diagram of biotic succession beginning on bare rock and leading to a climax forest that eventually acquires old-growth characteristics.

FIGURE 3.3

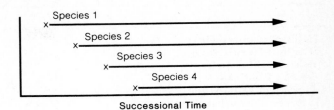

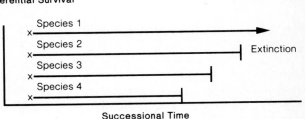

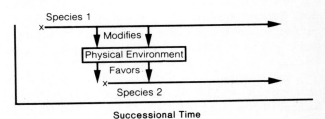

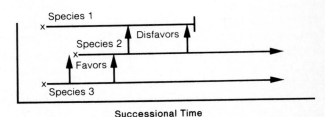

■ Diagrammatic representation of the four major mechanisms of biotic succession.

mechanisms arrive later. As species with different dispersal capabilities establish themselves, the composition of the community changes. A second mechanism is **differential survival.** Many kinds of species may initially reach and colonize a new area of habitat. In time, however, only some of these are able to tolerate the range of habitat conditions that occurs. As species drop out, the composition of the community changes. Third, some species modify the conditions of the habitat in a way that favors the establishment of other species, a process termed **facilitation.** Pioneer species may facilitate the **physical and chemical weathering** of terrestrial substrates and aquatic sediments, aiding purely abiotic processes of weathering. Facilitation was suggested as a major mechanism in the example of succession on bare rock given at the beginning of the chapter. Fourth, direct positive or negative **biotic interactions** may favor the establishment or hasten the disappearance of species. Establishment of a certain plant, for example, will favor herbivores that feed on it. The appearance of a predator may cause the decline of prey species unable to tolerate heavy predation. The key feature of biotic succession is that its mechanisms are primarily internal ecosystem processes, rather than changes in external controls (See chapter 1), such as changes in climate, physiography, or regional biota.

In real situations, not surprisingly, several or all of these mechanisms operate, and external controlling factors also show change. In a lake ecosystem, for example, successional processes are active, but in addition, geological processes act to transport sediment into the lake and reduce its depth, eventually filling it. The changes that occur in lake ecosystems through their geological history are not purely the result of biotic succession.

Biotic succession varies in pattern and duration in different ecosystems (MacMahon 1981). In chaparral shrublands and subarctic conifer forests, for example, the successional sequence may be short and the dominant plants of late succession may be the same species that resprout or germinate immediately after disturbance. Species of the same growth form, such as woody shrubs, may dominate the entire sequence, as well. In deciduous forests, on the other hand, succession may be an extended sequence with several distinct successional communities. In this case, early successional communities may be dominated by species of life forms quite different from those late in succession.

TABLE 3.1

Characteristics of Early and Late Successional Species.

	EARLY SUCCESSIONAL	LATE SUCCESSIONAL
Habitat Requirements	Generalized	Specialized
Resource Requirements	Generalized	Specialized
Behavioral Plasticity	High	Low
Genetic Plasticity	High	Low
Reproductive Potential	High	Low
Impact of Disturbance	Often Beneficial	Usually Detrimental
Exploitation Potential	High	Low

EARLY AND LATE SUCCESSIONAL SPECIES

Where succession involves a slow progression of distinct stages, and leads to climax communities that are likely to endure for a long time, both plants and animals tend to be adapted to particular stages in the successional spectrum (Leopold 1966). For simplicity, we can consider most species to be either **early successional** or **late successional** species. Basic patterns of adaptation, response to natural or human disturbance, and potential for exploitation of these kinds of species all vary with the stage of succession to which a species is adjusted (Table 3.1).

In the terrestrial environment, early successional habitats are usually variable and changing in their physical and biotic conditions. Extremes of physical conditions are greater, for example, on a bare rock outcrop than inside a mature forest. The specific kinds of plants and animals that exist in a weedy field, too, are more variable than the species making up the mature forest. Thus, early successional species must be broadly tolerant of habitat conditions and generalized in their patterns of adjustment to their habitat. They must also be able to use a wide variety of foods, either plant or animal, since the biotic makeup of early successional situations is so variable.

These generalized patterns of adaptation to habitats and resources usually include high behavioral adaptability—individuals can vary their responses in different situations. At the population level, species show high genetic plasticity; natural selection is able to adjust the local population to local environmental conditions. This adjustment, known as **ecogenetic variation,** is seen in almost all widespread early successional species. Taxonomists often recognize such local forms as subspecies. Another important population characteristic is usually high reproductive potential, which enables early successional species to colonize new areas rapidly, and also offsets the high mortality rates that usually occur under variable and unpredictable early successional conditions.

For such species, disturbance is often beneficial because it sets back the successional process that tends to eliminate the conditions to which they are adapted. Also, because of their plasticity and high reproductive potential, early successional species are often well suited to be exploited by humans—harvests of portions of their populations can be replaced quickly.

Late successional species, on the other hand, are adapted to more equitable and predictable environments, where much more specialized habitat and resource requirements can be satisfied. Because such requirements can be counted on, a lesser premium exists on behavioral and genetic plasticity. High reproductive potential is less important than extensive and prolonged parental care of a few offspring. When disturbance occurs, however, such species are usually unable to survive. Likewise, unregulated exploitation by humans usually cannot be tolerated because of the low reproductive potential, even when habitat conditions remain favorable.

AN EARLY SUCCESSIONAL SPECIES

The mule deer, *Odocoileus hemionus* (Fig. 3.4a), is an excellent example of an animal dependent on early successional habitats (Leopold 1966). This species occurs throughout much of western North America, from southern Alaska far south into Mexico. It occupies habitats ranging from hot southwestern deserts to the borders of the cool, wet rain forests of the Pacific Northwest. Even in a single region, its food plants are varied, and throughout its range the number of important food plant species is far into the hundreds. Local populations differ in behavior; some are permanent residents, others exhibit extensive seasonal migrations. Some 11 subspecies of mule deer are recognized, attesting to the genetic variation of this species from place to place.

The classic study by Taber and Dasmann (1957) of the Columbian black-tailed deer (*O. h. columbianus*), a subspecies of the mule deer, in northern California shows how disturbance can be used as a beneficial management tool for this early successional species. In the foothills of the coast range north of San Francisco, chaparral brushlands are one of the important vegetation types used by these deer. However, when protected from fire for periods of several decades, the chaparral shrubs become crowded and tall, reducing the growth of herbaceous plants at the ground surface and raising most

FIGURE 3.4

■ a. Mule deer, a representative early successional species. b. The woodland caribou, a representative late successional species.

TABLE 3.2

Responses of Columbian Black-Tailed Deer to Conversion of Mature Chaparral to Open Shrubland by Controlled Burning in Northern California.

	CHAPARRAL	SHRUBLAND	CHANGE
Deer Diet			
Shrub Browse Available (lbs/mi²)	113,000	317,000	+2.8 X
Shrub Browse Consumed (lbs/mi²)	17,888	29,692	+1.6 X
Consumption of Available (%)	15.2%	9.3%	−0.6 X
Shrub Browse in Diet (%)	92%	54%	−0.6 X
Herbaceous Plants in Diet (%)	8%	46%	+6.0 X
Deer Population Dynamics			
Pre-Fawning Adults (per 100 mi²)	2,498	5,153	+2.1 X
Fawns Born (per 100 mi²)	950	3,852	+4.0 X
Fawns Born per Female	0.59	1.10	+1.9 X

Source: Data from R. D. Tabor and R. F. Dasmann. "The Black-Tailed Deer of the Chaparral" in *California Department of Fish Game Wildlife Bulletin*, No. 8, 163, 1957.

of their foliage above the height that deer can browse. This, of course, is a natural successional trend. Taber and Dasmann showed that by controlled burning, this unfavorably mature chaparral community could be transformed into a more open shrubland with low, resprouting shrubs, a much greater growth of herbaceous plants, and scattered unburned patches of chaparral for cover.

Data on the use of unburned chaparral and the newly created shrubland by deer showed major dietary shifts (Table 3.2). In the shrubland, during the first year after fire, the amount of shrub browse available in the height range that deer could reach actually increased. Although the total amount of shrub forage taken by deer increased, due largely to an in-crease in deer numbers, the percentage of the available browse that was taken by deer declined. Since they took a smaller percentage of what was available, they could be more selective, choosing the preferred plant species and parts. Moreover, they were able to take a much higher percentage of herbaceous plants, in addition to shrub browse. The result of this dietary shift was almost a doubling of the number of fawns produced per female, and a four-fold increase in reproduction for the population as a whole. Thus, setting back succession by burning—clearly a "disturbance"—benefitted the population, and increased the potential hunting harvest from the population.

A LATE SUCCESSIONAL SPECIES

The woodland caribou (a name applied to the subspecies *R. t. caribou* and *R. t. sylvestris* of the caribou, *Rangifer tarandus*) is an ungulate comparable in general characteristics to the mule deer (Fig. 3.4*b*). In North America, woodland caribou once occupied near-climax coniferous forests from British Columbia east to Nova Scotia and south into the northern Rocky Mountains, upper Great Lakes region, and New England. Since European settlement, however, woodland caribou have declined greatly in abundance, and are restricted to scattered locations from Alberta to Quebec in Canada, and to one small population in northern Idaho in the United States. The total population of the woodland caribou is perhaps 25,000 animals.

Hunting by early settlers contributed to the decline of the species, but protection from hunting did not halt the decline. Changing climate, disease, competition from moose and deer (early successional species), or predation cannot account for this decline. Evidently, the major cause has been the change in structure of coniferous forests due to logging, and how this change influences the availability of critical foods (Cringan 1957). In winter, the critical food for woodland caribou is tree lichens—pale green, beard-like lichens that grow on tree branches much like Spanish moss in the Gulf Coast states. In a mature coniferous forest, the biomass of these slow-growing lichens is great, and every winter falling trees and branches make a certain amount available to caribou. In such forests, tree lichens are a dependable food that the animals themselves cannot overexploit. Logging of the northern forests, however, not only destroyed the old-growth trees and their lichen supplies, but created a young, even-aged successional forest that is incapable of supplying a dependable supply of tree lichens. Thus, the decline of the woodland caribou seems to be largely the result of the high specialization of these two subspecies for a food resource available only in near-climax forests.

OTHER EXAMPLES OF EARLY AND LATE SUCCESSIONAL SPECIES

Good examples of early and late successional species in North America are also provided by the ruffed grouse, *Bonasa umbellus,* and the spruce grouse, *Canachites canadensis,* respectively. The ruffed grouse, with 12 subspecies distributed from Alaska to Georgia, is a highly adaptable bird with a varied diet and an optimal habitat consisting of a mosaic of aspen stands of varying age—the products of succession after disturbance of mature forests (Gullion 1977). The spruce grouse, even though occurring from Alaska to Labrador and south beyond the Canadian border, shows less evolutionary differentiation (only 4 subspecies), and is restricted to mature coniferous forests, where it feeds almost exclusively on spruce buds and needles. The ruffed grouse is an important game species; the spruce grouse is not.

The widespread, behaviorally adaptable, prolific mourning dove, *Zenaida macroura,* is another good example of an early successional species, contrasting with the extinct passenger pigeon, *Ectopistes migratorius* (See chapter 2). The mourning dove is a familiar bird of farmland, urban areas, and other disturbed habitats. It typically produces a clutch of two eggs, and usually renests two or more times each breeding season. The passenger pigeon, a bird of mature oak forests, was specialized in its colonial breeding pattern and its dependence on tree fruits and nuts. Its reproductive potential was low, a pair nesting once a year and producing only a single egg.

Today, some of our most endangered species are those associated with late successional stages of forest ecosystems. Special efforts must be made to preserve the unique features of ecosystems that these forms require. The red-cockaded woodpecker, *Picoides borealis,* for example, was once common in pine forests from Maryland to Texas. This species requires large pines, 75 to 95 years in age, that have become infected by "red heart disease"—a fungal rot of the heartwood—for excavation of its large nest chambers, which serve as a communal nest for several females. Although pine forests are probably more abundant than ever within the woodpecker's range, the intensive forest management practices adopted in recent decades do not favor the growth of old trees infected by heartwood disease. Red-cockaded woodpeckers can profitably use large areas of younger forest for foraging, however. Thus, lengthening the harvest cycle of pine forests to permit the continued growth of old trees, coupled with a harvesting design that leaves small clusters of such trees scattered through younger stands, may meet the needs of the red-cockaded woodpecker without major loss of timber production (Seagle et al. 1987).

A BALANCE OF SUCCESSIONAL STAGES

The key strategy in management of temperate forest, woodland, and shrubland ecosystems is to maintain the full range of successional stages. This strategy often implies the deliberate use of disturbance such as fire to create early successional habitats. It also implies protection of areas of late successional communities. A balance of areas of different stages must be maintained, such that species adapted to all successional stages are able to survive, in effect guaranteeing that biotic succession continues to be an active ecological process.

In some places, however, early or mid-successional stages of succession are disappearing. In eastern North America, for example, modern land use policy tends to favor the extremes of forest succession: intensively farmed or grazed lands of early successional nature, and protected forests and woodlots that are becoming more and more mature. In this landscape, successional shrublands and the youngest forest stages may be represented inadequately. As a result, many forest parks and preserves have initiated "controlled succession" plots, on which areas of mature forest are cut periodically and succession is allowed to proceed.

Elsewhere, particularly in western North America, well-meaning protection of forests from fire and other disturbances has led to major changes in their characteristics through succession. In northern California, coast redwood stands on river floodplains were originally exposed to both flooding and fire. Redwoods are tolerant of both forms of disturbance, which essentially weed their groves of competitors and create favorable conditions for germination and growth of young redwoods (Stone 1965). Now protected from fire, many of these groves are being invaded by other tree species. In the Sierra Nevada of California, protection of sequoia groves from fire has fostered a dense understory of shrubs and other trees, increasing the danger of severe damage when a fire inevitably does come (Biswell 1989). The same is true of many other forest and shrubland communities on western North America.

The recognition that fire is a natural, and often beneficial, ecological factor has led to policies of allowing natural fires to burn in some areas (the so-called "Let-Burn Policy"), as well as gradually·reintroducing fire to forests by prescribed burning (Kilgore 1973). In sequoia groves, these efforts have resulted in the reduction of crown fire danger (Kilgore and Sando 1975).

OLD-GROWTH FORESTS: ULTRA-CLIMAX ECOSYSTEMS

When climax forests survive for centuries in regions with little disturbance, they acquire special ecological characteristics. These so-called **old-growth forests** (Fig. 3.5) typically contain trees of very large size, exhibit strong vertical differentiation of vegetational strata, show considerable patchiness due to succession in tree-fall gaps, and possess large standing dead trees ("snags") and decaying logs (Thomas et al. 1988, Feeney 1989). Old-growth forests are also home to many specialized plants and animals. Thus, from an aesthetic and conservational point of view, these forests are extraordinary and valuable. From a forestry standpoint they are over-mature, in the sense that no net increase in wood biomass is occurring. Often, in fact, wood biomass is less than that in younger, "mature" forest stands. In the Pacific Northwest, old-growth redwood and Douglas-fir forests exhibit these qualities.

Most of the original old-growth forest has been cut. Only about 17 percent of the original old-growth Douglas fir forest of the Pacific Northwest remains (Thomas et al. 1988), the bulk of this being in the United States National Forests. Old-growth Douglas fir has become the center of the controversy over the area of such habitat that should be preserved in the United States. Timber interests view these forests as decadent, and argue that their cutting and replacement by young stands is both economically justified and ecologically beneficial. Replacing old-growth forests by vigorous young stands, for example, has been claimed to be a way to counteract the greenhouse effect by removing CO_2 from the atmosphere. On the other hand, many species of birds, mammals, and amphibians are closely associated with this habitat, including the marbled murrelet, northern goshawk, northern

FIGURE 3.5

■ Old-growth forests of the Pacific Northwest, like this coast redwood stand in Muir Woods, northern California, require 350 to 750 years to grow. These stands are characterized by very large trees, distinct vertical strata of foliage, strong patchiness due to succession in tree-fall gaps, and numerous standing dead trees and decaying logs.

spotted owl, Vaux's swift, silver-haired bat, red tree vole, and northern flying squirrel (Simberloff 1987, Marshall 1989). Furthermore, analysis of the CO_2 releases and uptakes resulting from forest conversion shows that net storage does not result, and the greenhouse effect is not counteracted (Harmon et al. 1990).

Much of the recent controversy about old-growth forest in the Pacific Northwest has centered on the spotted owl (Fig. 3.6). The total population of the northern spotted owl is estimated to be 2,500 to 3,000 pairs (Simberloff 1987, Doak 1989). These birds occupy home ranges of 800 to 2,000 hectares (ha) in old-growth forests, and appear to feed heavily on flying squirrels. The value of old-growth timber is about $10,000 per ha, so that an 800–ha home range of a pair of owls contains about $8 million worth of timber. A decision about preserving habitat for a difference of 500 pairs of owls, by extension, involves $4 billion worth of timber. Modeling

FIGURE 3.6

■ The northern spotted owl is largely restricted to old-growth coniferous forests of the Pacific Northwest.

studies suggest, however, that continued harvesting of old-growth forest, as proposed by the United States Forest Service, will lead to increased fragmentation of old-growth stands and the severe decline of the species (Doak 1989). In this case, in June, 1990, under strong pressure from conservation groups, the United States Fish and Wildlife Service eventually listed the spotted owl as a threatened species, giving strong backing to the preservation of much of its old-growth habitat. Similar controversies exist for old-growth forests in other areas, such as coastal British Columbia and the southern panhandle of Alaska.

CYCLICAL PATTERNS OF SUCCESSION

In certain regions, succession leads to a community that seems to have climax qualities, but is highly vulnerable to destruction by fire or insect outbreaks. Many of the conifer forests of western North America show this characteristic (Heinselman 1981), as do shrublands such as the California chaparral (Hanes 1977). The vegetation in such regions often consists of a mosaic of communities of different successional stages, only a fraction of which are near-climax in structure. In such regions, many animals are adapted to a landscape mosaic with a certain mix of communities of different successional status. The California black-tailed deer and ruffed grouse discussed earlier show this pattern of adaptation quite clearly. In such ecosystems, too much disturbance or too little disturbance can both be detrimental to conservation objectives, and management of successional processes must be designed very carefully.

In the summer of 1988, for example, forest fires burned about 36 percent of Yellowstone National Park and surrounding areas of National Forest (Fig. 3.7). The stage for this event was set by a series of dry winters leading up to the driest summer since the dust bowl years of the 1930s. When fires of natural origin began, the park followed a policy designed to restore fire as a natural factor in the park's ecology. This policy called for natural fires that occurred within specified weather and fuel conditions to be classified as "prescribed natural fire" and allowed to run their course, unless they threatened buildings or other facilities. In 1988, however, fires that were initially classed as prescribed natural fires were converted by dry, windy weather into virtually uncontrollable fire storms. Although $125 million was spent in efforts to halt the fires, they were ultimately stopped only by autumn snows (Fig. 3.8). Successional recovery of burned areas began in the following summer at lower elevations with a flush of growth of herbaceous plants and tree seedlings. At many higher elevations, however, recovery proceeded very slowly (Fig. 3.9). The fires caused little mortality of large animals, and will probably benefit populations of elk and deer (Fig. 3.10), but their aesthetic impacts raised great controversy over fire policy in western parks.

Analysis of forest history of Yellowstone Park suggests that fires of comparable scale have occurred at intervals of 200 to 300 years (Romme and Despain 1989). Following fire, single-species stands of lodgepole pine tend to develop, due to adaptations of this species for post-fire seed dispersal and germination. In about 150 years, these stands reach maturity and some large trees begin to die, creating openings that are invaded by Engelmann spruce and subalpine fir. From then on, death of pines and growth of young spruces and firs creates a forest condition highly vulnerable to fire. Fallen dead trees create a heavy fuel load on the forest floor, and living trees of varying sizes provide a route for fire to climb into the canopy. In the 1980s, these conditions had developed over much of Yellowstone Park, largely as a result of natural successional processes.

Many ecologists argue, however, that although such fires may be part of the original natural order, they now represent an excessive disturbance of the small areas of natural landscape that exist in parks and national forests. A review of federal policies after the 1988 fires, furthermore, concluded that the fire management plan for Yellowstone National Park, developed to implement the policy of restoring fire as a natural force, was inadequate. In particular, its criteria for designating fires as prescribed natural fires that should be allowed to burn, as opposed to wildfires that should be controlled, were inadequate (Wakimoto 1990). Thus, efforts to use prescribed burning as a tool to manage the successional state of forests are sound, but must be designed to prevent large fractions of the landscape from becoming highly vulnerable to fires like those of Yellowstone National Park in 1988.

FIGURE 3.7

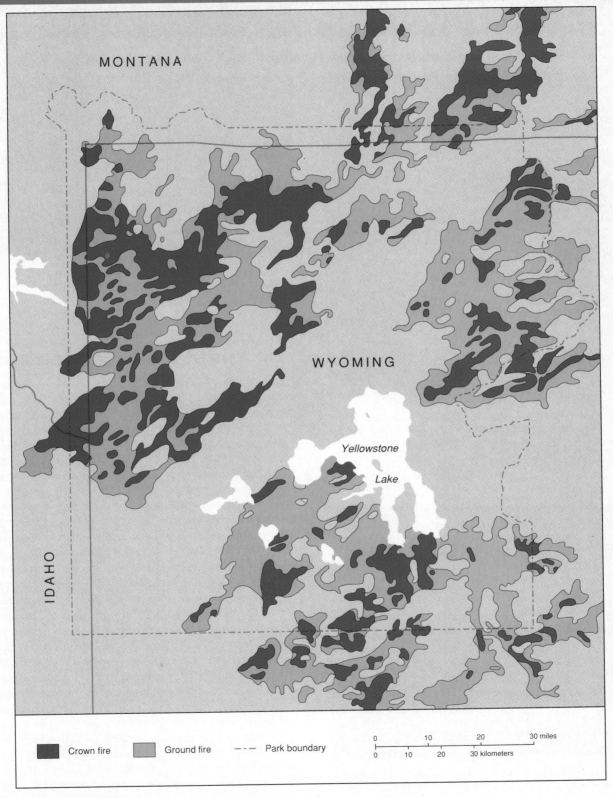

MONTANA

WYOMING

Yellowstone

Lake

IDAHO

| | Crown fire | | Ground fire | – – – Park boundary |

0   10   20   30 miles
0   10   20   30 kilometers

■ The fires in Yellowstone National Park in the summer of 1988
burned about 35.7 percent of the park area. (Courtesy of D. Despain,
Yellowstone National Park U.S. Park Service.)

FIGURE 3.8

■ In the summer of 1988, dry, windy conditions in Yellowstone National Park transformed fires initially classed as prescribed natural fire into wildfires that jumped fire lines and were ultimately controlled only by autumn rain and snow.

FIGURE 3.9

■ By the summer of 1990, successional recovery of many stands burned at higher elevations in Yellowstone National Park in 1988 had scarcely begun.

FIGURE 3.10

■ The Yellowstone National Park fires of 1988 caused little mortality of elk and other large animals, which began to forage in burned areas as soon as new plant growth appeared.

FRAGMENTATION OF FOREST AND SHRUBLAND ECOSYSTEMS

In many areas, forests and shrublands remain only as relict fragments of natural habitat. In North America, for example, much of the eastern deciduous forest has been transformed into island-like woodlots surrounded by farmland and suburban residential areas (Fig. 3.11). Transformation of continuous forest into scattered woodlots has a major impact on the forest fauna (Saunders et al. 1991). The microclimate of the forest is greatly modified, especially near the woodlot edge. Isolation from other woodlots reduces the interchange of forest species, and increases the likelihood of extinction of populations in individual woodlots. Small, isolated forest patches often lack the resources necessary to support viable populations of forest species, and make it difficult for these species to recolonize if they do accidentally disappear from areas that are minimally adequate. Wilcove (1985) and Yahner and Scott (1988) have shown that nests of birds in small woodlots in the eastern United States are subject to much heavier predation than those in larger forest tracts. Predators from adjacent open habitats, such as skunks, crows, jays, and grackles, are able to penetrate into the edges of forest areas, and in the case of small woodlots, the entire area of forest may become accessible to such species. As a result, many studies have shown that the number of species of birds declines much faster as one goes from large to small woodlots than from large to small areas within a region of continuous forest (Galli et al. 1976, Ambuel and Temple 1983, Lynch and Whigham 1984, Blake and Karr 1984). This decline is greatest among species typical of forest interior habitats, among insectivorous and predatory

FIGURE 3.11

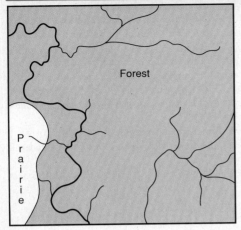

1831

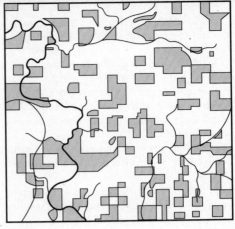

1882

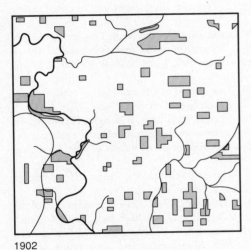

1902

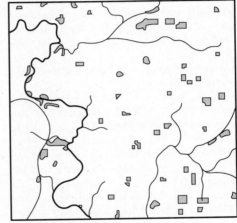

1950

■ From the time of settlement of Cadiz Township, Green County, southern Wisconsin to the mid-1900s, the forest was reduced from a nearly continuous cover to isolated woodlots covering less than 1 percent of the original area (Modified from Curtis 1956).

species, and among species that are long distance migrants than among other kinds of birds (Foreman et al. 1976, Whitcomb 1977, Whitcomb et al. 1981, Blake 1983, 1986).

In California, Soulé et al. (1988) showed a similar pattern for native chaparral communities in canyons isolated in an urban development. The number of chaparral bird species declined with decreasing canyon size and with increasing time since isolation. In addition, canyons that were still visited by coyotes retained more native chaparral birds than those that were not. Apparently, coyotes control the abundance of small predators such as foxes, skunks, and domestic cats, favoring the survival of species such as the California quail that are vulnerable to these animals. When coyotes disappear, these "mesopredators" increase in numbers, increasing predation on birds.

KEY CONSERVATION STRATEGIES FOR TEMPERATE FOREST AND SHRUBLAND ECOSYSTEMS

1. Maintain the full range of successional and climax communities in an appropriate balance and distribution pattern.

2. Provide adequate protection from disturbance for late successional ecosystems that contain highly specialized species.

3. Utilize active manipulation to maintain early successional ecosystems that contain species with high exploitation potential.

LITERATURE CITED

Ambuel, B. and S. A. Temple. 1982. Area-dependent changes in the bird communities and vegetation of southern Wisconsin forests. *Ecology* **64**:1057–1068.

Biswell, H. H. 1989. *Prescribed burning in California wildlands vegetation management.* Univ. Calif. Press, Berkeley, CA.

Blake, J. G. 1983. Trophic structure of bird communities in forest patches in east-central Illinois. *Wilson Bulletin* **95**:416–430.

Blake, J. G. 1986. Species-area relationship of migrants in isolated woodlots. *Wilson Bulletin* **98**:291–296.

Blake, J. G. and J. R. Karr. 1984. Species composition of bird communities and the conservation benefit of large versus small forests. *Biol. Cons.* **30**:173–187.

Connell, J. H. and R. O. Slatyer. 1977. Mechanisms of succession in natural communities and their role in community stability and organization. *Amer. Nat.* **111**:1119–1144.

Cringan, A. T. 1957. History, food habits and range requirements of the woodland caribou of continental North America. *Trans. N. A. Wildl. Conf.* **22**:485–501.

Curtis, J. T. 1956. The modification of mid-latitude grasslands and forests by man. Pp. 721–726 *in* W. L. Thomas (Ed.), *Man's role in changing the face of the earth.* Univ. of Chicago Press, Chicago, IL.

Doak, D. 1989. Spotted owls and old growth logging in the Pacific Northwest. *Cons. Biol.* **3**:389–396.

Feeney, A. 1989. The Pacific Northwest's ancient forests: Ecosystems under siege. Pp. 93–153 *in* W. J. Chandler (Ed.), *Audubon wildlife report 1989/1990.* Academic Press, San Diego, CA.

Foreman, R. T. T., A. E. Galli, and C. F. Leck. 1976. Forest size and avian diversity in New Jersey woodlots with some land use implications. *Oecologia* **26**:1–8.

Galli, A. E., C. F. Leck, and R. T. T. Forman. 1976. Avian distribution patterns in forest islands of different sizes in central New Jersey. *Auk* **93**:356–364.

Gullion, G. W. 1977. Forest manipulation for ruffed grouse. *Trans. N. A. Wildl. Conf.* **42**:449–458.

Hanes, T. L. 1977. Chaparral. Pp. 417–469 *in* M. G. Barbour and J. Major (Eds.), *Terrestrial vegetation of California.* John Wiley and Sons, New York.

Harmon, M. E., W. K. Ferrell, and J. F. Franklin. 1990. Effects on carbon storage of conversion of old-growth to young forests. *Science* **247**:699–702.

Heinselman, M. L. 1981. Fire and succession in the conifer forests of northern North America. Pp. 374–405 *in* D. C. West, H. H. Shugart, and D. B. Bodkin (Eds.), *Forest succession: Concepts and application.* Springer-Verlag, New York.

Kilgore, B. M. 1973. The ecological role of fire in Sierran conifer forests: Its application to national park management. *J. Quaternary Res.* **3**:496–513.

Kilgore, B. M. and R. W. Sando. 1975. Crown-fire potential in a sequoia forest after prescribed burning. *Forest Sci.* **21**:83–87.

Leopold, A. S. 1966. Adaptability of animals to habitat change. Pp. 66–75 *in* F. F. Darling and J. P. Milton (Eds.), *Future environments of North America.* Doubleday and Co., New York.

Lynch, J. F. and D. F. Whigham. 1984. Effects of forest fragmentation on breeding bird communities in Maryland, USA. *Biol. Cons.* **28**:287–324.

MacMahon, J. A. 1981. Successional processes: Comparisons among biomes with special reference to probable roles and influences on animals. Pp. 277–304 *in* D. C. West, H. H. Shugart, and D. B. Bodkin (Eds.), *Forest succession: Concepts and application.* Springer-Verlag, NY.

Marshall, D. B. 1989. The marbled murrelet. Pp. 434–455 *in* W. J. Chandler (Ed.), *Audubon wildlife report 1989/1990.* Academic Press, San Diego, CA.

Pickett, S. T. A., S. L. Collins, and J. J. Armesto. 1987. Models, mechanisms and pathways of succession. *Bot. Rev.* **53**:335–371.

Romme, W. H. and D. G. Despain. 1989. The Yellowstone fires. *Sci. Amer.* **261(5)**:37–46.

Saunders, D. A., R. J. Hobbs, and C. R. Margules. 1991. Biological consequences of ecosystem fragmentation. *Cons. Biol.* **5**:18–32.

Seagle, S. W., R. A. Lancia, D. A. Adams, M. R. Lennartz, and H. A. Devine. 1987. Integrating timber and red-cockaded woodpecker habitat management. *Trans. N. A. Wildl. Nat. Res. Conf.* **52**:41–52.

Simberloff, D. 1987. The spotted owl fracas: Mixing academic, applied, and political ecology. *Ecology* **68**:766–772.

Soulé, M. E., D. T. Bolger, A. C. Alberts, J. Wright, M. Sorice, and S. Hill. 1988. Reconstructed dynamics of rapid extinctions of chaparral-requiring birds in urban habitat islands. *Cons. Biol.* **2**:75–92.

Stone, E. C. 1965. Preserving vegetation in parks and wilderness. *Science* **150**:1261–1267.

Taber, R. D. and R. F. Dasmann. 1957. The black-tailed deer of the chaparral. *Calif. Dept. Fish Game Wildl. Bull. No. 8,* 163 pp.

Thomas, J. W., L. F. Ruggiero, R. W. Mannan, J. W. Schoen, and R. A. Lancia. 1988. Management and conservation of old-growth forests in the United States. *Wildl. Soc. Bull.* **16**:252–262.

Wakimoto, R. H. 1990. The Yellowstone fires of 1988: Natural process and natural policy. *Northwest Science* **64**:239–242.

Whitcomb, R. F. 1977. Island biogeography and "habitat islands" of eastern forest. *Amer. Birds* **31**:3–5.

Whitcomb, R. F., C. S. Robbins, J. F. Lynch, B. L. Whitcomb, M. K. Klimkiewicz, and D. Bystrak. 1981. Effects of forest fragmentation on avifauna of the eastern deciduous forest. Pp. 125–205 *in* R. L. Burgess and D. M. Sharpe (Eds.), *Forest island dynamics in man-dominated landscapes.* Springer-Verlag, New York.

Wilcove, D. S. 1985. Nest predation in forest tracts and the decline of migratory songbirds. *Ecology* **66**:1211–1214.

Yahner, R. H. and D. P. Scott. 1988. Effects of forest fragmentation on depredation of artificial nests. *J. Wildl. Manag.* **52**:158–161.

Grasslands and Tundra

Generally, grassland and tundra ecosystems lie between forests and deserts—specifically the broad-leaved forests and the arid mid-latitude deserts border the temperate grasslands and the coniferous forests and the polar or alpine wastelands border the tundras. Both grasslands and tundras are dominated by grasses, or by grass-like plants such as sedges, together with broad-leaved herbaceous plants and low shrubs. The faunas of grasslands and tundras are also quite similar, and these ecosystem types share many of the same conservation problems.

In this chapter we shall first examine the distribution and some of the important ecological characteristics of these ecosystems. Then we shall consider how human activities have affected the plant and animal life of grasslands and tundras. Finally, we shall identify key strategies for the restoration and conservation management of these systems.

DISTRIBUTION OF GRASSLANDS AND TUNDRA

Temperate grasslands occur in regions with strongly seasonal precipitation varying roughly between 25 and 75 cm per year (Fig. 4.1). Grassland climates nearly always exhibit one or more dry seasons severe enough to exclude most kinds of trees and shrubs. In some regions the dry season is summer, in others winter, and in still others spring and fall. These dry seasons also enable fire to be an influential environmental factor in many places. Frequent fire eliminates all but the most resilient species of trees and shrubs, and in many places prevents the replacement of grasslands by communities dominated by woody plants.

North America has several important grassland regions. In all of these areas the original grasslands were dominated by perennial **bunch grasses,** species that grow in dense clumps or large patches, rather than as a continuous turf. The most extensive grassland region occupies the Great Plains, where precipitation tends to occur both in summer and winter, but severe dry periods are frequent in spring and fall. In the Great Plains, total precipitation declines from east to west and from south to north. In the eastern and southern plains, tall grasses predominate, so that this region is known as the **tall-grass prairie** (Fig. 4.2). The leafy stems of many of these bunch grasses grow to heights of 1–2 m, and their flowering stalks even higher. The tall-grass prairie is a fire-dependent ecosystem, originally having been maintained and stimulated by frequent fire (Collins and Wallace 1990). Farther to the west and north lies the **mid-grass prairie,** where the dominant species tend to be of intermediate height. The dry western plains are home to the **short-grass prairie,** dominated by low-growing species of bunch grasses. Both the species of grasses and the luxuriance of growth of individual species change along this gradient.

Grasslands also occur on the Columbia Plateau of eastern Washington, Oregon, and western Idaho, a region known as the **palouse prairie,** and in the **shrub steppe** region of the northern Great Basin. In California, **valley grassland** originally covered most of the central valley and the coastal zone in the south. In these regions, most precipitation comes in the winter. Finally, in southern Arizona, southern New Mexico, and northern Mexico lies a region of **desert grassland,** again with a combination of winter and summer rainfall, but dry conditions in spring and fall.

Similar grasslands exist on other continents. These include the steppes of Asia, the pampas of southern South America, the grassveld of southern Africa, and grasslands of southern Australia. These areas share basic features of grassland ecology, although the species that make up these grassland ecosystems are almost entirely different (Fig. 4.3).

FIGURE 4.1

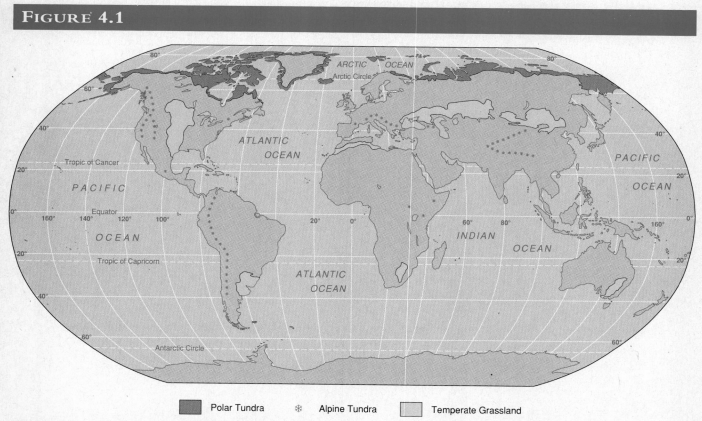

Polar Tundra   ❋ Alpine Tundra   Temperate Grassland

■ The distribution of temperate grasslands and arctic and alpine tundra in North America and other regions of the world.

FIGURE 4.2

■ The tall-grass prairies of the eastern and southern Great Plains were dominated by bunch grasses with flowering stalks that reached more than two m in height.

FIGURE 4.3

■ The temperate grasslands of South America possessed a distinctive fauna of large herbivores, such as guanacos and rheas.

Tundra ecosystems occur where severe cold, strong wind, and in some cases a permanently frozen subsoil zone, known as **permafrost,** combine to exclude large woody plants (Billings 1973). **Arctic tundra** (polar tundra) occurs throughout the lowlands and hills of northernmost North America and Eurasia (Fig. 4.1). Arctic tundra areas experience a very short summer with long days, and a long, cold, dark winter. Permafrost is a common feature of arctic tundra areas, contributing to cold, wet soil conditions and poor drainage. Ecosystems similar to those of arctic tundra also exist at high latitudes in some locations in the southern hemisphere. **Alpine tundra** occurs at high elevations in mountains at all latitudes. Day length and seasonal change in temperature and precipi-

FIGURE 4.4

■ The Afro-alpine vegetation of Mt. Elgon in western Kenya is dominated by tussock grasses and giant groundsels of the genus *Senecio,* of tropical affinity, rather than by arctic plants.

tation vary with latitude and other factors, and are often quite different from conditions of the arctic tundra. Permafrost is rare in alpine areas, and wind, rather than cold, is sometimes the critical factor that excludes larger woody plants.

Arctic tundra vegetation is dominated by perennial herbaceous plants such as sedges, low-growing forbs, and sprawling or creeping woody plants. In the northern hemisphere, the organisms inhabiting alpine tundra areas are often closely related to those of the arctic tundra, but in high mountain areas of the tropics and southern hemisphere, they are derived mostly from lower elevation groups of organisms characteristic of those regions (Fig. 4.4).

POPULATION CYCLES IN GRASSLANDS AND TUNDRA

In grassland and tundra ecosystems, plant production is restricted to part of the year, usually less than half, by highly seasonal temperatures or rainfall. In arctic tundra areas, the growing season is often less than three months. In both ecosystems, grazing or browsing herbivores, ranging in size from meadow voles to bison and caribou, are prominent, and their feeding activities are year-round (Fig. 4.5), as are those of their important predators. Climatic variation from year to year is also great. As Oksanen (1990) has shown, the strong seasonality of primary production predisposes these ecosystems to major fluctuations of herbivore populations. One of the striking phenomena of tundra ecology, shown also to a significant degree in grasslands, is "cycles" of abundance of many vertebrate herbivores and their predators. These fluctuations have attracted the attention of ecologists for more than a century, and several distinct theories of their cause have been developed.

Most ecologists recognize two types of cycles, differing in the average interval between peak populations. In North

FIGURE 4.5

■ Barren-ground caribou are the most common large herbivore of the arctic tundra of North America.

America, cycles of three to four years between peaks are shown by various species of lemmings and meadow voles (a group of mammals collectively known as *microtine* rodents), together with predators such as the arctic fox and the snowy owl that feed on such rodents. Cycles of meadow voles of the genus *Microtus* occur in many grassland areas as well as in the arctic tundra. In Eurasia, some kinds of grouse and ptarmigan are also reported to show cycles of three to four years duration. Cycles with a period of nine to ten years are also shown in North America by northern hares, particularly the snowshoe hare, and several species of grouse and ptarmigan, as well as by their predators, particularly the Canadian lynx and red fox (Keith 1963). In northern grassland regions, species such as the sharp-tailed grouse and the black-tailed jackrabbit show nine to ten-year cycles (Keith 1963, Anderson and Shumar 1986). Vertebrate population cycles of nine to ten years duration are unknown in the Old World.

Larger tundra animals may show even longer cycles, although this is somewhat speculative. Caribou, for example, were abundant in Alaska in the mid-1800s, when early prospectors and settlers first appeared in numbers. In the 1880s the populations of caribou declined. By the 1930s, however, caribou numbers had recovered, but once again, in the 1970s, serious declines occurred. In western Alaska, for example, the caribou herd that migrates between a summer range on the arctic slope and a winter range south of the Brooks Range declined from 242,000 in 1970 to 75,000 in 1976. In the 1980s Alaskan caribou populations showed substantial recovery, with the western Alaska herd reaching 230,000 by 1986 (Morgan 1988). Whether or not caribou show true cycles is uncertain, however, due to the small number of peaks and lows that have been observed and the poor quality of information prior to the 1930s.

Some biologists doubt that fluctuations of smaller species are true cycles, with regularity of period and amplitude.

Some of these populations of arctic lemmings, in fact, may show little more than chaotic fluctuations of numbers (Oksanen 1990). Nevertheless, these fluctuations are real, and are an important aspect of tundra and grassland ecology.

Many theories about the cause of three to four and nine to ten-year cycles have been proposed, and several of these have effectively been discarded by ecologists. Some of the latter include cycles of favorable and unfavorable weather, periodic episodes of mass mortality caused by stress-induced failure of the endocrine system, and shifting selective pressures that periodically allow genotypes vulnerable to catastrophic mortality to become predominant in the population. Still another theory is that cycles are simply a statistical result of factors acting randomly on populations, together with the fact that populations in one year have a biological (and mathematical) correlation with those of the next (being the next year's breeding unit). Most biologists have concluded that the regularities of the three to four and nine to ten-year fluctuations are greater than can be accounted for by this last hypothesis, however.

Three major theories of three to four and nine to ten-year vertebrate population cycles are now active. The first, which can be termed the **herbivory hypothesis,** suggests that the interaction between herbivores and their plant foods is the determining relationship, and that populations of predators are simply following the abundance of their prey. Lemmings, northern hares, and caribou are active year-round, but their food plants grow for only about three months. In temperate grasslands, plants grow only during periods of warmth and moisture, again only a few months of the year. Thus, if herbivore populations reach high levels, they can overexploit plant foods for a long time before recovery can even begin. Oksanen (1990) has suggested that this general mechanism accounts for population fluctuations of herbivores with short generation times in environments of low primary productivity, such as high arctic tundra.

For three to four-year cycles of lemmings and their predators, for example, many ecologists favor the **nutrient recovery hypothesis,** a specific version of the herbivory hypothesis. This hypothesis proposes that as the lemming population grows and feeds on the tundra vegetation, the nutrients vital to basic plant productivity increasingly become tied up in lemming (and other animal) tissues and wastes. The tie-up results simply from the slowness of decomposition in the arctic environment. At some point, depletion of the preferred food plants, in both quantity and nutrient quality, leaves the lemming population with inadequate food. Starvation, predation, and death by exposure lead to a population crash. According to this hypothesis, populations of predators mainly track those of their prey, the assumption being that the lemmings would cycle whether or not the predators were present.

A version of the herbivory hypothesis also has been applied to the nine to ten-year cycles of snowshoe hares and lynx. Fox and Bryant (1984) noted that during the winter, hares feed primarily on the shoots of shrubs and small trees

TABLE 4.1

Major Categories of Human Impact on Grassland and Tundra Ecosystems.

	GRASSLANDS	TUNDRAS
Conversion to Cropland	Extensive	Negligible
Grazing by Domestic Animals	Extensive	Slight
Physical Disturbance of Soil	Severe	Locally Severe
Change in Fire Regime	Reduced Frequency	Increased Frequency
Introduction of Exotic Plants	Extensive	Slight
Hunting and Animal Control	Extensive	Extensive

such as willows, birches, alders, and poplars. They postulated that when hare populations are low, the light browsing in many cases simply stimulates these plants to produce more shoots, increasing the supply of food for hares the next winter. Later, as the intensity of browsing by hares increases, some of these plants undergo a physiological shift, and begin to produce shoots that are defended against browsing by dense pubescence (plant hairs), surface resins, or distasteful chemicals. These defenses reduce the palatability and digestibility of the shoots to the point that the animals die if forced to feed only on such materials. At some point, therefore, growing populations of snowshoe hares may exhaust the supply of palatable food plants, triggering population crashes due to starvation, exposure, and inability of the animals to escape predators. Once developed, the plant defenses are retained for two to three years. This retention may create a lag in the recovery of hare populations, and this lag, together with the population growth capacity of the hare translates into the nine to ten-year periodicity of the cycle. Under this hypothesis, populations of lynx are simply tracking the abundance of their prey, and contributing to a crash that would occur regardless of their presence or absence.

Other biologists support the **food supply-predation hypothesis** as the cause of these cycles. Keith (1981), for example, proposed that populations of snowshoe hares grow until they are halted by shortage of winter food. Populations of predators, which have also been growing because of the increased availability of snowshoe hares, are then able to catch up with the hare population. Starvation of hares and predation then reduce hare numbers to levels that permit plant recovery. To test this hypothesis, Keith et al. (1984) studied the ecology of snowshoe hares during the decline phase of a population cycle in Alberta, Canada. This study showed that plant foods were in short supply, and that hares showed clear signs of malnutrition. Actual death of hares by starvation probably occurred only for a short time after the population had peaked, however, and the actual cause of death in 80 to 90 percent of all cases was predation. Keith et al. (1984) interpreted this to support the food supply–predation hypothesis, since both malnutrition and predation interacted in the decline. Theoretical analyses by Oksanen (1990) suggest that truly cyclical fluctuations of herbivores may occur by this mechanism in

productive, highly seasonal environments where specialized predator-prey interactions exist.

Still other biologists favor the **predation hypothesis,** which suggests that food shortage rarely affects herbivores such as lemmings and snowshoe hares, and that the basic cause of cycles is predation. In other words, without the action of predators, no cycles would occur. Trostel et al. (1987), for example, supported this hypothesis for snowshoe hare cycles. Long-term experimental studies in the Yukon Territory of Canada, designed to manipulate food availability and predation intensity, suggest that several relationships may be at work. These studies suggests that relative shortages of food may occur, predisposing hares to predation by forcing them to move about more in search of food (Smith et al. 1988). Responses of shrub defenses to hare browsing do not seem to correlate with the population cycle, however, and crashes occur even when hare populations receive supplemental food, suggesting that predation is the key cause (Sinclair et al. 1988).

The fluctuations of populations of large animals, such as caribou, are likely to be the result of more complex interactions, but they may also be a natural phenomenon in tundra and other northern ecosystems. As suggested by the difference in cycle length for lemming and snowshoe hares, the size of the herbivore may be a strong determinant of the length of a population cycle. For mammals, it has been hypothesized that cycle length is approximately equal to 8.15 times the body mass (in kg) to the exponent 0.26 (Peterson et al. 1984). For an animal of 100 kg, about the size of a caribou, this suggests a cycle length of 27 years, or possibly longer.

HUMAN IMPACTS ON GRASSLAND ECOSYSTEMS

Humans have made an impact on grassland ecosystems in several different ways, most of them directly or indirectly related to ranching and farming (Table 4.1). In North America, most of the tall-grass prairie and much of the palouse prairie have been converted to rain-fed or irrigated farming. In Illinois, for example, only fragments of the original grasslands now exist. Surviving remnants tend to occur on the thin soils of bluffs (hill prairies), along railroad rights of way (cinder prairies), and in cemeteries. Elsewhere in North America, particularly in central California, most of the original grasslands have been replaced by irrigated farming operations. In North America, the conversion of grasslands to croplands began in the early 1800s. In some other world regions, this conversion has been less extensive and more recent. In the Argentinean pampas, for example, the conversion has accelerated greatly during the 1980s and 1990s.

In North America, livestock grazing preceded intensive farming in many grassland regions, such as the western Great Plains, the Columbia Plateau, the California grassland, and the desert grassland. The grazing patterns of cattle, sheep, and other livestock differ from those of native large herbivores, such as bison, pronghorn antelope, and elk. The overall grazing pressure under livestock tends to be more selective. As a result, many of the preferred species of plants decrease in abundance, and the non-preferred species increase. Range managers term these kinds of plants **decreasers** and **increasers.** Plants that increase under heavy grazing include woody species such as mesquite, juniper, and many species of shrubs and cacti. In some regions, such as the palouse prairies of the Columbia Plateau, hoofed animals were originally low in abundance, and the dominant grasses were not adapted to heavy trampling, as they appear to be in the Great Plains (Mack and Thompson 1982). With the introduction of livestock to the Columbia Plateau, the native perennial grasses have largely disappeared in many areas.

In addition, many kinds of weedy plants, especially species from the Old World that evolved in close association with early agriculture over the past 12,000 years or so, are able to invade disturbed and overgrazed grasslands. Range managers call these kinds of non-native species **invaders.** On the Columbia Plateau and neighboring regions, cheatgrass, *Bromus tectorum,* was introduced accidentally in the late 1800s, and spread through the region by 1928 (Mack 1981). This aggressive annual grass, native to Eurasia, has replaced native grassland species over vast areas. In California, the original valley grassland was so quickly and so completely replaced by Mediterranean annual grasses and forbs that its original composition is very uncertain.

Grassland areas are also subject to extended droughts. During such periods a major shift in the dominant species of grasses can occur, with species adapted to lower rainfall becoming more abundant. Heavy grazing under such conditions can severely damage rangeland, as demonstrated during the droughts of the 1930s and 1950s in the Great Plains (Albertson et al. 1957). During these dry years, heavily grazed ranges experienced a much higher decline in total plant cover, and a much greater invasion by exotic annual weeds. Whereas well-managed ranges were able to recover from drought in about five years, heavily grazed areas required 20 years or more to recover.

Overgrazing may lead to major changes in grassland animal life, some of which may further promote degradation. Plant species typical of disturbed and early successional habitats tend to be more vulnerable to herbivores in general (Cates and Orians 1975), suggesting that outbreaks of insect and small herbivorous mammals in grasslands may be due, at least in part, to disturbance. In areas of tall-grass prairie, Koford (1958) noted, for example, that prairie dog towns almost never occurred on rangeland in good to excellent condition. In short-grass prairie, on the other hand, Hansen and Gold (1977), found that blacktail prairie dogs tended to reduce primary productivity and encourage the invasion of desert cottontails. Prairie dogs and desert cottontails consumed more than 20 percent of the plant productivity that did occur.

The original role of fire has also been modified in most grassland regions. As the land was settled, the common Indian practice of burning grasslands was eliminated, and natural wildfires were deliberately controlled. In many locations, the reduced frequency of fire favored the invasion of grasslands by woody species, to the extent that many former grasslands have now become forests, shrublands, or desert scrub. In southern Arizona, for example, the introduction of livestock grazing and the reduction of grassland fire has led to the invasion of much of the original desert grassland by mesquite, cholla cactus, and creosote bush (Humphrey and Mehrhoff 1958).

Finally, grassland areas have been the sites of some of the most intensive hunting and deliberate control of animal populations. In the 1800s, the large native grazing animals of the Great Plains, such as bison, elk, and pronghorn, were almost completely eliminated by hunting. The original bison population of North America is estimated to have been 30 to 60 million; all but a few hundred were killed for their skins, meat, or simply to destroy the basic food on which groups of Plains Indians depended. More recent decades have seen massive campaigns to control populations of prairie dogs (Fig. 4.6), jackrabbits, and coyotes. Widespread control of prairie dogs was a major cause of the decline of the black-footed ferret (Fig. 4.7) to the edge of extinction. The last known wild population of this species lived near Meeteetse, Wyoming, where about 129 individuals were present in 1984. The devastation of the local prairie dog population by plague, and the resulting crash of the ferret population, forced the remaining animals to be brought into captive breeding (See chapter 25).

HUMAN IMPACTS ON TUNDRA ECOSYSTEMS

Tundra ecosystems have experienced some of the same kinds of impacts. Agricultural activities, however, have been a very minor cause of disturbance in tundra ecosystems. Crop cultivation is negligible in tundra regions. In northern Scandinavia, semi-domesticated reindeer are herded in tundra areas. Local efforts have been made to introduce reindeer herding to North America, and to develop ways of managing herds of musk oxen on tundra range. Some summer grazing of livestock also occurs in alpine tundra areas in the Temperate Zone.

Severe local impacts have resulted from disturbance of the tundra surface, both in arctic and alpine areas. Studies of

FIGURE 4.6

■ Control programs for prairie dogs have greatly reduced their numbers, and some populations of the white-tailed prairie dog are now formally designated as threatened.

FIGURE 4.7

■ The black-footed ferret, a specialist predator on prairie dogs, has been reduced to the brink of extinction by control of prairie dog populations on rangeland in western North America.

the impacts of petroleum exploration activities show that the arctic tundra is sensitive to disturbance and very slow to heal. On the Alaskan north slope, recolonization of disturbed areas by native species is relatively rapid if the peaty surface layer of the soil remains intact, with 90 to 100 percent plant cover often being achieved in 5 to 10 years (Van Cleave 1977). Where the organic surface layer is destroyed, succession may be much slower. In sites heavily disturbed by an exploratory drilling operation in 1949, effects were still severe 28 years later (Lawson et al. 1978). Where petroleum spills had occurred, or where increased thawing of the permafrost had been triggered by disturbance, very little recovery had occurred. In some places thawing of the permafrost led to surface subsidence and extensive erosion. In general, where disturbance of the tundra surface has led to permafrost melting, more than 30 years are required to restore surface stability (Lawson 1986). Even seismic surveys carried out in winter, when the tundra surface is frozen and snow-covered, crush and compact the vegetated surface, often inducing permafrost melting and subsidence (Felix and Reynolds 1989). Again, these effects are slow to heal (Fig. 4.8). These effects suggest that major environmental damage may accompany the proposed exploitation of oil and gas reserves in the Arctic National Wildlife Refuge (Laycock 1988).

Alpine tundra is likewise sensitive to physical impacts. Trampling by visitors on foot in alpine areas in the Rocky Mountains can cause severe impacts on the vegetation (Willard and Marr 1970, Billings 1973). Where foot traffic becomes concentrated along regular routes, tundra plants quickly die, and the soil surface begins to erode.

Fire in the arctic tundra seems to have been increased by human activities, contrary to the pattern in temperate grasslands. Although it might seem that cold, wet conditions

FIGURE 4.8

■ The passage of vehicles conducting winter seismic surveys for evaluation of petroleum reserves on the Arctic National Wildlife Refuge creates trails that require many years for natural processes to heal.

CHAPTER FOUR GRASSLANDS AND TUNDRA

in the arctic would make the vegetation hard to burn, the truth is that many tundra plants possess large quantities of resinous materials on their surfaces, making them highly flammable.

Hunting of animals such as caribou, and control of populations of predators such as the wolf, are additional human impacts of significance. The declines of caribou populations in the 1970s raised major conservation concern. This was a period of cultural and economic change in many of the native villages of arctic Alaska and Canada. The winter hunting capabilities of Inuit and Indian inhabitants of villages that depended heavily on wild game had increased considerably by the introduction of snowmobiles and better high-powered rifles, and some instances of wasteful killing occurred. Tundra fires, caused both by lightning and by human activity, were more frequent and extensive than usual, too. Although fire may be beneficial in the long term, by increasing habitat diversity along the border of forest and tundra, its short-term effect is the destruction of ground lichens and other caribou browse (Klein 1982). With decline in caribou numbers, as well, predation by wolves and other predators became relatively greater in importance (See chapter 9). Beyond these factors, however, the high caribou populations that built up between the 1930s and 1960s may also have caused deterioration of range quality, somewhat in the manner postulated for lemmings and snowshoe hares.

GRASSLAND AND TUNDRA RESTORATION

Some of the earliest ecosystem restoration efforts were directed at recreating examples of tall-grass prairie in the midwestern United States, where these ecosystems had been almost completely destroyed (Kline and Howell 1987). In several locations, tall-grass prairies have been recreated on sites from which all native prairie species had been eliminated by farming activities. The techniques used in restoration have been derived largely from agronomy. Herbicides and mechanical cultivation have usually been used to eliminate existing non-prairie plants. Dominant prairie grasses and forbs are then seeded into the cultivated soil, or transplanted individually onto the site.

Once a healthy stand of prairie dominants is created, efforts turn to adding typical, but less common, prairie species and to eliminating exotic herbaceous plants and native shrubs and trees that often tend to invade aggressively. Prescribed burning is often effective in reducing these non-prairie species (Fig. 4.9), but complete elimination of exotics is difficult to achieve. This is particularly true on small prairie plots where the use of fire is often complicated by restrictions on weather conditions under which burning can be done, and where seed

FIGURE 4.9

■ Prescribed burning is used to reestablish the natural role of fire in tall-grass prairie on the Konza Prairie Research Natural Area in the Flint Hills of northeastern Kansas. Fires are ignited around the periphery of a designated burn unit, and extinguish themselves when they reach the center.

sources for exotics may be abundant in neighboring areas. Most of these restoration efforts have been limited to recreation of the plant community, with the assumption that success with plants will lead to colonization by the prairie animals that can exist on areas of the size in question.

At the University of Wisconsin, prairies 24 and 16 ha in area have been created, the first efforts beginning in 1934 (Cottam 1987). These sites now contain over 300 species of native prairie plants. Periodic observations have shown that major changes in composition patterns occur as newly introduced native species find the microhabitats that are optimal for them, and as short-term climatic cycles influence microhabitat conditions within the landscape.

In a few locations, such as Grasslands National Park in Saskatchewan (McCrea 1981), Konza Prairie Research Natural Area in Kansas (Reichman 1987), the Niobrara Valley Preserve in Nebraska (Harrison 1980), and the Tallgrass Prairie Preserve under development by the Nature Conservancy in Oklahoma (Madson 1990), larger areas of native grasslands are being restored. These areas have been used for livestock grazing but not extensively disturbed by cultivation. In these areas, the objectives are to establish prairie ecosystems containing the full complement of native animals, such as bison, pronghorn antelope, and elk (Fig. 4.10).

Little success has been achieved in restoration of tundra ecosystems. In the arctic, deliberate efforts to revegetate disturbed sites, such as the thick gravel pads on which roads, camps, and work sites are placed in order to prevent melting

FIGURE 4.10

■ Bison, the dominant original herbivore of Great Plains grasslands, are being reintroduced from areas such as the Wichita Mountains National Wildlife Refuge to the Tallgrass Prairie Preserve in the Osage region of Oklahoma and Konza Prairie in the Flint Hills of Kansas.

of the underlying permafrost, have not been very successful. Seeding such areas with non-native perennial grasses can create a temporary cover that helps stabilize the disturbed soil surface (Van Cleave 1977), but colonization by native species is very slow (Bishop and Chapin 1989a). Current efforts involve planting native species, together with fertilization of the site and its surroundings, to encourage natural succession to occur as fast as possible (Bishop and Chapin 1989a, b).

In alpine tundra areas restoration is somewhat more successful. A number of alpine grasses and a few forbs can be seeded onto disturbed areas, and many others can be reintroduced by transplantation (Brown et al. 1978). Seeding or transplanting is usually done in the late fall so that the species are able to begin growth as early the next spring as possible. Mulching and fertilization can aid the establishment of such

species, as well. When fertilization is stopped, however, such plantings often tend to decline, indicating that restoration must consider ecological processes beyond those of initial establishment.

KEY MANAGEMENT STRATEGIES FOR GRASSLAND AND TUNDRA ECOSYSTEMS

1. Maintain grazing and browsing pressure of livestock and wildlife within the capacity of production of the preferred food plants.

2. Maintain a fire regime that favors native species and maintains an optimal combination of productivity and habitat diversity.

Literature Cited

Albertson, F. W., G. W. Tomanek, and A. Reigel. 1957. Ecology of drought cycles and grazing intensity of grasslands of central Great Plains. *Ecol. Monogr.* **27:**27–44.

Anderson, J. E. and M. L. Shumar. 1986. Impacts of black-tailed jackrabbits at peak population densities on sagebrush-steppe vegetation. *J. Range Manag.* **39:**152–156.

Billings, W. D. 1973. Arctic and alpine vegetations: Similarities, differences, and susceptibility to disturbance. *BioScience* **23:**697–704.

Bishop, S. C. and F. S. Chapin III. 1989*a*. Patterns of natural revegetation on abandoned gravel pads in arctic Alaska. *J. Ecol.* **26:**1073–1081.

Bishop, S. C. and F. S. Chapin III. 1989*b*. Establishment of *Salix alaxensis* on a gravel pad in arctic Alaska. *J. Ecol.* **26:**575–583.

Brown, R. W., R. S. Johnston, and D. A. Johnson. 1978. Rehabilitation of alpine tundra disturbances. *J. Soil Water Cons.* **33:**154–160.

Cates, R. G. and G. H. Orians. 1975. Successional status and the palatability of plants to generalized herbivores. *Ecology* **56:**410–418.

Collins, S. L. and L. L. Wallace (Eds.). 1990. *Fire in North American tallgrass prairies*. Univ. Oklahoma Press, Norman.

Cottam, G. 1987. Community dynamics on an artificial prairie. Pp. 257–270 *in* W. R. Jordan III, M. E. Gilpin, and J. D. Aber (Eds.), *Restoration ecology: A synthetic approach to ecological research*. Cambridge Univ. Press, New York.

Felix, N. A. and M. K. Reynolds. 1989. The effects of winter seismic trails on tundra vegetation in northeastern Alaska, U.S.A. *Arctic Alpine Res.* **21:**188–202.

Fox, J. F. and J. P. Bryant. 1984. Instability of the snowshoe hare and woody plant interaction. *Oecologia* **63:**128–135.

Hansen, R. M. and I. K. Gold. 1977. Blacktail prairie dogs, desert cottontails, and cattle trophic relations on shortgrass range. *J Range Manag.* **30:**210–214.

Harrison, A. T. 1980. *The Niobrara Valley Preserve: Its biogeographical importance and description of its biotic communities*. The Nature Conservancy, Minneapolis, MN.

Humphrey, R. R. and L. A. Mehrhoff. 1958. Vegetation changes on a southern Arizona grassland range. *Ecology* **39:**720–726.

Keith, L. B. 1963. *Wildlife's ten-year cycle*. Univ. Wisc. Press, Madison.

Keith, L. B. 1981. The role of food in hare population cycles. *Oikos* **40:**385–395.

Keith, L. B., J. R. Cary, O. J. Rongstad, and M. C. Brittingham. 1984. Demography and ecology of a declining snowshoe hare population. *Wildl. Monogr.* **90:**1–43.

Klein, D. R. 1982. Fire, lichens, and caribou. *J. Range Manag.* **35:**390–395.

Kline, V. M. and E. A. Howell. 1987. Prairies. Pp. 75–83 *in* W. R. Jordan III, M. E. Gilpin, and J. D. Aber (Eds.), *Restoration ecology: A synthetic approach to ecological research*. Cambridge Univ. Press, New York.

Koford, C. B. 1958. Prairie dogs, whitefaces, and blue grama. *Wildl. Monogr.* **3:**1–78.

Lawson, D. E. 1986. Response of permafrost terrain to disturbance: A synthesis of observations from northern Alaska, U.S.A. *Arctic Alpine Res.* **18:**1–17.

Lawson, D. E., J. Brown, K. R. Everett, A. W. Johnson, V. Komarkova, B. M. Murray, D. F. Murray, and P. J. Webber. 1978. Tundra disturbance and recovery following the 1949 exploratory drilling, Fish Creek, northern Alaska. *U.S. Army Cold Regions Res. Eng. Lab. Rep.* 78–28.

Laycock, G. 1988. Wilderness by the barrel. *Audubon* **90(3):**100–123.

Mack, R. N. 1981. Invasion of *Bromus tectorum* L. into western North America: an ecological chronicle. *Agro-Ecosystems* **7:**145–165.

Mack, R. N. 1981. Invasion of *Bromus tectorum* L. into western North America: An ecological chronicle. *Agro-Ecosystems* **7:**145–165.

Madson, J. 1990. On the Osage. *Nature Conservancy* **40(3):**6–15.

McCrea, J. 1981. Canada's new Grasslands National Park. *Parks* **6(3):**15–16.

Morgan, S. O. (Ed.). 1988. *Caribou*. Alaska Dept. Fish and Game, Juneau, AK.

Oksanen, L. 1990. Exploitation ecosystems in seasonal environments. *Oikos* **57:**14–24.

Peterson, R. O., R. E. Page, and K. M. Dodge. 1984. Wolves, moose, and the allometry of population cycles. *Science* **224:**1350–1352.

Reichman, O. J. 1987. *Konza Prairie: A tallgrass natural history*. Univ. Press of Kansas, Lawrence, KS.

Smith, J. N. M., C. J. Krebs, A. R. E. Sinclair, and R. Boonstra. 1988. Population biology of snowshoe hares. II. Interactions with winter food plants. *J. Anim. Ecol.* **57:**269–286.

Trostel, K., A. R. E. Sinclair, C. J. Walters, and C. J. Krebs. 1987. Can predation cause the 10-year hare cycle? *Oecologia* **74:**185–192.

Van Cleave, K. 1977. Recovery of disturbed tundra and taiga surfaces in Alaska. Pp. 422–455 *in* J. Cairns, Jr., K. L. Dickson, and E. E. Herricks (Eds.), *Recovery and restoration of damaged ecosystems*. Univ. of Virginia Press, Charlottesville.

Willard, B. E. and J. W. Marr. 1970. Effects of human activities on alpine tundra ecosystems in Rocky Mountain National Park, Colorado. *Biol. Cons.* **2:**257–265.

CHAPTER 5

Deserts

Desert ecosystems exhibit perhaps the most challenging combination of physical conditions in the terrestrial environment: extremes of both temperature and moisture. In many deserts, temperatures can reach extremes at either end of the scale, and although long periods of drought are frequent, cloudbursts and violent floods can also occur. Despite these challenges, the diversity of plant and animal life in deserts is great, and patterns of adaptation to conditions of the physical environment are more strikingly developed than in any other terrestrial ecosystem.

Despite the intrinsic stability of desert ecosystems in the face of climatic variability—their ability to tolerate extremes of temperature, moisture, and wind—these systems are easily damaged by factors that disturb the soil surface. After such disturbance, furthermore, they show low resilience. Human impacts have degraded large areas of desert, and such areas may take decades or centuries to recover, even when protected (Webb et al. 1983). Restoration of disturbed areas is very difficult, and usually requires special techniques to create favorable microsites for plant growth and to provide supplemental moisture until perennial plants become fully established (Bainbridge and Virginia 1990). Because of the low productivity of desert environments and the sparse populations of plants and animals, large areas of land must be protected to guarantee the survival of desert forms (Louw and Seely 1982). Not surprisingly, some of the most serious problems of conservation ecology relate to desert ecosystems.

ORIGIN AND DISTRIBUTION OF DESERTS

Deserts are ecosystems with an annual precipitation of less than 25 cm, coupled with an evaporation potential from the soil surface and from plants well in excess of precipitation. Furthermore, the amount of precipitation is highly variable from year to year. As in grasslands, the precipitation that does occur tends to be highly seasonal. The wet season varies greatly in length and intensity in different desert areas. In some cases, rain falls mainly in winter, in other cases largely in summer; still other deserts receive a little precipitation in both seasons. In some of the most extreme deserts, rains may fall only at intervals of several years, and seasonality becomes irrelevant. Temperature, too, varies among deserts, and ecologists commonly distinguish **hot deserts,** where freezing conditions are rare, from **cool deserts,** where winter temperatures regularly fall below freezing. **Cold deserts** also occur at very high latitudes (polar deserts) and at high altitudes, where precipitation is low and sub-freezing conditions further limit water availability.

Three kinds of relationships contribute to the occurrence of deserts: 1) prevailing high atmospheric pressure, 2) rain shadows, and 3) cold coastal waters. At latitudes of 25 to 30 degrees N and S, subtropical high pressure patterns of the earth's atmospheric circulation system dominate most of the year. In these regions, in the upper atmosphere, air flows converge from higher and lower latitudes and the air sinks downward toward the earth's surface. As it sinks, it warms by compression and declines in relative humidity—the actual amount of moisture in the air stays constant, but the capacity of the air to hold moisture increases. At the surface, the air is almost always hot and dry, and only very unusual weather systems disturb this pressure belt and bring rain. An analogous system of high pressure dominates high arctic latitudes, creating desert-like patterns of low precipitation. The second mechanism operates where weather systems cross major mountain ranges. On the windward slope of the mountains, air is forced upward in elevation, and the expansion and cooling of the air leads to condensation of moisture, and thus heavy rains or snows. When the air has crossed the mountains, it descends in elevation, warming and decreasing in relative humidity. This **rain shadow** is usually severe enough to create desert conditions. Finally, where cold ocean currents exist along the western edge of continents, weather systems moving across these waters tend to cool and lose their moisture over the water. When they move onto the warmer land, however, the air warms and declines in relative humidity, thus creating dry conditions.

These mechanisms of aridity have produced a complex distribution of deserts, which cover about 12 percent of the continents (Fig. 5.1). In North America, the belt of subtropical high pressure contributes to the desert regions of northern Mexico and the southwestern United States—the more southern portions of the Chihuahuan and Sonoran Deserts lie in this latitudinal zone. These are hot deserts, with the Chihuahuan experiencing most of its precipitation in the summer, and the Sonoran Desert (Fig. 5.2) experiencing either winter

FIGURE 5.1

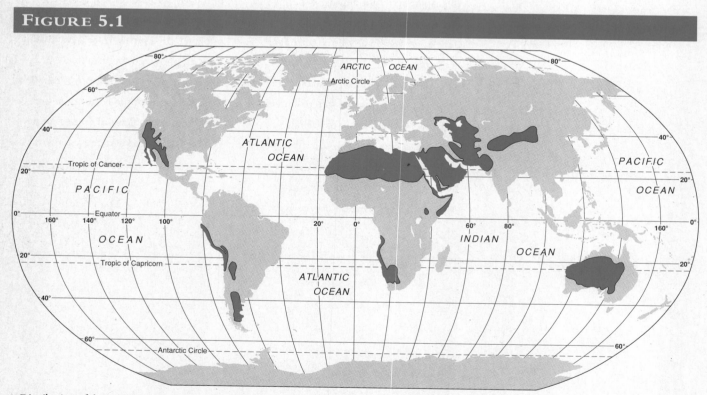

■ Distribution of deserts in North America and other world regions.

FIGURE 5.2

■ The Sonoran Desert receives a combination of both summer and winter rain, and has one of the most diverse biotas of any desert region in the world. This scene is in spring in Organ Pipe Cactus National Monument, Arizona.

FIGURE 5.3

■ In the Namib Desert of coastal Namibia (southwest coast of Africa), fog formed over the cold Benguela Current offshore often blows inland at night, depositing moisture that is important to the plant and animal life of this nearly rainless desert region.

FIGURE 5.4

■ Major features of desert dune fields, such as the major sand ridges of the Namib Desert, are stable, although their surface details are reworked constantly by wind.

precipitation or a combination of winter and summer precipitation. The belt of subtropical high pressure is also responsible for the vast deserts of North Africa and the Middle East, together with the interior deserts of Australia.

The more northern parts of the Sonoran Desert, together with the Mojave and Great Basin Deserts, on the other hand, lie largely in the rain shadow of the Sierra Nevadas, the Cascade Mountains, and the interior ranges of Nevada and Utah. Although the Mojave Desert is a hot desert and the Great Basin Desert is a cool desert, both experience primarily winter precipitation. The desert regions of Argentina and inner Asia are also examples of rain shadow deserts.

Cold upwelling oceans bordering the west coast of continents create the deserts of the Pacific Coast of Baja California in North America, the Atacama Desert of coastal Chile and Peru, and the Namib Desert of southern Africa (Fig. 5.3). In these deserts, much of the moisture on which plant and animal life depends comes in the form of fog that blows inland, mostly at night, from the cold ocean waters. In these deserts one can find the seemingly contradictory association of cacti bearing epiphytic lichens and bromeliads, both largely dependent on moisture deposited by fog.

SUBSTRATE STABILITY IN DESERT ECOSYSTEMS

Substrate conditions, both physical and chemical, are a key determinant of the structure and function of desert ecosystems. Rocky hillslopes, eroded badlands, sandy washes, stony alluvial fans, fine-textured alluvial and lake bed deposits, and active sand dunes are all common desert substrates. Extreme conditions of salinity or alkalinity may develop in desert basins or poorly drained desert lowlands. Chemical leaching of carbonates may carry these minerals downward in the soil to a certain depth, where they are deposited to form a cemented

horizon of material known as **caliche,** a formidable physical feature of the subsoil.

Despite the fact that thinly vegetated substrates of these varied types are exposed to fluctuating temperatures, frequent strong winds, and occasional violent rainstorms, desert landscapes undisturbed by humans develop a high degree of geomorphic stability (Wilshire 1983). Major sand dune systems, although continually being reshaped in detail by wind, are stable (Fig. 5.4). Their large-scale structure is in equilibrium with the long-term pattern of prevailing winds and surrounding topography. Some dune plants and animals are dependent on stability of the major dune masses and interdunal flats, others depend on the ephemeral microhabitats created by the small-scale reshaping of the sands by wind. Desert washes are similar in their structure, varying in the pattern of small channels that change with each storm flood, but maintaining a stable position on the landscape. The vegetation of the floor of washes is often destroyed by the scouring of intense floods, but is quickly reestablished by plants whose seeds are stimulated to germinate by the physical abrasion that occurs during the flood itself (Fig. 5.5).

Many desert soils also tend to develop stabilized surface layers or crusts that reduce the vulnerability of the soil to wind and water erosion (Wilshire 1983). The fine roots of trees, shrubs, and herbaceous plants form a shallow, soil-holding network below the surface. In many places, removal of fine soil by wind erosion over thousands of years has concentrated pebbles and stones to form a **desert pavement**—a dense surface layer that protects the underlying soil from further erosion. More delicate soil crusts are also formed by several processes. Algae, fungi, and lichens can form a fine network of organic tissues at the soil surface. Silt and clay particles can form a physically cemented surface layer in some soils. Precipitated salts may also form a protective crust over the surfaces of playas and alluvial lowlands. All of these layers and crusts are highly vulnerable to physical disruption by hoofed animals and motor vehicles.

HUMAN IMPACTS AND DESERTIFICATION

The low primary productivity and thin plant cover make arid and semiarid lands especially vulnerable to two types of human impacts: overgrazing and physical disturbance of the soil surface. Heavy grazing by domestic animals such as cattle, sheep, goats and camels, and by feral animals such as burros, can severely reduce the total cover and diversity of plants in desert ecosystems. Thinning of the plant cover and disturbance of the soil surface by trampling increase the vulnerability of the soil to erosion by wind and water. These changes also increase the patchiness of soil moisture and nutrients, which favors desert shrubs over grasses (Schlesinger et al. 1990). Once established, desert shrubs tend to reinforce this patchiness of resources and suppress the recovery of semiarid communities such as desert grasslands. Other types of physical disturbance, such as cultivation, mining, and vehicular activity, cause even more serious degradation of desert landscapes.

A third major impact often occurs where irrigation is practiced. Desert soils, which lack a history of intense leaching by water, often contain large quantities of salts. Many desert regions also lack good drainage. Consequently, when irrigated farming is undertaken without the construction of adequate drainage systems, the result is often waterlogged and salinized soil. In many areas, this has forced once-productive irrigated lands to be abandoned. Thus, the original desert ecosystems and the productive agricultural systems that replaced them have both been lost.

The reduction of the productivity of arid and semiarid lands as a result of such impacts is known as **desertification** (Dregne 1983, Mabbutt 1984). The effects of desertification

FIGURE 5.5

■ Desert washes are stable in their general location, although individual channels are reworked by each flood.

are serious for both wildlife and human populations. In some severely desertified regions, such as the sub-Saharan zone of Africa known as the Sahel, famine has been a chronic problem since the late 1960s. Desertification has thus become a problem of major concern to international organizations, particularly the United Nations Environment Program (UNEP) and the United Nations Educational, Scientific, and Cultural Organization (UNESCO). World attention was focussed on this problem at the United Nations Conference on Desertification, held in 1977 in Nairobi, Kenya.

Desertification is not restricted to Africa, however, and recent surveys (Dregne 1983) indicate that almost 48 percent of the world's arid lands have suffered moderate to very severe desertification (Fig. 5.6). Slightly over 5 percent of this affected area consists of rain-fed cropland, whereas almost 94 percent consists of rangeland. About 77 percent of the total rain-fed cropland in arid regions has experienced at least a 10 percent loss of crop productivity and 82 percent of rangeland is in fair to worse condition. About 675 million people inhabit regions that have become moderately to severely desertified; of these about 450 million have probably suffered substantial impairment of their livelihood (Dregne 1983). Unfortunately, global climatic warming is likely to exacerbate the desertification problem, expanding the area of land with desert climates perhaps 17 percent with a doubling of atmospheric CO_2 (Emanuel et al. 1985).

Desertification is extensive even in North America. About 80 percent of the arid rangeland and 50 percent of the arid, rain-fed cropland in the western United States are considered by Dregne (1983) to be at least moderately desertified. Along the edges of the warmer southwestern deserts, large areas of productive grasslands have been replaced by stands of desert shrubs (Schlesinger et al. 1990). In some of the cooler deserts, overgrazing has led to extensive erosion by wind and water. The Navajo Indian Reservation of northern Arizona and New Mexico, in fact, is considered the largest area of severely desertified land in the world (Dregne 1983).

FIGURE 5.6

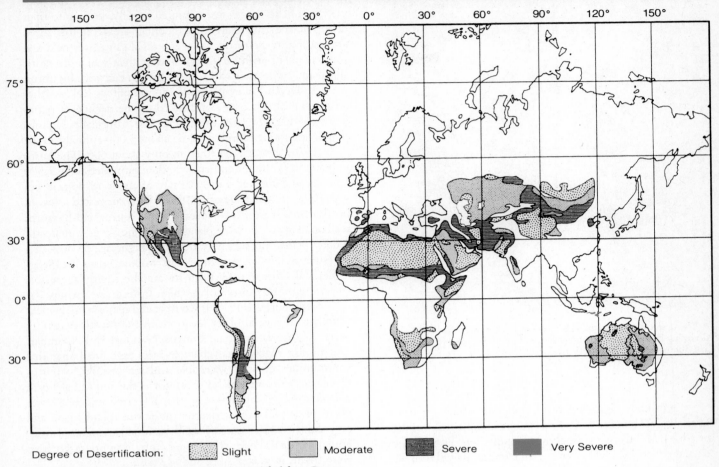

Degree of Desertification: Slight   Moderate   Severe   Very Severe

■ Extent of desertification of world arid lands (Modified from Dregne 1983).

CONSERVATION PROBLEMS IN NORTH AMERICAN DESERTS

Overgrazing by Livestock and Feral Ungulates

Livestock grazing affects desert ecosystems in several ways. Large animals such as cattle and burros impact these systems by grazing, trampling, dispersing weedy shrubs and cacti. In North America, these effects have contributed to the decline of several native desert animals.

In the Mojave Desert of California, for example, cattle grazing has been responsible for much of the decline of populations of the desert tortoise (Berry 1978, Campbell 1988). Other contributing factors have been vehicle mortality, predation on young tortoises by ravens, the shooting and removal of animals (for pets) by humans, and, most recently, a viral respiratory disease that may have been introduced into wild populations by release of pet animals.

Desert tortoises (Fig. 5.7) are long-lived animals, requiring 10 to 20 years to reach reproductive maturity and often living to an age of 50 or perhaps even 100 years. This tortoise, one of four species of tortoises in North America, occurs in scattered areas of the Mojave and Sonoran Deserts, including southeastern California, southwestern Arizona, southern Nevada, and extreme southwestern Utah. It is herbivorous, feeding on herbaceous plants that grow during the wet season and shrubs and cacti at other times. The home ranges of individuals vary in size, some as small as a few hectares, others more than 200 hectares in extent. In the warmer southern deserts, the animals use shallow burrows or depressions (pallets) built in shaded situations to avoid the extremes of heat and drought. In the cooler northern deserts, tortoises use shallow burrows in the summer, but construct deep winter dens in the banks of stream channels and similar sites. In the spring, mature females lay 2 to 14 eggs, which hatch after an incubation period of 90 to 120 days.

FIGURE 5.7

■ The desert tortoise of the southwestern United States has been declining in abundance due to a variety of impacts related to human activity in deserts.

Normally, desert tortoise populations show a sex ratio of about 150 females to 100 males, and about equal numbers of adults and juveniles. Declining populations, however, are characterized by fewer females and fewer juveniles. These patterns indicate high female mortality, and poor reproduction—both likely the result of inadequate nutrition of adult females. In many parts of the Mojave Desert, cattle are trucked in during the spring, when new, tender shrub foliage and herbaceous plants stimulated by winter rains are most abundant. Cattle grazing can quickly remove much of the food on which tortoises depend. Cattle also cause mortality of tortoises by trampling them, or by defoliating shrubs and collapsing the shallow burrows that provide shelter from heat and drought.

Recently, high levels of predation on young tortoises by ravens have been noted (Campbell 1988). Increased human activity in desert regions has led to increased numbers of ravens, which feed at garbage dumps and cattle ranches, and nest in trees or on utility poles. Predation by ravens may cause as much as 30 percent of total mortality of tortoises, leading to proposals for artificial reduction of raven populations as a means of increasing the survival rates of young tortoises.

The desert bighorn sheep is a second species of special concern in North America. Originally, 500,000 to 1 million desert bighorns probably occurred in the southwestern United States and northern Mexico (Cooperrider 1985). Today, total numbers are about 8,500 to 9,100, or about 1 to 2 percent of the original population. Between 1850 and 1900, the populations of this species were hunted heavily, leading to their disappearance from many small mountain areas. Bighorn sheep are also highly sensitive to several diseases of livestock, and some populations have disappeared in areas where domestic sheep and cattle have come into close contact with bighorns.

In Death Valley National Monument, as well as other areas, feral burros have contributed to the decline of populations of desert bighorn sheep (Fig. 5.8). Originally, bighorn populations were estimated to be about 5,000 animals; in the early 1970s they had declined to only about 600. The burro population of the monument was then estimated to be 1,500 animals, and showed a growth rate of about 18 percent per year (Norment and Douglas 1977). This suggested a loss of nearly three bighorns for every burro. Burros and bighorns feed on many of the same plants. Burros, however, can browse woody plants more heavily than can bighorns, because of a digestive system adapted to process woodier materials (Seegmiller and Ohmart 1981). Burros are also highly aggressive toward bighorns, and displace them from desert springs.

Feral burros (Fig. 5.9) have created major management problems in several other locations in the North American deserts, particularly Grand Canyon National Park. Comparison of the vegetational characteristics and small mammal communities of areas accessible and inaccessible to burros (Table 5.1) shows just how extensive the impact of burro grazing can be (Carothers et al. 1976). Grazed areas had fewer plant species, less than a quarter the normal plant cover, and only about a quarter as many small mammals as ungrazed areas. Some mammal species that were common in ungrazed areas were absent in grazed sites, and *vice versa.*

Overgrazing has long been recognized as a problem in the Grand Canyon area. Prior to 1969, however, populations of these animals were periodically reduced by "burro hunts." In 1971, the killing of feral burros and horses on many federal lands was made a felony under the Wild Horse and Burro Act (Federal Law 92–195). The result was that feral burro and horse populations on federal land increased from about 17,000 in 1971 to 64,000 in 1985, and overgrazing has become a serious problem in many areas.

Recent policy has been to reduce burro populations by capturing them alive and putting them up for "adoption." Adopting a burro requires paying a fee of $75, and agreeing to conditions relating to ownership and care. This practice, which has resulted in the adoption of over 50,000 animals, has not been adequate to reduce feral animal populations in the wild. The "willingness to adopt" has largely been exhausted, and large numbers of burros and horses must be maintained in captivity at an expense of about $730 per animal-year. In 1985, the federal budget for the maintenance and adoption program was $16.7 million, based on the handling of 17,000 animals.

FIGURE 5.8

■ Desert bighorn sheep in many mountain ranges of the western
United States have been decimated by diseases and competition from
livestock and feral burros.

FIGURE 5.9

■ Feral burros are strong competitors with desert bighorn sheep because
of their ability to browse woody plants heavily and their aggressive
displacement of sheep from water holes.

TABLE 5.1

Characteristics of Areas Inaccessible and Accessible to Burros in Grand Canyon National Park.

	INACCESSIBLE TO BURROS	ACCESSIBLE TO BURROS
Number of Plant Species	28	19
Total Plant Cover	80%	19%
Small Mammal Abundance	51.8	13.2
(No./ha)		
Canyon Mouse	Absent	Common
Rock Pocket Mouse	Common	Absent

Source: Data from S. W. Carothers, et. al. "Feral Asses on Public Lands: An Analysis of Biotic Impact, Legal Considerations and Management Alternatives" in *North American Wildlife Conference,* 41: 396–405, 1976.

Physical and Biotic Impacts of Off-Road Vehicles

The use of **off-road vehicles** (ORV's), such as trail bikes, all-terrain vehicles (ATV's), dune buggies, and 4–wheel drive wagons and trucks, for outdoor recreation has caused serious impacts on several types of ecosystems, especially sandy coastal ecosystems and deserts (Fig. 5.10). In the United States alone, there are probably 10 to 20 million motorcycles and ATV's, about half of which are used regularly for off-road recreation. A large fraction of the 5 million or so 4–wheel drive wagons and trucks are probably used for such activities, as well. These vehicles severely disturb desert soils, and cause direct damage to plant and animal populations.

ORV activity disrupts the integrity of surface protective layers, compacts the soil (Adams et al. 1982, Webb 1983), and increases its vulnerability to erosion (Iverson 1980). Soil crusts are easily disrupted by the passage of vehicles. The weight of ORV's also tends to compact the soil (Iverson et al. 1981, Hinckley et al. 1983). The first few passes of an ORV do most of the compaction, however (Table 5.2), with the first pass causing more compaction than any other. Loamy and gravelly soils with a mixture of textural classes (sand, silt, clay) are most susceptible to compaction, whereas dune sands and the fine alluvial soils of playas are least susceptible. Compaction is coupled with reduction in the pore space within the soil (Table 5.2), as well.

These effects promote erosion. Disruption of the soil crust exposes loose underlying particles to wind or water erosion. Compaction means that infiltration of water is slower, and that runoff, and the erosion that accompanies it, is greater. Experimental measurements of erosion caused by simulated rainfall on ORV recreational areas indicate that the general increase in erosion is 10 to 20-fold (Hinckley et al. 1983). On steep slopes, ORV trails promote severe gully erosion and occasionally cause massive **debris flows,** the mass slippage of destabilized, saturated soil.

Damage to vegetation is both direct and indirect. In the Mojave Desert, for example, the density of the dominant desert shrub, the creosote bush, declined by almost two-thirds on sites used as staging areas for vehicular recreation (Table 5.3). Moreover, when the fraction of the branches of individual bushes that were alive was taken into account, even moderate use areas experienced a decline of almost half in the amount of live shrub foliage, and staging areas showed only about 5 percent as much foliage as control (undisturbed) areas. Indirect effects, such as soil compaction, also tend to reduce the germination and growth of native annuals. Disturbance of the soil by ORV activity, on the other hand, favors the establishment of exotic weedy plants, such as the Russian thistle.

ORV activity causes a marked decline in the abundance and diversity of animals (Table 5.4). Many small animals are crushed and their burrows collapsed by ORV passage. Among small vertebrates, ORV noise may cause hearing loss, reduced ability to detect predators, or unnatural behavior (Brattstrom

FIGURE 5.10

■ Off-road vehicle activity is destructive to both physical and biotic features of desert ecosystems.

TABLE 5.2

Effects of Motorcycle Passes on the Density and the Large Pore Volume of a Loamy Sand Soil in the Mojave Desert, California.

NUMBER OF VEHICLE PASSES	SOIL DENSITY (g/cm³)	LARGE PORE VOLUME* (cm³/g)
0	1.52	0.21
1	1.60	0.19
10	1.68	0.17
100	1.77	0.15
200	1.78	0.14

*pores > 0.045 mm in diameter
Source: Data from B. M. Iverson, et. al. "Physical Effects of Vehicular Disturbance on Arid Landscapes" in *Science*, 212: 915–917, 1981, American Association for the Advancement of Science. Washington D.C.

TABLE 5.3

Effects of ORV Activity on the Abundance and Condition of the Creosote Bush, the Dominant Shrub Species, in the Mojave Desert of California.

ORV ACTIVITY LEVEL	SHRUB DENSITY (No./ha)	CONDITION INDEX* (Per ha)
None	240	216
Moderate	236	112
Heavy	145	56
Very Heavy ("Pit" Areas)	84	10

*Sum of decimal fraction of living branches for all shrubs present.
Source: R. B. Bury, et. al., "Effects of Off-Road Vehicles on Vertebrates in the California Desert." U.S. Department of the Interior, Fish and Wildlife Service, *Wildl. Res. Rep.* No. 8, 1977.

DISTRIBUTION OF MOIST TROPICAL FORESTS

Moist tropical forests comprise many distinct forest types. Indeed, the richness of species in humid tropical regions is paralleled by the richness of the types of communities they form. In lowland areas, forest communities range from **tropical rain forests** (Fig. 6.1), in which abundant rain occurs in every month, to **tropical deciduous forests,** which have a dry season long enough that many or most of the species lose their leaves for several months. In major river basins, distinct forest types occur in areas that are seasonally inundated by the river flood. In Amazonia, these forest types include the **varzea forest** (Fig. 6.2), bordering whitewater rivers, such as the Amazon proper, that carry heavy sediment loads, and the **igapo forests,** bordering low-sediment, blackwater rivers such as the Rio Negro. These forests may be deeply submerged for several months, and possess faunas intimately adjusted to the seasonal cycle of inundation. With increase in elevation, a variety of **tropical montane forests** can be recognized, culminating in high-elevation **cloud forests** (Fig. 6.3) that are bathed in mist much of the time. In a topographically varied tropical region such as Costa Rica or Peru, many more specific types of forest can be distinguished on the basis of the predominant tree species. All of these forest types fall into the general category of moist tropical forests.

FIGURE 6.1

■ Interior of a tropical rain forest in Costa Rica.

FIGURE 6.2

■ The rise and fall of the Amazon River creates an extensive, periodically flooded forest zone known as the varzea.

FIGURE 6.3

■ Tropical cloud forests rich in tree ferns and other moisture-loving plants occur where mountains create almost continual damp, foggy weather.

FIGURE 6.4

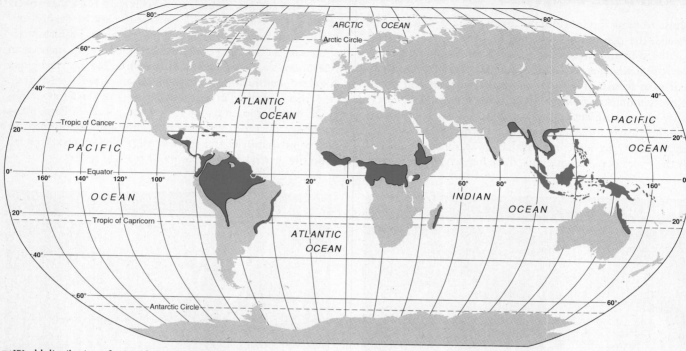

■ World distribution of original tropical rain forests.

TABLE 6.1

Original and Present-Day (1985) Moist Tropical Forests, and their Estimated Rates of Disappearance in Different Regions.

REGION	ORIGINAL AREA (10⁶ KM²)	PRESENT-DAY FORESTS (10⁶ KM²)	PERCENT LOST	PERCENTAGE OF ANNUAL LOSS
New World	8.03	6.57	18.2	0.61
Africa	3.62	2.09	42.3	0.53
Asia/Pacific	4.35	2.96	32.0	0.56
TOTAL	16.00	11.62	27.4	0.57

Source: Data based on the Global Assessment of Tropical Forest Resources, GEMS PAC Information Series No. 3 (Nairobi, UNEP, 1982).

Tropical forests can also be characterized as primary or secondary. Mature stands that are free from signs of disturbance and exhibit high biomass and diversity are termed a **primary forest,** whereas stands recovering from disturbances such as fire, hurricane impacts, or cutting by humans are termed a **secondary forest.** Secondary forests vary greatly in age, but by 60 to 80 years after disturbance they begin to approach primary forests in general appearance (Brown and Lugo 1990).

Moist tropical forests cover about 23 percent of the land area within the tropics, or about 7 percent of the total con-tinental area of the earth. Major areas of moist tropical forests occur in the three main regions of the tropics: the New World, Africa, and the Southeast Asia/Pacific region (Fig. 6.4, Table 6.1). The largest area of such forests, both now and originally, is in the New World tropics, the smallest in Africa. More significantly, about 80 percent of the remaining area of moist tropical forests occurs in only nine countries: Brazil (31 percent by itself), Indonesia, Zaire, Malaysia, Gabon, Venezuela, Colombia, Peru, and Bolivia (Myers 1980). This means that the decisions made by these few countries will ultimately determine the fate of most of the world's moist tropical forests.

TABLE 6.2

Biomass, Net Primary Productivity, and Quantities of Nitrogen and Calcium in Various Ecosystem Compartments for Tropical Rain Forest and Temperate Forest Ecosystems.*

CHARACTERISTIC	TROPICAL RAIN FOREST	TEMPERATE DECIDUOUS FOREST	TEMPERATE CONIFEROUS FOREST
Biomass (Mg/ha)	410.0	195.0	200.0
Canopy Tree Leaves	9.0	3.0	10.0
Canopy Tree Trunks	360.0	142.0	114.0
Understory	4.0	3.0	2.0
Roots	33.0	37.0	38.0
Litter	4.0	10.0	36.0
Net Production (Mg/ha/yr)	50.0	15.0	10.0
Total Nitrogen (Mg/ha)	8.0	8.0	6.0
Live Biomass	1.8	1.0	0.7
Litter	0.2	0.5	1.8
Soil	6.0 (75%)	6.5 (81%)	3.5 (58%)
Exchangeable Ca (kg/ha)			
Live Biomass	745	1000	640
Litter	5	10	15
Soil	40 (5%)	1100 (52%)	575 (47%)

*Values are Rough Approximations Derived from Several Sources.

CHARACTERISTICS OF MOIST TROPICAL FORESTS

Primary tropical forests are the richest and most productive of the earth's terrestrial ecosystems (Collins 1990). Of the conservatively estimated 4.8 million species of organisms living on earth (See chapter 2), about half can probably be found in tropical forests. Because tropical forests are the least explored tropical environment, most of these species are as yet undescribed, the estimate of their number being the sum of informed guesses based on the rate of discovery of new species. Local endemism, or the restriction of species to relatively small areas of distinctive habitat, is also high among tropical forest species (Gentry 1986). Some taxonomists believe that the number of undescribed species—mostly invertebrates and microorganisms—inhabiting the canopy of tropical forests, has been underestimated many-fold. Collection of insects from the rain forest canopy by insecticidal fogging has revealed an extraordinarily rich insect fauna, with many host-specific forms and a high degree of restriction of species to particular types of forest (Erwin 1983).

The richness of life in tropical forests is overwhelming to anyone who tries to learn the species of even a small unit of forest. A single hectare of primary tropical forest commonly contains 100 to 250 or more species of trees, a richness an order of magnitude greater than that of temperate forests. In a single hectare of wet, upper Amazonian forest, for example, Gentry (1988) found 283 tree species. In Malaysia, 835 species of trees were recorded on a 50-ha plot—the world's record for tree diversity. Animal life is likewise diverse. In a 10-square kilometer area of lowland rain forest, about 150 species of butterflies, 60 species of amphibians, 100 species of reptiles, 125 species of mammals, and 400 species of birds can typically be found (National Research Council 1982). Erwin (1983) estimated that more than 41,000 species of insects may occur on a single hectare of tropical lowland forest. This richness—often undescribed richness—often forces biologists interested in ecological or evolutionary studies to do descriptive taxonomy before they can attempt other studies.

The biomass structure of primary tropical forests is likewise complex (Table 6.2). The total biomass of moist tropical forests is considerably greater than that of most temperate deciduous and coniferous forests. Maximum tree height is related to annual precipitation, reaching 70 to 80 m in wet lowland forests. Wet lowland forests have several understory strata of smaller trees, and an abundance of **lianas,** or massive vines that expose their foliage in the forest canopy. Wet forests, especially the high mountain cloud forests, also possess a rich flora of **epiphytes** (Fig. 6.5), plants that grow on their trunks and branches. Epiphytes obtain their moisture from rainfall, and their nutrients from rainwater and the decomposition of organic matter in the epiphytic environment. In many cases, the foliage of lianas and epiphytes exceeds that of the host tree. The surfaces of leaves are often covered with **epiphylls,** a thin layer of algae, lichens, and mosses, which meet their nutrient needs by capturing nutrients from rainwater or fixing nitrogen from the air (Jordan 1985).

FIGURE 6.5

■ Epiphytes are abundant in cloud forests, sometimes covering the limbs of trees with a dense layer of bromeliads, ferns, orchids, and other plants.

Moist tropical forests are the most productive terrestrial ecosystem (Table 6.2), and are estimated to carry out about 29 percent of the total primary production of the biosphere. The productivity of nectar and fruits is especially high in these forests, and many birds and mammals depend on these resources. The biotic relationships that center on such producers can be very complex. Howe (1977) and Terborgh (1986) have shown that some trees function as **keystone mutualists** that are essential to the survival of many species of frugivorous birds and mammals. A relatively few species of figs and other fruiting trees, nut-bearing palms, and nectar-producing woody plants are vitally important to the survival of many forest mammals and birds (Terborgh 1986). These plants, in turn, are often dependent on animals that serve as **mobile links** by carrying pollen from individual to individual, or seeds from mature tree to potential establishment sites (Gilbert 1980). As for other forests, however, the bulk of the net primary production is processed through detritus food chains based on the forest floor. A rich invertebrate and decomposer biota, operating under warm, moist conditions, rapidly breaks down litter, so that litter biomass is typically small. Detritus food chains culminate in a rich community of forest floor insects, amphibians, and lizards that, in turn, support a diverse array of predators. These predators range from army ants, and the ant birds that follow them and feed on insects they flush out, to larger reptiles, mammals and birds.

From the diversity and productivity of the forests, one might conclude that their soils must be highly fertile, and that the potential for converting the land to permanent agriculture is great. Although some tropical soils are indeed very fertile, especially those of volcanic highlands, vast areas of the forested tropics possess very poor soils. Many of these soils are ancient and highly weathered. Their low organic content (due to high decomposition rates), the dominance of highly weathered clay minerals with poor nutrient-holding capacities, and the abundant rainfall make such soils prone to loss of nutrients by leaching. In some locations, such as the **caatinga forest** of the northern Amazon basin, the soils are little more than highly leached, quartzitic sands that do not retain appreciable quantities of nutrients in any form (Herrera 1985). Leaching is particularly severe for mineral cations such as calcium (Table 6.2), magnesium, potassium, and sodium. For these nutrients, the vast bulk of the active nutrient pool is in the living biomass of the forest, contrary to the case in tem-

FIGURE 6.6

FIGURE 6.7

■ Mycorrhizal root mat in a tropical rain forest in Costa Rica.

■ Plank buttresses of trees in tropical rain forests are believed to be a mechanism to brace these shallow-rooted trees against toppling by wind.

perate forests (Jordan 1985). For nitrogen and phosphorus, concentrations of available nutrients in the soil may not be much less than for temperate forests, except in caatinga forests, but a greater fraction of the active nutrient capital still tends to occur in living material than is true for temperate forests because of the larger biomass of the mature forest (Table 6.2).

The high productivity and biotic richness of the forests in areas of poor soils result from an efficient mechanism of retention and recycling of nutrients by the mature forest ecosystem (Jordan 1985). This mechanism is primarily a high concentration of mycorrhizal roots of trees in the surface soil, where nutrients released by decomposition can be taken up immediately (Fig. 6.6). **Mycorrhizae** are symbiotic associations between fungi and the fine roots of higher plants. In these associations, the filaments of the fungi extend outward into the soil or litter from the rootlets themselves. In some mycorrhizae the fungal filaments simply form a sheath around the rootlet; in others they penetrate into the interior of the root. In both cases, the fungal associate of the root assists in the uptake of water, nutrients, or both. The intimate association of some mycorrhizae with decomposing organic matter results in **direct nutrient cycling,** in which nutrients pass directly from the dead matter into fungal filaments and then into plant rootlets without entering the mineral soil.

Mycorrhizal root mats are highly effective in uptake of nutrients released by decomposition of litter. In forests on

leached quartzitic sands at San Carlos de Rio Negro, Venezuela, where these mats are 15 to 40 cm thick, Stark and Jordan (1978) found that 100 percent of the calcium and phosphorus in leaves placed on the ground surface was taken up by the root mat; none passed into deeper soil layers by leaching. Because of the high concentration of absorbing roots close to the soil surface, most trees of mature tropical forests have few deep roots. A common feature of many of these trees are **plank buttresses,** flanged extensions of the trunk base that connect with lateral roots (Fig. 6.7). Plank buttresses are thought to provide support for the trunk in place of deep roots.

When mature tropical forests are cleared, and the root systems of the trees killed, this nutrient retention and recycling system is destroyed. Nutrients then are leached into deeper soil layers, or flushed out of the soil altogether and carried into streams. This dispersion of nutrients means that successional plants must seek nutrients in deeper soil layers, and gradually re-concentrate them over many years. Successional trees in tropical forests thus tend to have widely spreading, deeply penetrating roots, rather than shallow root systems with surface root mats.

Leaching of nutrients is one of the factors that forces indigenous farming peoples throughout the tropics to practice **shifting cultivation** (Fig. 6.8). In these farming systems, which vary greatly in detail, a plot is cleared by felling and burning the woody vegetation. Burning releases large quantities of nutrients in the ash. Crops are grown for two to several seasons, and the plot is then abandoned to succession. Loss of fertility by leaching from the surface soil, together with weed and insect pest buildup and, often, an increase in soil acidity that causes aluminum ions to exert toxic effects, lead to a rapid decline in yields of crops in successive growing seasons (Fig. 6.9). Without heavy artificial fertilization it soon becomes unprofitable to farm the plot.

Shifting cultivation is practiced by about 240 million people in tropical forest regions (National Research Council 1982). When the interval between use of the land for farming is 70 or more years, shifting cultivation is probably an ecological stable strategy that does not degrade the tropical forest ecosystem. By initiating successional communities at scattered locations, in fact, shifting cultivation may promote biotic diversity by increasing the assortment of stands of different age and successional status within the forest. As long as the disturbance from shifting cultivation is small in scale and short in duration, nutrients are retained in deeper layers of the soil, and eventual recovery by succession is rapid because of the rich source of propagules in the surrounding forest. In addition, some shifting cultivators, such as the Kayapo Indians of Brazil, actively manage their farm plots to encourage successional recovery after cropping is ended (Hecht and Cockburn 1989). With growing human populations, however, the number of subsistence farmers in forest environments increases. Consequently, the period of cultivation lengthens, the fallow interval between cropping periods shortens, and farm plots are spaced closer together. With such changes, the nutrient capital of the landscape at large is depleted by leaching and export of harvested materials. Thus, a crisis of fertility looms even in areas where the basic forested character of the landscape is retained.

Recovery of moist tropical forests to a mature state is very slow. Many of the trees and other plants of the mature forest have specialized seed dispersal mechanisms, usually involving fruit-eating animals. How quickly such species reinvade an area depends on the distance from a source of seeds and the abundance of the animal dispersal species. In Costa Rica, Opler et al. (1977) estimated that about 1,000 years is required for full successional recovery of primary lowland forest, even when sources of seed were close at hand.

TROPICAL DEFORESTATION

Moist tropical forests are being converted to other vegetation types at an alarming rate, although actual conversion rates are still poorly known. Most estimates are based on a survey by the United Nations Food and Agriculture Organization in 1980. From these and other data, Melillo et al. (1985) estimated that annually about 59,000 to 75,000 square kilometers of moist tropical forest, or about 0.56 percent of the total forest remaining (Table 6.1), are cleared permanently or introduced into a shifting cultivation cycle by clearing. Of this, about 45,000 square kilometers are permanently converted to non-forest land annually. Brown and Lugo (1990) estimated that about 77,000 square kilometers of primary forest are converted to secondary forest or agricultural land annually. The area of secondary forest has thus increased as a result of disturbance. More than 31 percent of the remaining moist tropical forest is now estimated to be secondary forest. More recent estimates suggest that rates of deforestation are now 79 to 100 percent higher than those for 1980 (Myers 1989, Collins 1990, World Resources Institute 1990). Myers (1989) estimated that 142,200 square kilometers of tropical forest disappeared in 1989. As the remaining forest area shrinks, and as human populations in the tropics grow, of course, the rate of forest conversion is likely to increase.

The total loss of moist tropical forest has been least in the New World, where the original area was greatest (Table 6.1). In Africa, where the original area of forest was least, the loss has been greatest. In Madagascar, one of the tropical forest areas richest in endemic species, about 66 percent of the original moist forest has disappeared (Green and Sussman 1990).

The ultimate causes of tropical deforestation are complex, and are deeply rooted in international trade. Chiefly among them are growth of human populations and efforts for economic development in tropical forest countries, combined with the appetite for tropical products such as timber and beef in developed countries. These forces drive several deforestation processes. The transformation of shifting cultivation to permanent farming due to the growth of human populations is a major process, particularly in Africa and Latin America. Clearing of forest for cattle ranching is a major process in Latin America, as well. In some cases, most of the meat produced by these ranching operations is exported to developed countries for use in fast foods; this relationship has been termed the "**hamburger connection**" (Myers 1981). Often, clearing of forest for ranching is promoted by international developmental organizations (Fearnside 1987) and by governmental incentives and subsidies, based on the legal definition of such action as "improvement" of the land (Hecht 1989). In Southeast Asia, clearcutting of tropical hardwood forests for timber destined for export has been the major cause of forest destruction. The rapid depletion of suitable timber in this region is now forcing timber harvesters to shift their efforts to other parts of the tropics.

FIGURE 6.8

■ In Brazilian Amazonia, an Indian farmer clears a plot of forest for shifting cultivation of maize, manioc, and other crops.

FIGURE 6.9

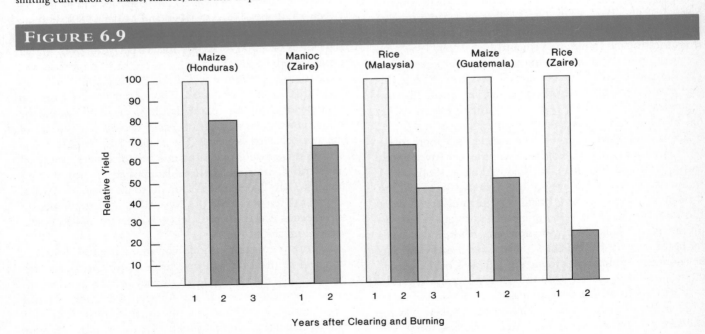

■ Decline of productivity of crops in shifting agriculture (Modified from Cox and Atkins 1979).

TABLE 6.3

Estimated Percentages of Tropical Forest Species that have Disappeared and are Projected to Disappear During Various Time Periods Due to Decline in Total Forest Area.

		AT CURRENT DEFORESTATION RATE			AT 2X CURRENT RATE	
WORLD REGION	LOST PRIOR TO 1990	1990–2000	1990–2010	1990–2020	1990–2020	With Loss of All But Currently Protected Areas
Africa	10–22	1–2	2–4	3–7	6–14	35–63
Asia/Pacific	12–26	2–5	4–10	7–17	22–44	29–56
New World	4–10	1–3	3–6	4–9	9–21	42–72

Source: Data from W. V. Reid and K. R. Miller, *Keeping Options Alive: The Scientific Basis for Conserving Biodiversity,* 1989 World Resources Institute, Washington, D.C.

The thinly populated Amazon Basin, where the largest area of unspoiled tropical forest remains, is viewed by several nations as their last frontier. Brazil, in particular, considers Amazonia a vast underdeveloped area that must be integrated into the national economy and culture (Lovejoy 1985), lest it be settled through default by people from other countries. Brazil has sponsored several efforts to promote such integration by the construction of highways, by agricultural settlement schemes, and by efforts to encourage major corporations to undertake large-scale development schemes (Hecht and Cockburn 1989). The Trans-Amazon Highway, crossing the Amazon Basin from east to west, opened up much of the southern part of the basin to settlement. In the eastern Amazon State of Para, the Grande Carajas Program has promoted mining and smelting of metal ores, using charcoal derived from the tropical forest as a principal fuel (Oren 1987). To the west, in the State of Rondonia, the Polonoroeste Project has promoted settlement on small farm plots, giving this region the most rapid deforestation rate in Brazil. The unsuitability of the land for permanent farming in this region has kept the settlers in a state of impoverishment and forced many to abandon their original land and clear still more forest.

Another major development was the Jari River Forest and Agricultural enterprize, initiated in the late 1960s on a 1 million ha site on a tributary of the Amazon River 350 km west of the port of Belem, Brazil. The plan of operation at this location was to harvest and clear the original forest, replacing it with pulpwood plantations, rice fields, and pastures for beef cattle. By 1981 about $1 billion had been invested in development. Pulpwood production fell well below expectations, however, and in 1982 the entire operation was sold to a consortium of Brazilian companies for $280 million, or $720 million less than the cumulative investment. Having effectively destroyed an extensive area of tropical forest, the project still operates, but at a low level of profit (Russell 1987).

IMPACTS OF DEFORESTATION

Destruction of tropical forests threatens enormous numbers of species with extinction, many before they are even described and named. Lovejoy (1980), for example, estimated that 15 to 20 percent of forest species would disappear between 1980 and 2000 A.D. due to forest loss. Simberloff (1986) estimated that 12 percent of plant species and 15 percent of bird species in the Amazon Basin would disappear by 2000 A.D. due to deforestation. Raven (1988) projected a loss of 25 percent of tropical forests species by 2015 A.D. The most comprehensive estimate (Table 6.3) is that between five percent and 15 percent of tropical forest species will be lost between 1990 and 2020 A.D. (Reid and Miller 1989). If about 2.4 million species still exist in tropical forests, this translates into losses of 4,000 to 12,000 species per year. Even higher losses are projected if rates of forest destruction increase, and a catastrophic pattern of loss if the only areas to survive are those that are now protected. These estimates are based on mathematical species-area relationships, or, simply stated, the rate at which total number of species decreases as the size of the area occupied decreases (See chapter 26). From these estimates it is obvious that large numbers of species, nearly all undescribed, must already have disappeared in tropical forest regions: 12 to 26 percent in Asia, 10 to 22 percent in Africa, and four to ten percent in the New World. Disappearance of these species is an immense scientific, aesthetic, and economic loss to humankind.

Most extinctions due to continued tropical deforestation will be of plants and invertebrates, particularly insects. Nevertheless, the vertebrates of tropical forest regions are increasingly under threat. In the Amazon Basin, for example, 18 mammals, seven birds, and three reptiles are designated as rare or endangered (Barrett 1980). Of the mammals, six are monkeys, highlighting the fact that deforestation is a particular threat to forest primates throughout the tropics. In the Atlantic forest region of eastern Brazil, where the original

THE SAVANNA AND WOODLAND ENVIRONMENT

Savannas and woodlands possess a highly seasonal climate with one or two dry and wet seasons annually. Total annual rainfall ranges from about 30 cm to 160 cm or more, but is often patchy and highly variable from year to year. Temporal and spatial heterogeneity thus play a major role in the ecology of these ecosystems.

The pattern of wet and dry seasons is determined by the seasonal shift of the **intertropical convergence,** the belt of most intense solar heating of the land, and the zone into which the trade winds of the northern and southern hemispheres flow (Fig. 7.2). The position of the intertropical convergence at a given time is close to the latitude where the sun is directly overhead, and thus most intense in its radiation. This overhead position migrates seasonally, passing northward over the equator on March 21, reaching the Tropic of Cancer on June 21, moving southward and passing over the equator again on September 21, and reaching the Tropic of Capricorn on December 21.

The zone of maximum solar heating is also a zone of convectional rainfall—thunderstorms created by condensation of moisture in rising masses of heated air. Proximity of the intertropical convergence thus means a rainy season. Near the equator, therefore, rainy seasons tend to occur around March and September, and dry seasons around December and June (although secondary influences tend to shift these periods somewhat). At the limits of the tropics, a single, short rainy season occurs, when the intertropical convergence is nearby, and the rest of the year is dry. Thus, basic climatic relationships define a complex pattern of dry and wet seasons within much of the tropics.

The pronounced seasonal changes in moisture, combined with soil nutrient conditions, directly influence the growth of the vegetation. Grasses and forbs grow actively during the wet seasons, and their shoots die during the dry seasons, creating conditions favorable to fire. Fire is thus an important environmental factor in almost all savanna and woodland areas. Frequent fire tends to kill woody plants and stimulate perennial bunch grasses. The chance of fire is, in turn, influenced by the harvest of plant material by herbivores. Thus, moisture supply, nutrient availability, herbivore activity, and fire are the major determinants of savanna and woodland ecology. Savannas in some areas, including parts of northern Australia, Central America, and northern South America, may actually be of human origin, the result of fires set by aboriginal peoples (Bourliere and Hadley 1983, Cole 1986).

The seasonality and variability of rainfall make migration one of the key patterns of adaptation to savannas by both wildlife and humans. The life cycles of migratory ungulates are closely keyed to the seasonal conditions of different areas. During the wet seasons, the forage that appears in the drier

FIGURE 7.1

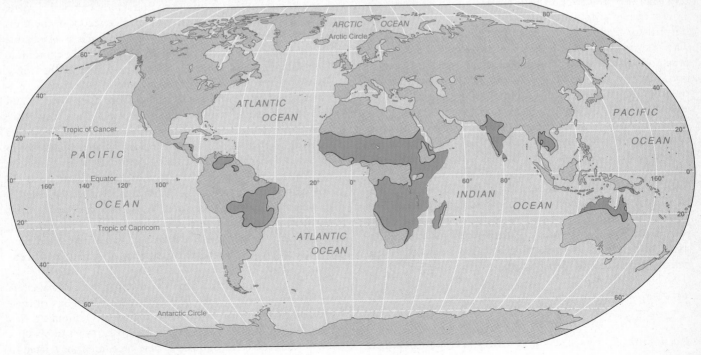

■ World distribution of tropical savannas.

FIGURE 7.2

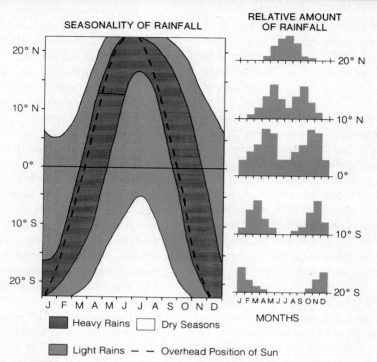

SEASONALITY OF RAINFALL

RELATIVE AMOUNT OF RAINFALL

MONTHS

■ Heavy Rains ☐ Dry Seasons

■ Light Rains — — Overhead Position of Sun

■ Wet and dry seasons within the tropics are determined by the seasonal movement in the overhead position of the sun, which determines the latitude of most intense heating of the earth's surface. Periods of heaviest rainfall, corresponding to the location of the intertropical convergence, tend to occur shortly after the direct overhead position of the sun has occurred at a given location (Modified from Niewolt 1977).

portions of the migrational range is typically higher in protein than that in wetter areas. It is also higher in many mineral nutrients (McNaughton 1990). These areas are thus optimal for feeding by animals in later stages of pregnancy. The drier areas may also be more open, making it difficult for predators to stalk and kill ungulates at the time the young are born. During the dry season, of course, the wetter areas of the migrational range form a drought refuge, with forage adequate for adult survival.

The native human peoples of savanna regions in Africa and Asia largely lived by **pastoralism,** the herding of animals such as cattle, sheep, goats, and camels. These livestock provided humans with most of their food and clothing, in the form of milk, blood, meat, and skins. Like migratory ungulates, pastoralists moved their herds seasonally to exploit grazing habitats that were optimal at different seasons. Many pastoralist peoples still retain a fundamentally subsistence lifestyle, while others are becoming integrated into the commercial economies of their countries.

BASIC TYPES OF SAVANNAS AND WOODLANDS

The dominant plants of savanna ecosystems are varied in their life forms. Trees, shrubs, succulents, grasses, and broad-leafed herbaceous plants are all important in the vegetation. Often,

communities dominated by plants of different life forms exist as a mosaic resulting from the complex interaction of climate, soils, fire, and animal influences (See chapter 1). In these interactions, some of the mega-herbivores, such as elephants, rhinoceroses, and hippopotamuses, may play a keystone role (McNaughton and Sabuni 1988, Owen-Smith 1988, 1989). By their intense feeding on woody plants, for example, elephants may create and maintain the more open savanna and savanna woodland habitats that are optimal for many other large herbivores. Elephants are capable of pushing over trees up to half a meter in diameter to obtain foliage. They also rip off strips of bark with their tusks, eventually killing trees by girdling them. Large, soft-wooded baobab trees can be killed when deep cavities are gouged into their trunks by elephants to obtain the moist xylem wood. With some tree species, such as the mopane of southern Africa, heavy elephant browsing can keep large areas in a resprouting, shrub-like condition, effectively preventing their maturation as normal trees (Owen-Smith 1988).

From a fertility standpoint, savannas can be characterized as **eutrophic,** having high soil nutrient supplies, or **dystrophic,** having low nutrient availability. Combining this feature with moisture availability, four basic types of savanna ecosystems can be recognized. These types are thus related to the dominating influences of climate and soils, and differ strikingly in their biotas, and especially in the role of herbivores in nutrient recycling (Ruess 1987). Even more importantly, they differ in their stability and resilience in the face of disturbance and stress imposed by human activities.

Dry eutrophic savannas are characterized by low rainfall, 30 to 70 cm annually, but possess soils rich in nutrients. These savannas occur where soils have not been depleted of nutrients by leaching over a long period of geological time. In some cases, these savannas occur on young volcanic soils in which weathering releases new quantities of mineral nutrients. In other cases, they exist on alluvial soils that receive nutrient inputs by occasional flooding. Although rainfall is low, the richness of the soil means that during wet seasons plant productivity is high. The nutrient quality of the plant material is likewise high. This type of savanna supports the greatest biomass and diversity of large animals. The high quality of the plant forage means that many kinds of vertebrate grazing and browsing specialists can satisfy their food needs. The importance of fire in this type of savanna depends on the intensity of herbivory, but often grazing is heavy enough that fuel to support extensive fires does not remain into the dry seasons. Likewise, the heavy consumption of plant material by large herbivores means that the dead matter available to termites is relatively small. Because of the high nutrient quality of plant material, decomposition is rapid during the wet season, and nutrients are quickly recycled.

Dry eutrophic savannas are not the most stable of savanna types, as they are subject to droughts that may cause heavy mortality of plant and animal life. They are, however, highly resilient as a result of their fertility and of the adjustment of the vegetation to intense grazing. Savannas of this type are among the most important wildlife ecosystems on earth, and are exemplified by the Mara-Serengeti Ecosystem of Kenya and Tanzania. We shall discuss this ecosystem in detail later in the chapter.

Moist dystrophic savannas contrast sharply with those described earlier. These savannas receive 60 to 160 cm of rain annually, but occur on soils that are ancient, and have been leached of their nutrients over millions of years. The high rainfall permits a high level of primary production, but the bulk of the plant tissue produced is low in nutrient quality and high in fiber. Specialist herbivores consequently are often unable to meet their food needs, and the large herbivore fauna tends to be dominated by animals such as elephants and cape buffalo, which are able to process large quantities of plant material and extract the small fraction of digestible material present. In this type of savanna, therefore, the principal herbivores are frequently termites, which create distinctive landscapes dotted with their large nests (Fig. 7.3). Termites may consume a third or more of the total primary productivity in these savannas (Josens 1983). Many of these termites process plant material by using it as a substrate for growth of certain fungi, which they then consume. The lower abundance of large herbivores means that much plant material remains in the dry season, making fire a nearly annual event in savannas of this type. Some of these savannas, such as those of the Gran Sabana of Venezuela, may owe their origin to burning (Folster

FIGURE 7.3

■ A dystrophic savanna near Darwin, Australia, with large nests of fungus-gardening termites.

1986). Because of the low nutrient quality of the plant material, decomposition is slow, and nutrient recycling is much slower than in dry eutrophic savannas.

Moist dystrophic savannas and woodlands are probably somewhat more stable than the drier eutrophic savannas and woodlands. Ecosystems of this type are widespread in central and western Africa in the climatic belt immediately north of the forests of the Congo Basin and West African coastline. The extensive *miombo* woodlands of central and southern Africa exemplify this type of ecosystem. Many of the savannas of Brazil and northern Australia also fall in this category.

Dry dystrophic savannas combine the less desirable features of the first two savanna types. These savannas occur where shallow soils and less than average rainfall severely limit moisture availability to plants during the dry seasons. They have a low, highly variable productivity and a low biomass of vertebrate herbivores. Like moist dystrophic savannas, termites are major herbivores and fire is frequent. Both stability and resilience are low, and the stresses of drought or overgrazing are likely to convert these systems to desert-like ecosystems. Savannas of this type occur in parts of Africa and South America.

Moist eutrophic savannas occur where basic climatic and soils conditions might permit forest or dense woodland types to exist, but where fire, animal, or possibly human influences have resulted in the conversion of these vegetation types to open woodland and savanna. Plant productivity and nutrient quality are high, and thus diversity and abundance of large herbivores are great. Decomposition and nutrient cycling also are rapid. Stability of this type of savanna is high, but resilience is limited. Excesses of fire or animal influence can transform the plant community to one dominated by a different life form. In fact, the diversity of animal and plant life in this type of system often depends on a delicate balance of soils-animal-fire influences that create a mosaic of communities dominated by plants of different life forms. Savannas of this sort also occur in parts of East Africa.

Lying just south of the equator in East Africa (Fig. 7.4), the Serengeti-Mara ecosystem contains the world's greatest concentration of large mammals (Sinclair 1979). Covering an area of over 25,000 square kilometers, this natural ecosystem unit centers on the Serengeti National Park in Tanzania and the neighboring Mara National Park in Kenya. The northern and western parts of this region, where total annual rainfall averages up to 120 centimeters, are covered by savanna and woodland. The southeastern portion of the region is plains grassland, with an annual rainfall of about 35 cm. Rainfall is strongly seasonal, coming mainly from December through April, but the fertile volcanic ash soils permit high plant production when moisture is available.

The Serengeti-Mara ecosystem dramatically illustrates many of the major kinds of interactions (Fig. 7.5) that occur in ecosystems (Sinclair 1979). Among these are interactions between organisms and the physical environment. Conditions of the physical environment clearly influence organisms. Moisture is a key factor to plant production in semiarid environments like the Serengeti-Mara region. In the 1970s, dry season rainfall increased from about 15 cm to about 25 cm, perhaps not permanently but as a long, wet cycle. Increased moisture led to increased plant productivity, which favored the increase in populations of several large herbivores during this period.

More than 3 million large herbivorous animals of 27 species inhabit the Serengeti-Mara region (Fig. 7.6), the most abundant being the wildebeest (1.4 million), the Thomson's gazelle (600,000), the plains zebra (200,000), the topi (60,000), and the Cape buffalo (52,000). Of these, the wildebeest, Thomson's gazelle, and plains zebra are mostly migratory; individuals of the remaining species live in the same location throughout the year (Fig. 7.7). Together they consume an astounding 66 percent of the aboveground net primary production of the ecosystem—a fraction probably greater than for any other terrestrial ecosystem (McNaughton 1985). However, this intense grazing, rather than degrading the productivity of the system, may actually stimulate it (Belsky 1986). Experiments suggest that areas subjected to natural grazing show

FIGURE 7.4

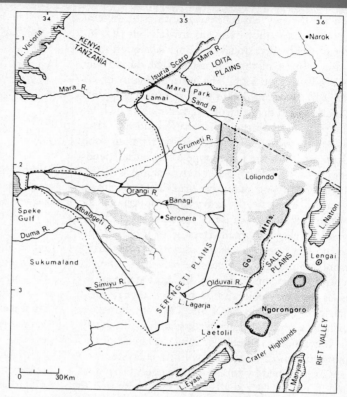

■ The Mara-Serengeti ecosystem. The solid line indicates the boundary of Serengeti National Park, and the dashed line shows the total area used by migratory ungulates (Modified from Sinclair 1979).

about twice the net primary production of protected areas (McNaughton 1985). In part, this reflects the fact that grazing, with return of all elements to the soil of the system itself in urine, feces, and decomposed carcasses, stimulates rapid recycling of nutrients. In some places, high grazing intensity creates **grazing lawns** (Fig. 7.8), which are dense carpets of actively growing grasses that are kept in an actively growing, juvenile state by this mechanism (McNaughton 1985).

The migratory wildebeest, plains zebras, and Thomson's gazelles spend the wettest portion of the year, typically from March through May, on the southern plains region of the Serengeti (Bell 1971). As the forage begins to dry and disappear,

these species move northwest into the western corridor, where they tend to remain from June through October. Finally, from November through February, they concentrate in the northern part of the system, near the Kenya border. Utilization of the vegetation by the major migratory species also appears to be complementary (Bell 1971). Zebras are usually the first animals to move into new areas. They feed in the tallest grasses, and consume the coarsest components of the vegetation. The wildebeest follow, foraging in the shortened stands of grasses and selecting primarily the more tender leaves. Finally, when the stands of grasses have been decimated and trampled by these species, the Thomson's gazelles appear to feed on re-

FIGURE 7.5

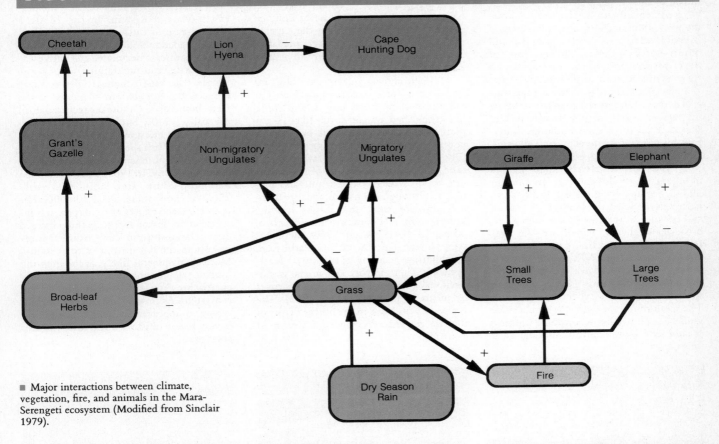

■ Major interactions between climate, vegetation, fire, and animals in the Mara-Serengeti ecosystem (Modified from Sinclair 1979).

FIGURE 7.6

■ Wildebeest and plains zebra are the most numerous migratory ungulates in the Mara-Serengeti ecosystem.

FIGURE 7.7

■ About 60,000 topi, one of several non-migratory ungulates, occupy the Mara-Serengeti ecosystem.

FIGURE 7.8

■ Intense grazing by animals such as wildebeest can maintain grasses in a low, actively growing state known as a grazing lawn.

sprouting grasses and the leaves and fruits of low broad-leafed herbs exposed by removal of the grasses.

An equally impressive group of carnivores—lions, leopards, cheetahs, spotted hyenas, wild dogs, several jackals, and others—prey on these herbivores. Most of the carnivores are permanent residents, and their numbers seem to be limited by the abundance of resident, rather than migrant, prey. One of the major benefits of migration, in fact, may be that it enables these species to escape population limitation by resident carnivores (Fryxell et al. 1988).

Interactions between major groups of organisms are also conspicuous (Sinclair 1979). Feeding by the large grass-eating herbivores reduces the biomass of grasses, and permits an increase in the biomass of broad-leafed herbs, which are the preferred foods of gazelles, particularly the Grant's gazelle (Fig. 7.9). As the numbers of large grass-eaters have grown, so have the numbers of Grant's gazelles and their specialist predator, the cheetah (Fig. 7.10). The continuing growth in numbers of wildebeest, however, seems to be limiting population growth of plains zebra and Cape

buffalo by competition. Competition also seems to occur among some of the predators. As populations of resident herbivores have increased, so have those of lion and spotted hyena, in this case at the expense of the wild dog, which the larger predators are able to frighten away from their own kills.

Plant-animal interactions also involve the savanna trees (Sinclair 1979). Giraffes browse on small acacias, tending to keep them hedged and shrub-like. Heavy giraffe browsing thus delays or prevents young acacias from maturing into trees (Fig. 7.11). This, of course, favors the giraffe, since all of the foliage of small trees can be browsed by these animals. Elephants, although not present in large numbers, browse trees and shrubs heavily, killing some and stunting the growth of others. In the 1950s and 1960s, elephants were forced into Serengeti Park from neighboring areas by agricultural development. Their impact on acacia woodlands was great, reducing tree canopy cover by up to 50 percent (Pellew 1983). The greatest change was in the northern part of the park where acacia woodlands were most extensive.

An interaction between a disease organism and its hosts has exerted one of the most long-reaching influences on the Serengeti-Mara ecosystem. In the 1890s, a virus disease of cattle, known as **rinderpest,** invaded sub-Saharan Africa. In a few years this disease decimated the native African cattle, causing famine among the people who raised them. It also severely reduced the populations of many of the large herbivores. Outbreaks recurred until the early 1940s, but vaccination of domestic cattle has now virtually eliminated the disease. With the disappearance of rinderpest, the numbers of large herbivores climbed steadily. In 1961, wildebeest in the Serengeti-Mara area numbered only 250,000; now there are 1,400,000. Release from the effects of rinderpest and increase in dry season rain have thus changed basic relationships in this ecosystem enormously in the last quarter of the century. More change seems likely, as the growing human population of East Africa is now encroaching on this remarkable natural ecosystem. As in so many areas of the biosphere, it appears that human beings will have a major impact on the future of this ecosystem.

FIGURE 7.9

■ The Grant's gazelle is a specialist feeder on broad-leaf herbaceous plants.

FIGURE 7.10

■ The cheetah is a specialist predator on the Grant's gazelle.

FIGURE 7.11

■ Heavy browsing by giraffes can cause severe hedging of preferred woody plants, greatly slowing their growth to tree size.

Human Populations and Savanna Preserves

Rapidly growing human populations, especially in Africa, Asia, and South America, are placing greater pressures on savanna areas that were once the exclusive domain of wildlife and subsistence pastoralists. Increasing conflict between wildlife and human interests is the inevitable result.

One of the most serious general effects is increasing constraints on the ability of migratory species to move between their dry- and wet-season ranges. Nairobi National Park, for example, lies immediately adjacent to the city of Nairobi, Kenya. The park itself, only 114 square kilometers in area, is continuous with drier plains to the south that are utilized and administered by the Masai tribe, whose traditional way of life is pastoralism, their livestock mostly cattle. Like the Mara-Serengeti ecosystem, the Nairobi Park region possesses migratory ungulates, mostly plains zebra and wildebeest. During the wet seasons, zebra and wildebeest disperse onto the plains south of the park, and during the dry seasons they return to the park, where grazing conditions remain more favorable. Growth of the Nairobi urban area, population growth of the Masai people themselves, and the gradual transition of subsistence pastoralism to more intensive ranching and farming activity are placing increased pressure on this overall ecosystem. These changes are tending to interrupt migration corridors and to confine wildlife more fully to the park itself. Similar changes are occurring for Amboseli National Park on the border of Tanzania.

Growing populations of pastoralist peoples and their herds of cattle and other livestock are placing increased pressure on grazing lands in and near many African wildlife areas. Traditionally, pastoralists have lived in relative harmony with wildlife, except for livestock predators such as lions, since they did not hunt wild animals for food. Many ecologists have suggested that pastoralists' herds often increase to levels beyond those which can be sustained in dry years, and that the result is range degradation (Lamprey 1983), with wildlife also experiencing the effects of range degradation. Others (Ellis and Swift 1988) doubt that pastoralism substantially contributes to range degradation. Nevertheless, increasing numbers of people and livestock, together with changing cultural, social, and economic values, will require major adjustments to enable large migratory wildlife species to survive in many parts of Africa.

A major objective of conservationists in such areas is to find ways to involve local peoples in park management and wildlife conservation, so that they share both responsibility and economic benefit of these activities. In Kenya, for example, ways are being explored to enable both Masai pastoralists and wildlife to utilize the entire geographical area, both within and outside formal parks, on which migratory animals depend (D. Western, personal communication). Specific proposals include the sharing of park income with residents of areas used by wildlife outside the parks, developing wildlife-

FIGURE 7.12

■ Waterbuck in Lake Nakuru National Park, Kenya, experienced a population explosion after large predators had disappeared and the park was fully fenced.

related tourism in areas outside park boundaries, and perhaps even permitting limited exploitation of park lands for livestock grazing.

Fencing of all or part of the borders of wildlife areas in some cases has had disastrous effects on migratory savanna animals. In Botswana, for example, long fences have been constructed to separate cattle ranching and wildlife areas, the objective being to reduce transmission of certain diseases from wild ungulates to cattle. These fences have diverted dry-season migrations of wildebeest from the extensive surface water areas of the Okavango delta to Lake Xau, a single water area that itself is being used with increasing intensity by cattle ranchers (Williamson et al. 1988). Restriction of access to permanent water during the dry season has contributed to a severe decline in numbers of wildebeest.

In Kenya, several game parks lie close to major cities and centers of intensive farming. Fencing has become necessary to separate wildlife from both farming and residential areas. Lake Nakuru National Park, established in 1960, lies in an area of the Rift Valley that was settled by Europeans in the early 1900s. Killing of wildlife by these settlers eliminated several species of large mammals, including lions, cheetahs, and hyenas. The park, with a total area of 200 square kilometers, 38 square kilometers of which are covered by the lake itself, lies adjacent to the city of Nakuru, now a major commercial and industrial center. This park has become the first East African park to be completely fenced. Freed from predators, some of the remaining ungulates have increased enormously. The park now contains about 6,000 waterbuck and 4,000 impala, and severe overbrowsing of the woody vegetation is apparent (Fig. 7.12).

Fragmentation of Savanna Wildlife Habitat

Like forests, natural savanna ecosystems are suffering fragmentation, and the remaining parks are becoming islands within agriculturally developed and urbanized landscapes. How serious this isolation is for the survival of the wildlife of these parks is subject to widely differing views. Miller and Harris (1977) noted that Mkomazi Game Reserve in Tanzania had recently lost several species of large mammals, perhaps in part because of its isolation. When it was established in 1952, the reserve had 43 species of large mammals; by 1977 it had only 39 (Miller 1978). Projecting this rate of loss, Miller suggested that in about 114 years, the game reserve would only contain 22 species of large mammals.

Soulé et al (1979) offered an even grimmer scenario of probable losses from isolated preserves, based on an analysis that projected losses at a rate comparable to that on newly isolated oceanic islands. They predicted that losses of all species over 500 years might range from 34 percent, for the largest parks, to 65 percent, for the smallest preserves. Western and Ssemakula (1981), on the other hand, noted that the relation of number of species to park area was very weak, and that projection of losses based on faunal relaxation theory (See chapters 9, 26) was not well-justified. They also suggested that long coevolution of savanna animals greatly reduced the chance of competitive exclusion. They concluded that even small parks, such as Nairobi National Park, could be viable "faunal enclaves" as long as wildlife could continue to make reasonable use of areas outside the park boundaries. Nevertheless, Western and Ssemakula (1981) also found that if parks were reduced in size from the effective ecological units now being utilized by animals to the legally protected areas alone, significant losses would likely occur (Table 7.1). For Nairobi National Park in Kenya, for example, only 11 of the 21 species now present were predicted to survive.

East (1981, 1983) noted that many of the smaller savanna parks have only very small populations of many large mammals—less than 25 individuals in many cases. In these cases, it is obvious that local extinction can easily occur due to any of a multitude of factors, including drought, poaching, fire, disease, and competition from livestock. Problems of inbreeding may also arise in such small populations (See chapter 24). Thus, strong, active management on an overall regional basis will be required to maintain the existing species diversity of many of these savanna parks.

Restoring the Ecological Balance in Savanna Preserves

In spite of the wild character that still invests many of the regions where savanna parks exist, the ecosystems of many of these parks are missing some of their original members. In other cases, the confinement of large animals within park

TABLE 7.1

The Numbers of Large Herbivores Expected to Survive in Some East African National Parks if the Effective Area of Wildlife Habitat is Reduced to the Legally Protected Minimum Area.

PARK	PARK AREA KM²	NUMBER OF SPECIES	
		NOW	AFTER ISOLATION
Serengeti NP, Tanzania	14,504	31	30
Mara NP, Kenya	1,813	29	22
Meru NP, Kenya	1,021	26	20
Amboseli NP, Kenya	388	24	18
Samburu NP, Kenya	298	25	17
Nairobi NP, Kenya	114	21	11

Source: Data from D. Western and J. Ssemakula, "The Future of the Savannah Ecosystems: Ecological Islands or Faunal Enclaves?" in African Journal of Ecology, 19: 7–19, 1981, Blackwell Scientific Publications Ltd., Oxford, England.

boundaries has resulted in exaggerated impacts of these species on the park environment. Efforts are now being made to restore natural relationships in several locations.

In Natal Province, South Africa, elephants were eliminated from the region now occupied by Hluhluwe ("Shush-SLOO-wee") and Umfolozi National Parks in the late 1800s. Large predators such as lions, cheetahs, and wild dogs were apparently also eliminated. These two parks, connected by a broad corridor of wild land through which large animals move freely, have experienced major vegetational change since the 1930s, when they were established. This change is apparently due to absence of the elephant, a keystone herbivore, and of the several predators. Under protection, populations of many open-country herbivores reached high densities, exerting heavy pressure on the savanna vegetation. Woody vegetation replaced open grassland, in spite of efforts to maintain grassland by controlled burning. In turn, major declines of black rhinoceros and several open-country ungulates occurred, with local extinction of at least three species (Owen-Smith 1989).

Efforts to restore the biotic balance of Hluhluwe and Umfolozi Parks have included the reintroduction of cheetahs and hunting dogs in the 1970s. This effort was aided by the natural recolonization of the area by lions. In the early 1980s, elephants were reintroduced to the two parks, and by the late 1980s this population had grown to about 100 animals (Owen-Smith 1989).

In Amboseli National Park, Kenya, in contrast, poaching in surrounding areas has led to a high, year-round concentration of elephants within the park. This abnormal concentration has led to widespread destruction of trees and shrubs, converting woodlands and thickets to open grasslands (Western 1989). Meanwhile, outside the park, savanna areas without elephants are experiencing heavy brush invasion, reducing grazing forage for cattle and increasing the potential for livestock disease transmitted by the tsetse fly, which flourish in dense brushland. A secondary effect of the vegetational change has been to enable tourist vehicles to penetrate into almost every corner of the park. The impacts of tourist vehicles on park landscapes in East Africa have now become a serious problem in themselves. Consideration is thus being given to ways to encourage elephant use of areas outside the park, in the interest of improving both environments (D. Western, personal communication).

Efforts are also being made to reestablish several major animal species and restore a natural predator-prey structure in Nakuru National Park, Kenya. Giraffe and black rhinoceros, both native to the region, have both been reintroduced to the park in recent years. More recently, hyenas and lions have been reintroduced. Management of this park now faces the challenge of maintaining a balanced system of predators, prey, and productive vegetation in a relatively small, fully fenced environment.

KEY MANAGEMENT STRATEGIES FOR SAVANNA AND WOODLAND ECOSYSTEMS

1. Developing protection and management schemes for natural ecosystem units in which wildlife can carry out essential seasonal movements.

2. Regulation of grazing pressure and fire to maintain vegetational diversity and prevent loss of plant communities critical to the survival of particular wildlife.

LITERATURE CITED

Bell, R. H. V. 1971. A grazing system in the Serengeti. *Sci. Amer.* **225(1):**86–93.

Belsky, A. J. 1986. Does herbivory benefit plants? A review of the evidence. *Amer. Nat.* **127:**870–892.

Bourliere, F. and M. Hadley. 1970. The ecology of tropical savannas. *Ann. Rev. Ecol. Syst.* **1:**125–152.

Bourliere, F. and M. Hadley. 1983. Present-day savannas: An overview. Pp. 1–17 *in* F. Bourliere (Ed.), *Tropical savannas.* Elsevier, Amsterdam.

Cole, M. M. 1986. *The savannas.* Academic Press, San Diego, California.

East, R. 1981. Species-area curves and populations of large mammals in African savanna reserves. *Biol. Conserv.* **21:**111–126.

East, R. 1983. Application of species-area curves to African savannah reserves. *Afr. J. Ecol.* **21:**123–128.

Ellis, J. E. and D. M. Swift. 1988. Stability of African pastoral ecosystems: Alternate paradigms and implications for development. *J. Range Manag.* **41:**450–459.

Folster, H. 1986. Forest-savanna dynamics and desertification processes in the Gran Sabana. *Interciencia* **11:**311–316.

Fryxell, J. M., J. Greever, and A. R. E. Sinclair. 1988. Why are migratory ungulates so abundant? *Am. Nat.* **131:**781–798.

Josens, G. 1983. The soil fauna of tropical savannas. Pp. 505–524 *in* F. Bourliere (Ed.), *Tropical savannas.* Elsevier, Amsterdam.

Lamprey, H. F. 1983. Pastoralism yesterday and today: The overgrazing problem. Pp. 643–666 *in* F. Bourliere (Ed.), *Tropical savannas.* Elsevier, Amsterdam.

McNaughton, S. J. 1985. Ecology of a grazing ecosystem: The Serengeti. *Ecol. Monogr.* **55:**259–294.

McNaughton, S. J. 1990. Mineral nutrition and seasonal movements of African migratory ungulates. *Nature* **345:**613–615.

McNaughton, S. J. and G. A. Sabuni. 1988. Large African mammals as regulators of vegetation structure. Pp. 339–354 *in* M. J. A. Werger, P. J. M. van der Aart, and J. T. A. Verhoeven (Eds.), *Plant form and vegetation structure.* SPB Academic Publishing, The Hague, The Netherlands.

Miller, R. I. 1978. Applying island biogeographic theory to an East African reserve. *Env. Conserv.* **5:**191–195.

Miller, R. I. and L. D. Harris. 1977. Isolation and extirpations in wildlife reserves. *Biol. Conserv.* **12:**311–315.

Ojasti, J. 1983. Ungulates and large rodents of South America. Pp. 427–439 *in* F. Bourliere (Ed.), *Tropical savannas.* Elsevier, Amsterdam.

Owen-Smith, R. N. 1988. *Megaherbivores: the influence of very large size on ecology.* Cambridge Univ. Press, Cambridge, England.

Owen-Smith, R. N. 1988. *Megaherbivores: The influence of very large size on ecology.* Cambridge Univ. Press, Cambridge, England.

Pellew, R. A. P. 1983. Impacts of elephant, giraffe and fire upon the *Acacia tortilis* woodlands of the Serengeti. *Afr. J. Ecol.* **21:**41–74.

Ruess, R. W. 1987. Herbivory. The role of large herbivores in nutrient cycling of tropical savannas. Pp. 67–91 *in* B. H. Walker (Ed.), *Determinants of tropical savannas.* Int. Union. Biol. Sci., Paris, France.

Sarmiento, G. 1984. *The ecology of neotropical savannas.* Harvard Univ. Press, Cambridge, Mass.

Sinclair, A. R. E. 1979. Dynamics of the Serengeti ecosystem. Pp. 1–30 *in* A. R. E. Sinclair and M. Norton-Griffiths (Eds.), *Serengeti: Dynamics of an ecosystem.* Univ. of Chicago Press, Chicago.

Soulé, M. F., B. A. Wilcox, and C. Holtby. 1979. Benign neglect: A model of faunal collapse in the game reserves of East Africa. *Biol. Cons.* **15:**259–272.

Western, D. 1989. The ecological role of elephants in Africa. *Pachyderm* **12:**42–45.

Western, D. and J. Ssemakula. 1981. The future of the savannah ecosystems: Ecological islands or faunal enclaves? *Afr. J. Ecol.* **19:**7–19.

Williamson, D., J. Williamson, and K. T. Ngwamotsoko. 1988. Wildebeest migration in the Kalahari. *Afr. J. Ecol.* **26:**269–280.

CHAPTER 8

Coastal Ecosystems

Some of the most important terrestrial ecosystems for wildlife are those that border lakes and oceans. Beaches and coastal dunes, both on mainland areas and on barrier islands, and sea cliffs are the most important of these coastal systems. These environments are ecosystems in their own right, but they are also breeding or non-breeding habitats for many animals that are members of other ecosystems. These include waterfowl and other freshwater birds, many raptors and seabirds, and all species of marine turtles. Still another coastal ecosystem of great importance is the salt marsh. We shall consider salt marshes as integral parts of estuarine marine ecosystems (See chapter 17).

The immediate coastline also attracts human development, both commercial and residential. Because of their highly dynamic geological nature, however, some coastal environments that are vital to wildlife are poorly suited to intensive development. Sandy coastal environments, including barrier islands, fall into this category. Severe storms can reshape these areas overnight, destroying property worth millions of dollars. Unfortunately, many such areas have been developed, often with the aid of government subsidies for road and bridge construction. Following storm damage, many of these areas have received still other subsidies, in the form of governmental funds and low-interest disaster loans to repair damage. Understanding the dynamics of coastal environments, and planning appropriate human uses that are compatible with their wildlife values are major conservation challenges.

ECOSYSTEMS OF SANDY COASTLINES

Beaches and coastal sand dunes occur along most ocean coastlines, as well as along the shores of many large lakes. They are best developed where rivers discharge large volumes of sandy sediment into shallow coastal waters, or where rising sea levels have caused the erosion of headlands, creating large volumes of sandy sediment along the coastline itself. The combined influences of a sandy substrate and marine (or lake-influenced) climate usually make sandy coastal ecosystems sharply distinct in habitat conditions from the ecosystems lying inland. Often, too, these systems are discontinuous, being separated by stretches of rocky coastline. The distinctiveness and isolation of these ecosystems mean that they often contain isolated, endemic populations of plants and animals.

Sandy coastal ecosystems are highly dynamic. The beaches themselves change in form seasonally. During periods of frequent storms, intense wave action erodes the beach, carrying sand into deeper water and producing a long, flat profile across which the waves expend their energy. During storm-free periods, waves carry sand toward the shore, building up a short, steep beach leading up to a flat berm or beach terrace. These changes can usually be seen on lake and ocean beaches in the temperate zone from winter to summer.

Sandy areas inland from the beach are also highly dynamic. The first plant colonists of newly deposited sand are often grasses, such as sea oats and marram grass, that spread by rhizomes (Fig. 8.1). These grasses trap wind-blown sand and cause the formation of a **foredune ridge** (Fig. 8.2). They also begin to stabilize the sand surface enough to permit other plants to become established, facilitating biotic succession. Succession often leads to a dense vegetation of grasses, shrubs, and trees that tend to cover and stabilize the sandy substrate.

As long as the vegetation cover is undisturbed the landscape is now stable. Where this cover is disturbed, however, an even more dramatic example of the potential instability of sandy coastal environments may occur. Wind action may initiate a **blowout**—an erosion basin that gradually enlarges as wind sweeps out sand and carries it inland, where it is deposited as an advancing dune front (Fig. 8.3). The advancing dune may bury mature forests, highways, and residential developments. Blowout formation is a normal phenomenon of coastal dunelands, but, of course, a process that can also be triggered by disturbance of vegetation by human activity.

In North America, coastal sand dune systems occur in many locations along the Atlantic and Pacific Coasts, as well as along the shores of the Great Lakes. In Oregon, Washington, and northern California, coastal dunelands reach perhaps their greatest extent and size. Here, these ecosystems occupy about 42 percent of the coastline, often extending inland for 2 to 4 kilometers (Wiedemann 1984). These dune ecosystems have developed since the last glacial maximum, and are often perched on non-sandy marine terraces below

FIGURE 8.1

■ Marram grass and other colonists of sandy beaches spread by rhizomes that connect individual shoots and anchor them in the loose substrate.

FIGURE 8.2

■ Colonization of inland edge of the beach at Cape Hatteras, North Carolina, by marram grass results in the deposition of wind-blown sand and the formation of a foredune.

FIGURE 8.3

■ Disturbance of the plant cover in Indiana Dunes State Park, at the south end of Lake Michigan, has led to the formation of a blowout. Inland from the blowout, sand is deposited in an advancing dune front.

FIGURE 8.4

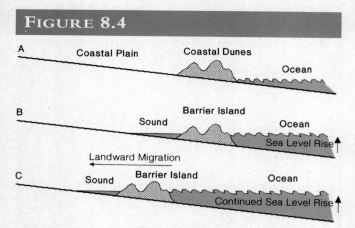

■ Barrier islands are believed to have originated by the isolation of beach and dune complexes by rising sea levels as continental glaciers melted. A. Formation of coastal dunes and beach ridges at a low interglacial sea level. B. Rising sea level caused by glacial melting allows water to flood in behind coastal dunes and ridges, isolating them as barrier islands. C. Continued sea level rise, coupled with the dynamics of transport of sand from the front to back of barrier islands, causes their landward migration.

FIGURE 8.5

■ At Canaveral National Seashore, Florida, wind transport, storm overwash, and inlet formation have all carried sand landward from the ocean side of the barrier island, at right, to create a mosaic of peninsulas, islets, and marshes on the inner side of the barrier.

sea level. Much of this coastal duneland has been covered and stabilized by vegetation, but spectacular, active blowouts and high dunes exist in many locations.

Barrier islands and beaches are also important types of sandy coastal ecosystems. **Barrier islands** are long, narrow islands formed from beach sediments and lying more or less parallel to the seacoast. **Barrier beaches** are similar, differing only in that they are connected to the mainland at one end. Barrier beaches and islands lie offshore from about 10 percent of the world's seacoasts. These landforms are well developed along the Atlantic and Gulf coasts of the United States, where about 295 barrier islands total 6,500 square kilometers in area.

Barrier beaches and islands are one of the most dynamic landforms on earth. Several mechanisms of origin of barrier islands have been suggested, but the most accepted is the **mainland beach detachment hypothesis** (Fig. 8.4). This hypothesis suggests that barrier islands originate at the end of periods of continental glaciation, when ocean levels are beginning to rise as a result of the melting of continental ice sheets (Dolan et al. 1980). At the height of the last Pleistocene glaciation, for example, sea level was about 120 m below its present level, and the coastline of North Carolina and other regions now bordered by barrier islands was far oceanward. As sea level rose, headlands were eroded, producing large volumes of sediment that contributed to the formation of extensive coastal beach and dune systems. The rising seas also flooded inland in river lowlands and embayments, spreading parallel to the coast behind the coastal dunes. Eventually, the rising sea completely cut the beach and dune complex from the mainland, forming a barrier island that shelters a **sound** on its inner side. Some barrier beaches and islands also form by the growth of sand spits parallel to the coast, beginning from a headland. Breaching of a long spit by one or more inlets then may create a barrier island system.

After barrier islands were formed in late glacial time, they began to migrate coastward. This migration results from the transport of sediment from the ocean side to the sound side of the island. Sediment transport occurs in several ways (Fig. 8.5). Wind carries sand inland from the beach, and blowout formation creates active dunes that advance toward the inland side of the island. During winter storms or hurricanes, heavy waves on top of a storm surge may break through the dune system and overflow the island, washing large quantities of sand to the back side of the island, and extending the island into the sound. Storm waves may also cut through the island, forming a new inlet or **pass** through which currents may carry sediment into the sound, depositing it as an underwater delta. Landward migration has enabled barrier islands to keep up with rising sea level, which would have inundated them in their original location.

Historical records document major changes in the position of island shorelines due to landward migration. Along much of the coastline of Virginia and North Carolina, the average rate of migration is about 1.2 m per year, but this rate is as great as 8.0 m per year in a few spots (Inman and Dolan 1989). The shells and other biotic relics that one finds on the ocean side of a barrier island also attest to island migration. These are often the shells of mollusks that live in the sheltered sounds on the opposite side of the island. The island has simply migrated across the site where these shells were deposited in sound sediments, and they have been exposed on the beach. In a number of locations tree stumps—relics of a forest once growing on the inland side of the island—are exposed on barrier island beaches.

The fate of barrier islands is thus eventually to fuse with the coastline. "Fossil" barrier islands from the last interglacial, such as the Ingleside Barrier along the Texas coast between Corpus Christi and Galveston, demonstrate this fate. The Ingleside Barrier is a sandy coastal formation fused with the continental coastline, but bordered oceanward by Matagorda Island and other modern barrier islands.

Another dynamic feature of barrier islands is the migration of the passes or inlets between islands. Longshore currents tend to deposit sediment on the up-flow side of passes and remove sediment on the down-flow side. Again, large changes in the measured position of passes are documented by historical records.

The dynamic nature of beach and dune systems, and especially of those that develop into barrier beaches and islands, means that these environments are poorly suited to urban development. Most parts of the United States Gulf and South Atlantic coastline have a 5 to 15 percent chance of being hit by a hurricane in a given year. Nevertheless, summer home developments and resorts exist on—in some cases completely cover—70 barrier islands from New Jersey to Texas. Almost annually, urbanized barrier islands somewhere experience massive damage as a result of the natural processes that reshape beaches and dunes, and slowly carry the islands landward. The damage in the United States from major hurricanes is now counted in billions of dollars. Much of this damage has led to massive expenditures of public funds to replace roads, bridges, and other public facilities that many feel should never have been built in the first place.

PLANT AND ANIMAL LIFE OF SANDY COASTAL ECOSYSTEMS

Plant Communities

From the beach inland, a transect of successional development of plant communities is usually evident. The colonists of the foredune region are perennial grasses, such as marram grass (*Aimophila breviligulata*) and sea oats (*Uniola paniculata*) that spread by rhizomes. Farther inland, other grasses, broad-leafed herbs, vines, and shrubs invade the pioneer stands of grasses. Along ocean coastlines, tolerance of salt spray is a requisite of all of these early successional species.

Succession usually leads to forest communities that often have a distinctive coastal duneland character. On Fire Island National Seashore, off the coast of Long Island, New York, for example, a 200- to 300-year-old forest of American holly, white sassafras, and shadbush has developed on the protected inner side of the barrier island (Art 1976). Known as the Sunken Forest (Fig. 8.6), this community has a nutrient budget dependent almost entirely on atmospheric inputs.

At the south end of Lake Michigan, succession leads through a complex sequence of cottonwood, pine, oak, and mixed hardwood forests. Between dune ridges paralleling the shore, pond and marsh communities also occur, and in drier locations tall-grass prairie occupies dune sands. These com-

FIGURE 8.6

The Sunken Forest in Fire Island National Seashore, New York, is a dwarf, barrier island forest of holly, sassafras, and shadbush that receives almost all of its nutrients in salt spray. The wooden walkway is designed to protect the fragile, sandy surface against erosion caused by trampling.

munities occur on beach ridges of Lake Michigan that were formed at different stages in glacial retreat, and are of differing age. These dunes, in fact, were the site at which the concept of plant succession was recognized. H.C. Cowles, who began his studies of dune succession in 1896, was the first of several prominent ecologists to study successional processes at this location. Indiana Dunes State Park, established in 1923, and Indiana Dunes National Lakeshore, created in 1972, preserve the full range of this remarkable successional sequence.

In the Pacific Northwest, the humid climate and complex microtopography of the coastal dunelands likewise result in a complex of communities, some 21 in all (Wiedemann 1984). The isolation and the distinctive nature of coastal dunes mean that they often contain endemic species. The Nipomo Dunes, in coastal Santa Barbara County, California, for example, contain 18 species of rare or endangered plants, several of them highly adapted to the dune environment (Smith 1976).

Terrestrial Vertebrates

Sandy coastal ecosystems, especially barrier islands, are breeding grounds or winter homes to many water birds. Many species of gulls, terns, plovers, herons, egrets, ibis, rails and other water birds nest on barrier islands or in the salt marshes bordering the sounds on their landward sides. Migratory shorebirds and waterfowl use many of the same areas as stopover points on migration, or as wintering habitats. Several

FIGURE 8.7

■ The piping plover, a nesting shorebird of lake and ocean beaches, is designated as endangered or threatened in all of its North American range.

FIGURE 8.8

■ The key deer, restricted to a few islands in the Florida Keys, is a small subspecies of the white-tailed deer with a total population of a few hundred individuals.

species of raptors, including the bald eagle, osprey, and peregrine falcon, are closely associated with coastal ecosystems, especially in winter.

Most immediately threatened by development and disturbance of sandy coasts are nesting species such as the least tern and the piping, snowy, and Wilson's plovers that nest on open sandy beaches (Fig. 8.7). Least terns require long stretches of bare sand for nesting, their protection against predators being isolation and camouflage of their nests and young. Increasing human use of beaches has led to the disappearance of these species from much of their original ranges.

Under the geographic isolation provided by barrier islands, many endemic forms of terrestrial vertebrates have evolved. Because of destruction of vegetation, many of these have become endangered. Along the coast of Georgia, the Cumberland Island pocket gopher, *Geomys cumberlandius,* a barrier island endemic species, is now considered endangered. Several endemic subspecies of beach mice, cotton rats, and rice rats on barrier islands along the Florida and Alabama coasts are also listed as threatened or endangered because of habitat disturbance.

The Florida Keys, a chain of islands of tropical reef origin, harbor many endemic reptiles and mammals, several of which are threatened or endangered (Schomer and Drew 1982). These include the endangered key deer, *Odocoileus virginianus clavium,* a diminutive subspecies of the white-tailed deer (Fig. 8.8). Reduced to less than 30 individuals in the late 1940s, this subspecies has recovered under protection to several hundred animals. Other endangered forms include endemic subspecies of the raccoon, marsh rabbit, and several rodents. Several tropical West Indian birds and reptiles also reach their northernmost distribution in the Florida Keys.

Sea Turtles

Sea turtles are intimately tied to sandy coastal ecosystems. Of the seven species of sea turtles, six are formally classified as threatened or endangered (Table 8.1). These species are very diverse in their habits, some being herbivores, others carnivores (Hendrickson 1980). Five species are widely distributed in tropical and subtropical oceans, but two are much more restricted. The Kemp's ridley is confined largely to the Caribbean and Gulf of Mexico, and the Australian flatback is limited to the northern coastal waters of Australia (Zangerl et al. 1988).

Sea turtles are tied to sandy coastal ecosystems by their nesting habits. The females of all species come ashore to deposit their eggs in nests excavated in beach sands (Fig. 8.9). From 50 to 250 eggs are laid at a time, with some females nesting several times in one season. The eggs incubate for 45 to 75 days by the heat of the sun-warmed sand. When the young hatch, they crawl immediately to the water and begin a juvenile life that may be 10 to 50 years in duration, during which time they may move thousands of kilometers from their home beach. During their first year, turtles of several species may live a pelagic life in association with accumulations of *Sargassum* seaweed accumulating in the centers of oceanic gyres and in drift lines formed where surface water masses converge and sink (Carr 1987). In the North Atlantic, the Sargasso Sea is such an area. Eventually they return as adults to mate and lay eggs. Some species nest at widely scattered locations, but

TABLE 8.1

Sea Turtles and their Ecological Characteristics. Endangered (E) and Threatened (T) Species are Designated.

SPECIES	HABITAT	FOOD	NESTING
Leathery Turtle (E) *Dermochelys coriacea*	Tropical open ocean	Jellyfish	Scattered tropical beaches (colonial)
Hawksbill Turtle (E) *Eremochelys imbricata*	Tropical oceans near coral reefs	Sponges	Coral reef beaches (solitary)
Kemp's Ridley Turtle (E) *Lepidochelys kempi*	Caribbean coastal waters	Crustaceans	Rancho Nuevo Beach, Tamaulipas, Mexico (colonial)
Olive Ridley (T) *Lepidochelys olivacea*	Tropical coastal waters except Caribbean	Crustaceans	Scattered tropical beaches (colonial)
Loggerhead Turtle (T) *Caretta caretta*	Subtropical to temperate coastal waters	Mollusks and crustaceans	Scattered beaches (colonial)
Green Sea Turtle (T) *Chelonia mydas*	Tropical and subtropical coastal waters	Sea grass, algae	Scattered beaches (colonial)
Flatback Turtle *Natator depressus*	Northern Australian coastal waters	Reef and benthic invertebrates	Tropical beaches (colonial)

FIGURE 8.9

■ The Kemp's ridley turtle nests exclusively on a barrier beach at Rancho Nuevo, Tamaulipas, Mexico. Because of the rapid decline of this species, eggs are collected for incubation in protected enclosures and for incubation and release at Padre Island National Seashore, Texas.

others concentrate their nesting on certain beaches year after year. At Tortuguero Beach on the Caribbean coast of Costa Rica, for example, green, hawksbill, and leatherback turtles nest in large numbers (Carr 1967). Adults of these species congregate offshore over a period of days or weeks, and then come ashore together in what is termed an **arribada** (arrival), concentrating their egg-laying into a period of a few nights.

Commercial exploitation of sea turtles for a variety of body parts and products has been a major cause of decline of all the endangered and threatened species (King 1981, Ross 1981). All are hunted for their meat, and the nests of all species are robbed of eggs for human consumption, especially in parts of Latin America where the eggs are believed to impart sexual vigor. The green turtle is hunted heavily for its **calipee,** the cartilaginous material lining the inner side of the ventral shell that is the essential ingredient of green turtle soup. Green turtle oil is used in certain cosmetics, and the skin and shell of the ridleys and the hawksbill are used for manufacture of leather goods and trinkets. Stuffed juveniles, mainly green and hawksbill turtles, are also a common tourist item. A 1987 analysis of turtle products sold in Japan indicated that this trade involved about 2 million animals annually, almost a million of which were stuffed juveniles (Gregg 1988).

Sea turtles have been hunted for centuries. Indeed, one of the earliest laws relating to wildlife exploitation in the New World was made to protect young sea turtles in the Bahamas (Carr 1967). Often, hunting is concentrated on or offshore from nesting beaches. This hunting, together with the raiding of nests by animals such as dogs and pigs that forage on the beaches, is certainly one of the factors responsible for the decline of sea turtle populations.

Recently, another major cause of mortality has been recognized: the capture of young turtles at sea in the nets of shrimp trawlers and in other fishing gear. Since turtles are air-breathing reptiles, being trapped in nets often results in

FIGURE 8.10

a.

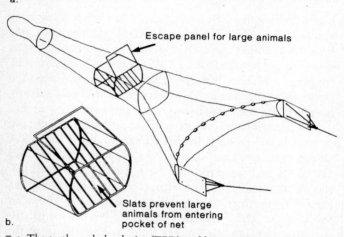

Escape panel for large animals

Slats prevent large
animals from entering
pocket of net

b.

■ a. The turtle excluder device (TED) enables animals the size of
young sea turtles to push open a trap door and escape from trawl nets
designed for shrimp. b. Diagram of TED structure and placement
(Modified from National Research Council 1990).

drowning if the animals are held submerged for more than
an hour or two. Analyses of population dynamics of animals
with a long juvenile life suggest that this cause of mortality
may be more serious than the more obvious mortality of adults
and eggs at nesting beaches (Crouse et al. 1987). By the late
1980s, annual shrimp trawler mortality of loggerhead turtles
along the Atlantic and Gulf Coasts of the United States was
estimated to be at least 5,000 animals, and possibly as much
as 50,000 individuals, with the mortality of Kemp's ridleys
being somewhere between 500 and 5,000 (National Research
Council 1990). Several kinds of **turtle excluder devices
(TED's),** essentially a trap-door arrangement that allows turtles and other large animals to escape trawl nets (Fig. 8.10),
have been developed and tested. In the United States, TED's
are now required for the larger offshore shrimp trawlers,
which make up about 10 percent of the entire shrimp fleet,

FIGURE 8.11

■ Newly hatched loggerhead turtles are released on a protected Florida
beach in an effort to establish new nesting beaches for the species.

but mandatory use by all trawlers has been resisted bitterly by
the fishing industry. Mexico has not required the use of TED's
on shrimp trawlers.

Special efforts are being made to encourage recovery of
the Kemp's ridley in the Gulf of Mexico, the most endangered of the sea turtles (Klima and McVey 1981). The Kemp's
ridley of the Caribbean and Gulf of Mexico nests almost entirely on a beach at Rancho Nuevo, Mexico. This population
has declined catastrophically in recent decades. In 1947, an
estimated 40,000 nesting female turtles were seen on the beach
on one night. Recent estimates are that only about 350 nesting
females remain (National Research Council 1990). During
the egg-laying season, the beaches at Rancho Nuevo, Tamaulipas, are now guarded against predators and poachers.
Clutches of newly laid eggs are collected (Fig. 8.9) and placed
inside predator-proof fencing or incubated artificially. Over
65,000 newly hatched turtles were released at Rancho Nuevo
in 1979 through this effort.

A second approach is **headstarting,** or raising young
turtles to the age of about a year and a weight of about
600 g, before their release (Klima and McVey 1981). These
turtles are then released in areas believed to be optimal habitat
for animals of this age. Although headstarting has been used
with some species for about 30 years, there is little evidence
that headstarted animals survive even as well as those entering
the ocean at hatching (National Research Council 1990). An
experimental headstarting program exists for the Kemp's
ridley, with releases being made along the west coast of Florida
and near the Florida Keys.

Efforts are also being made to introduce certain species
to new, more protected nesting beaches (Fig. 8.11). Beginning in 1978, an effort was begun to establish a new nesting
colony of Kemp's ridleys on Padre Island National Seashore,
Texas. Eggs taken from turtles at Rancho Nuevo have been
incubated in Padre Island sand, and the young released into
the ocean at Padre Island. Over 12,000 young have been released. Many of the headstarted turtles of this species have
also been imprinted on Padre Island sand, as well (Woody
1986). The juvenile life of this species is estimated to last about
15 years, so that in about 1993, adult females should begin
returning to Padre Island to nest, if this effort is successful.

The Outer Banks are a series of barrier beaches and islands fronting the Atlantic Coast for about 290 km from southern Virginia to central North Carolina. The barriers enclose sounds up to 45 km wide. Much of the northern portion of the barrier system is privately owned, but most of the southern portion has been preserved as Cape Hatteras and Cape Lookout National Seashores. Five National Wildlife Refuges border the sounds enclosed by the Outer Banks.

The Outer Banks show the dynamic nature of barrier island systems. Many inlets have formed, closed, reformed, and shifted. The tiny town of Currituck, lying inside Currituck Banks, was a bustling fishing port in the 1700s and early 1800s. Closing of the inlet serving the port in 1828 cut off access to the sea, freshening waters of that part of the sound and destroying the estuarine fisheries that were the port's mainstay.

Oregon Inlet separates Bodie and Pea Islands a short distance south of the community of Nags Head. Although an inlet has periodically existed at about this location at several times since settlement of the eastern seaboard, the present Oregon Inlet was formed during a hurricane in 1846. Since its formation, the inlet has migrated southward more than 2.8 km, shifting about 23 m per year (Inman and Dolan 1989). A bridge 5 km long was constructed across this inlet at a cost of several million dollars. The bridge has been an expensive victim of the shifting inlet, with several more millions of dollars being spent to protect and extend its southern end, while its northern end spans an increasing length of dry land.

Hurricanes and winter storms have struck the Outer Banks many times, destroying property in beach communities such as Kill Devil Hills and Nags Head. On Hatteras Island, the chance of a tropical storm in a given year is 18 percent, the chance of a hurricane 11 percent, and the chance of a high intensity hurricane 8 percent. Intense storms create a storm surge of 2 m, topped by storm waves that often breach the foredunes. Water flows across the islands, carrying beach sand and debris to be deposited as overwash fans. In severe storms, much of the islands are awash from the combination of surge and high waves. With each storm the beachline retreats a little and the inner side of the island advances landward.

The shoreward migration of the Outer Banks is evidenced by the Cape Hatteras lighthouse (Fig. 8.12). Built at a location about 1,000 m from the water in the late 1860s (Dolan et al 1973), it now stands at the edge of the beach. To save this lighthouse by moving it inland will cost about $8.7 million. If it is not moved, costly revetments or groins must be built to protect it as long as possible, an exercise in delaying the inevitable loss of the site to the sea.

The Outer Banks are immensely important to wildlife. Pea Island National Wildlife Refuge, south of Oregon Inlet, was established in 1938. About 265 species of birds breed in or regularly visit the refuge. It is a major wintering area for 25 species of ducks, Canada and snow geese, and tundra swans, as well as a breeding site for herons, ibis, rails, terns and many other water and land birds. It is also an important stopover area for many migrating shorebirds.

Human occupation of the Outer Banks began in the mid-1700s. On Shackleford Island, originally largely forested, settlers cleared much of the forest. Heavy livestock grazing disturbed the vegetation, fostering blowouts and actively migrating dunes that buried the remaining forest. The island was abandoned after a hurricane in 1889. Farther north, resort development began at Nags Head in the 1830s. With little concern about island dynamics, the foredunes were flattened and houses built immediately behind the beach line. This pattern continues to the present, in spite of several disastrous hurricanes. Elsewhere, attempts were made in the 1930s to stabilize the shorelines by planting marram grass behind the beaches. This effort resulted in an artificially high foredune system that some scientists believe has increased beach erosion by deflecting storm waves seaward. Again, this effort reflects failure to recognize the dynamics of barrier islands.

Today, the protected areas of the Outer Banks illustrate the valuable use to which these coastal environments can be put. More than 1 million people visit the national seashores and wildlife refuges of the Outer Banks annually.

FIGURE 8.12

■ The Cape Hatteras lighthouse, built about a kilometer from the ocean in the 1860s, now stands in imminent danger of destruction, testifying to the retreat of the barrier island shoreline.

PROTECTING BARRIER ISLAND ECOSYSTEMS

Sandy coastal ecosystems are particularly vulnerable to physical impacts on the vegetation or the soil surface. Trampling by humans is a serious problem in some areas, tending to promote the expansion of foredune vegetation into areas farther back from the beach, where other types of dune plants normally dominate (McDonnel 1981). Off-road vehicles can exert heavy damage to any type of dune vegetation, as well as causing mortality of animal life (Godfrey et al. 1980). At Sand Lake, along the central Oregon coast, for example, the activities of up to 5,000 ORV's on a single weekend have destroyed some of the finest examples of duneland vegetation (Wiedemann 1984). Even when ORV activity is confined to non-vegetated areas of the beach, repeated disruption of the sand surface promotes erosion and wind transport of sand inland, where excessive deposition can kill vegetation and lead to blowout development (Hosier and Eaton 1980).

Other impacts have resulted from the introduction of plants to coastal dunes, ostensibly to stabilize them. European beachgrass (*Ammophila arenaria*) and gorse (*Ulex europeus*), among other species, were introduced to dunes along the Pacific Coast (Wiedemann 1984). European beachgrass has spread widely, crowding out many of the native strand plants that fostered a distinctive hummocky dune topography. It also creates a strong foredune, unnatural for the region, that traps sand that formerly moved inland in greater quantity, feeding other duneland landscapes. Gorse, a shrub, has covered thousands of hectares of dunelands in southern Oregon.

The environmental unsuitability of barrier islands for urban development alone does not deter many developers, who nevertheless demand that public funds be used to provide roads, bridges, and other basic services, and to repair these services when they are damaged. When these funds are withdrawn, development often becomes impractical. In 1982, the United States Congress passed the **Coastal Barrier Resources Act,** which designated 186 barrier island units as a **Coastal Barrier Resource System.** These barrier islands extend along 751 miles of Atlantic and Gulf of Mexico shoreline. The key feature of this act is that these areas are ineligible for federal flood insurance or for federal road and bridge funds. Thus, public subsidization of development has been eliminated.

The importance of barrier islands to wildlife is evidenced by the location of national wildlife refuges and other wildlife preserves. About 35 national wildlife refuges and 10 national seashores lie on or inside barrier islands of the United States Atlantic and Gulf coasts. These refuges and parks represent appropriate uses of dynamic sandy coastal environments: wildlife conservation and recreation.

FIGURE 8.13

■ Thick-billed murres, one of the most abundant of the seabirds of the Auk Family, utilize narrow ledges of sea cliffs for nesting.

OTHER COASTAL ENVIRONMENTS

Cliffs, offshore rocks, and islands are major nesting areas for coastal water birds and certain raptors. More than 60 species of colonial water birds, including petrels, pelicans, cormorants, herons, egrets, ibis, gulls, terns, auks, and other birds, use such habitats in the United States (Fig. 8.13). In California, for example, some 23 species of seabirds nest at 260 locations along the coast. The total populations of these birds are about 700,000 individuals. Of these, about 89 percent belong to four cliff-nesting species: the common murre, Cassin's auklet, Brandt's cormorant, and western gull.

Colonial nesting makes many of these species vulnerable to exploitation. In the late 1800s and early 1900s, nesting colonies of many species, particularly egrets and ibis, were killed for their feathers, which were used for decorating women's hats. Until early in this century, sea bird colonies were also exploited for eggs almost everywhere (See chapter 18). This practice is still pursued in many parts of Latin America, Africa, and the Middle East (Croxall et al. 1984). Often, egg collectors destroy all the eggs present on an initial visit to a colony, and then return a few days later to harvest freshly laid eggs for sale in markets.

Many colonies of coastal water birds have also been decimated by the introduction of rats, cats, foxes, and other predatory animals to nesting islands (Nisbet 1979, Croxall et al. 1984). Increased numbers of opportunistic predators, such as ravens and gulls, their populations supported by human food sources such as garbage dumps and fish processing operations, have also contributed to the decline of some coastal birds.

Increasingly, the impacts of human visitation of nesting colonies is a serious cause of nest failure and mortality (Anderson and Keith 1980, Croxall et al. 1984). Recreational activities, tourism, and even scientific study can disturb colonies. Off-road vehicles or humans on foot can cause birds to flee from their nests, often spilling the eggs or young and leaving the nests open to predation by gulls, ravens, or other predators.

KEY MANAGEMENT STRATEGIES FOR COASTAL ECOSYSTEMS

1. Land use planning that limits urban development on sandy coastal areas, especially barrier islands, and designates these areas for recreation and wildlife use.

2. Protection of concentrated nesting areas of marine birds, mammals, and reptiles against disturbance and overexploitation.

LITERATURE CITED

Anderson, D. W. and J. O. Keith. 1980. The human influence on seabird nesting success: Conservation implications. *Cons. Biol.* **18:**65–80.

Art, H. W. 1976. *Ecological studies of the Sunken Forest, Fire Island National Seashore, New York.* U.S. National Park Service, Washington, D.C.

Carr, A. 1967. *So excellent a fishe.* Natural History Press, Garden City, New York.

Carr, A. 1987. New perspectives on the pelagic stage of sea turtle development. *Cons. Biol.* **1:**103–121.

Crouse, D. T., L. B. Crowder, and H. Caswell. 1987. A stage-based population model for loggerhead sea turtles and implications for conservation. *Ecology* **68:**1412–1423.

Croxall, J. P., P. G. H. Evans, and R. W. Schreiber. 1984. *Status and conservation of the world's seabirds.* Int. Council for Bird Preservation, Cambridge, England.

Dolan, R., P. J. Godfrey, and W. E. Odum. 1973. Man's impact on the barrier islands of North Carolina. *Amer. Sci.* **61:**152–162.

Dolan, R., B. Hayden, and H. Lins. 1980. Barrier islands. *Amer. Sci.* **68:**16–25.

Godfrey, P. J., S. P. Leatherman, and P. A. Buckley. 1980. ORV's and barrier beach degradation. *Parks* **5:**5–11.

Gregg, S. S. 1988. Of soup and survival: The plight of the sea turtles. *Sea Frontiers* **34:**297–302.

Hendrickson, J. R. 1980. The ecological strategies of sea turtles. *Amer. Zool.* **20:**597–608.

Hosier, P. E. and T. E. Eaton. 1980. The impact of vehicles on dune and grassland vegetation on a south-eastern North Carolina barrier beach. *J. Appl. Ecol.* **17:**173–182.

Inman, D. L. and R. Dolan. 1989. The Outer Banks of North Carolina: Budget of sediment and inlet dynamics along a migrating barrier system. *J. Coastal Res.* **5:**193–237.

King, F. W. 1981. Historical review of the decline of the Green Turtle and the Hawksbill. Pp. 183–188 *in* K. A. Bjorndal (Ed.), *Biology and conservation of sea turtles.* Smithsonian Institution Press, Washington, D.C.

Klima, E. F. and J. P. McVey. 1981. Headstarting the Kemp's Ridley Turtle, *Lepidochelys kempi.* Pp. 481–487 *in* K. A. Bjorndal (Ed.), *Biology and conservation of sea turtles.* Smithsonian Institution Press, Washington, D.C.

McDonnell, M. J. 1981. Trampling effects on coastal dune vegetation in the Parker River National Wildlife Refuge, Massachusetts, USA. *Biol. Cons.* **21:**289–301.

National Research Council. 1990. *Decline of sea turtles: Causes and prevention.* National Academy Press, Washington, D.C.

Nisbet, I. C. T. 1979. Conservation of marine birds of northern North America—a summary. Pp. 305–315 *in* J. C. Bartonek and D. N. Nettleship (Eds.), *Conservation of marine birds of northern North America.* U.S. Fish Wildl. Serv. Wildl. Res. Rep. 11.

Ross, J. P. 1981. Historical decline of Loggerhead, Ridley, and Leatherback Sea Turtles. Pp. 189–195 *in* K. A. Bjorndal (Ed.), *Biology and conservation of sea turtles.* Smithsonian Institution Press, Washington, D.C.

Schomer, N. S. and R. D. Drew. 1982. *An ecological characterization of the lower Everglades, Florida Bay, and the Florida Keys.* U.S. Fish and Wildl. Serv., Office of Biol. Services, Washington, D.C. FWS/OBS-82/58.1.

Smith, K. A. 1976. *The natural resources of the Nipomo Dunes and wetlands.* Coastal Wetlands Series #15, Calif. Dept. Fish and Game, Sacramento.

Wiedemann, A. M. 1984. *The ecology of the Pacific Northwest coastal sand dunes: A community profile.* U.S. Fish Wildl. Serv. FWS/OBS-84/04.

Woody, J. B. 1986. Kemp's Ridley Sea Turtle. Pp. 919–931 *in* R. L. Di Silvestro (Ed.), *Audubon wildlife report 1986.* National Audubon Society, New York.

Zangerl, R., L. P. Hendrickson, and J. R. Hendrickson. 1988. *A redescription of the Australian flatback sea turtle.* Bishop Museum Press, Honolulu, HI.

CHAPTER 9

T Islands

The plants and animals of ocean islands have suffered some of the most serious impacts of human disturbance. More species have become extinct on islands than in the vastly more extensive continental environment (See chapter 2). At the same time, island ecosystems contain some of the most remarkable plants and animals that exist on earth (Carlquist 1965). Preservation of the members of these island ecosystems is thus one of the major challenges of conservation ecology.

Islands are of two major types: oceanic and continental. **Oceanic islands** are those which have never been connected to a continent, good examples being the Hawaiian Islands and the Galápagos Islands (Fig. 9.1). Most oceanic islands are of volcanic origin, and many show extensive areas of recent volcanic activity (Fig 9.2). The flora and fauna of such islands are thus derived exclusively by overwater transport (including recent introductions by man). The biotas of oceanic islands are often very biased samples of the kinds of forms that exist on the nearest continents. **Continental islands** are those that once possessed a continental connection. Trinidad, for example, lying off the northern coast of South America, was connected with Venezuela during the last Pleistocene glacial period, when sea level was about 125 m lower than at present. Continental islands may possess a flora and fauna much like that of the continental areas to which they were once connected.

CHARACTERISTICS OF ISLAND BIOTAS

The isolation of island ecosystems profoundly influences biogeographic, ecological, and evolutionary processes (Mueller-Dombois et al. 1981). As a result, the organisms of islands, and the communities they form, possess many distinctive characteristics. These characteristics are the product of the biogeographic hurdles that organisms have to surmount to reach islands, and of the distinctive evolutionary environment they encounter once they arrive on islands.

Species diversity is low in island ecosystems. In the case of small oceanic islands, the number of species present is largely the outcome of two processes: colonization of the island by species from other areas, and extinction of the populations of species inhabiting the island. Isolation causes colonization rates to be low, but at the same time, the risk of extinction may be high because of the small area of the island and the small size of the population of an individual species. For archipelagos such as the Hawaiian Islands and the Galápagos Islands, speciation may be a second source of new species. Speciation requires that populations of an ancestral form colonize two or more islands, where the different populations can eventually diverge to become distinct daughter species. It is doubtful that speciation is accelerated on islands compared to continental areas, and, of course, the products of insular speciation are also affected by high extinction rates.

The low diversity of island biotas is illustrated by the native land birds of the Galápagos Islands, lying 1,000 km west of South America. The Galápagos Islands are one of the few archipelagos of substantial size and diversity for which we have an idea of the full biota prior to disturbance by humans. Only 28 land bird species occur on the Galápagos, the product of perhaps as few as 13 colonization events. Some of these species are undifferentiated from their relatives on the coast of Ecuador. Others, such as 13 species of finches of the sub-family Geospizinae, are forms that were derived by speciation within the archipelago (Grant 1986). An area of equivalent size and vegetational diversity on the South American continent, however, would probably have a bird fauna ten or twenty times as rich.

Even continental islands exhibit a reduced level of species diversity. When the continental connection existed, their biotas were presumably as rich as those of similar neighboring areas. But when the land connection is severed, the rate of colonization of the area drops, and the rate of extinction increases due to the fact that an isolated land area with populations of small size has been created. For some time, the rate of extinction exceeds the rate of colonization, and the number of species declines, a process known as **biotic relaxation.** In the Gulf of California, for example, lizards occur on islands that were connected to the mainland during the height of Pleistocene glaciation. At that time, the future island areas probably had faunas identical to those of the mainland at large. With isolation, however, the number of species on the islands has declined, and the longer the isolation, the greater this loss has been (Wilcox 1978).

A dramatic example of biotic relaxation is provided by forest birds on Barro Colorado Island, Panama (Karr 1982). This island, about 15.6 square kilometers in area, was created in 1913 by the flooding of Gatun Lake as part of the construction of the Panama Canal. Since the first faunal surveys of the island, some 23 species of forest birds are known to have disappeared. Comparisons of the present bird fauna of the island with that of comparable forests in nearby Parque Nacional Soberania, on the mainland, however, suggest that as many as 50 additional species were probably present on Barro Colorado Island at the time of isolation, and have since disappeared from the island. This loss amounts to over 27 percent of the probable original breeding bird fauna of the island.

A second characteristic of islands is high endemicity. **Endemic species** are those that have evolved in, and are restricted to, a particular area. All of the native land birds of the Hawaiian Islands, for example, are endemic. Over 40 percent of the plant species on well isolated oceanic islands are commonly endemic (Table 9.1). The isolation of the islands means that once a species becomes established, its evolution is likely to be independent of that occurring in its parental population, thus leading to divergence. Endemic groups of plants and animals on islands often show striking examples of **adaptive radiation,** the evolutionary diversification of species for different ways of life. The Hawaiian honeycreepers, birds of the endemic sub-family Drepanidinae, show remarkable patterns of specialization of beaks for feeding on insects, seeds, and nectar (Scott et al. 1988). The Galápagos finches, which form the endemic sub-family Geospizinae, have diversified from an ancestral seed-eater to become fruit-eaters, wood-excavators, bark-scalers, and foliage-gleaners (Grant 1986).

A third characteristic of island biotas, especially those of oceanic islands, is absence or poor representation of certain ecological groups. Large predators tend to be absent or poorly represented on islands compared to continental areas. In the Hawaiian Islands, for example, the Hawaiian hawk and the short-eared owl are the only predatory land birds; the hawk is restricted to one island, Hawaii. Flightless mammals, including browsing and grazing mammals as well as carnivores, tend to be absent. No native terrestrial mammals, for example, occur in the Hawaiian Islands. Early successional plants that tend to invade areas disturbed by animals and even other types of disturbance are often poorly represented. In the absence of large grazing animals that can disturb the vegetation, of large rivers that can cause flooding and erosion, and often of causes of fire, evolution has not favored adaptation of plants for disturbed sites.

FIGURE 9.1

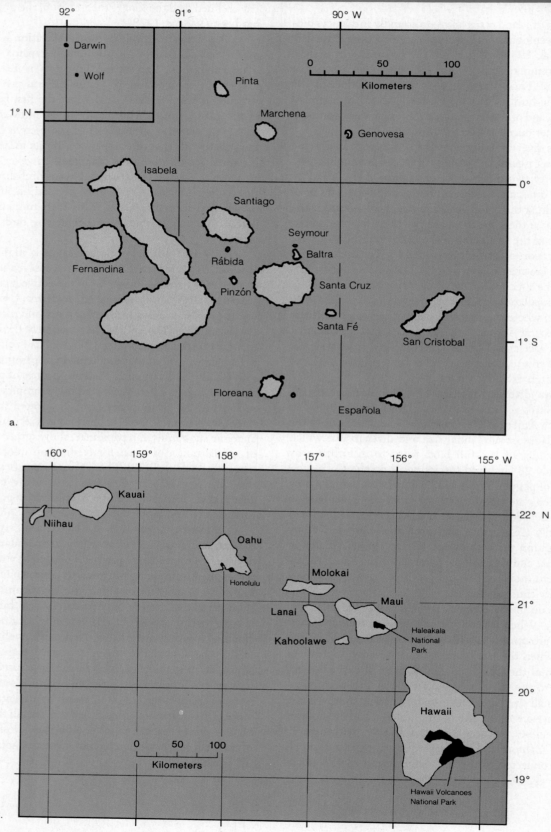

a.

b.

■ The Galápagos (a) and Hawaiian (b) archipelagos.

TABLE 9.1

Native and Endemic Plant Species on Various Islands and Archipelagos, and their Conservation Status.

| | NUMBER OF SPECIES | | PERCENT OF SPECIES | |
LOCATION	NATIVE	ENDEMIC	ENDEMIC	THREATENED ENDEMIC
North Atlantic				
Azores	600	55	9	42
Canary Islands		500		75
South Atlantic				
Ascension	25	11	44	82
St. Helena	60	50	83	8
North Pacific				
Hawaiian Islands	1,200 to 1,300	1,140 to 1,235	95	50
South Pacific				
Galápagos	541	229	42	
New Calidonia	3,250	2,474	76	6
Norfolk Island	174	48	28	94
New Zealand	2,000	1,620	81	8
Juan Fernandez	147	118	80	79
Indian Ocean				
Mauritius	850	280	33	42

Source: W. V. Reid and K. R. Miller, *Keeping Options Alive: The Scientific Basis for Conserving Biodiversity*. Copyright © 1989 World Resources Institute, Washington, DC; and W. L. Wagner, et al., "Status of the Native Flowering Plants of the Hawaiian Islands" in C. P. Stone and J. M. Scott, (Editors), *Hawaii's Terrestrial Ecosystems: Preservation and Management*, 1985, University of Hawaii, Honolulu; Data from Davis, et al., 1986.

FIGURE 9.2

■ The Galápagos Islands are oceanic islands of volcanic origin, as evident in this view of the eastern coast of Santa Cruz from the neighboring small island of Bartolome.

FIGURE 9.3

■ The flightless Galápagos cormorant is endemic to the Galápagos Islands.

A fourth characteristic is modified ecology and behavior. Flightlessness among birds (Fig. 9.3) and insects, and fearlessness of humans and other large animals are common features of island animals. Presumably, these characteristics reflect the absence of predators for these species. Sensitivity to grazing or browsing is a comparable characteristic of many island plants. The physical or chemical defenses, such as thorns or bad-tasting chemicals, by which continental plants defend themselves against herbivores have often been lost. Where grazing or browsing animals are absent, natural selection does not maintain energetically costly defenses of these sorts, favoring instead the diversion of energy to some adaptation that has an important function in the island situation.

Finally, a fifth characteristic is **ecological release:** the expansion of the ecological niche by occupation of more habitats and the use of more kinds of resources. In a sense, island species make up for the shortage of species by becoming generalists. In a comparison of bird communities in Panama and on the small Caribbean island of St. Kitts, for example, Cox and Ricklefs (1977) found that island species both occupied more kinds of habitats and were more abundant within each habitat than were birds in a continental area with a similar range of habitats. In some cases, species that show ecological release may evolve ecological races or sexual differences that are like the differences shown by different species in continental regions.

HUMAN IMPACTS ON ISLAND ECOSYSTEMS

The characteristics of island biotas have combined to make them highly vulnerable to the impacts of human activity (Mueller-Dombois and Loope 1990). The most serious human impacts have been due to hunting and harvesting of particular species, destruction of native vegetation, and the introduction of alien plants and animals.

Hunting has affected only a few selected species, but the impact in these instances has often been heavy because of the fearlessness of the animals, or their physical inability to escape humans. The Dodo, a large, flightless pigeon that inhabited the island of Mauritius, in the Indian Ocean, was hunted to extinction in the seventeenth century. Because of their fearlessness, dodos were captured for meat by the crews of ships that visited the islands. On many oceanic islands, such as in the Mascarene Islands and Seychelles in the Indian Ocean, populations of giant tortoises were hunted to extinction, or severely reduced in numbers, for the same reason. Polynesian colonists of the Hawaiian Islands probably contributed to elimination of many endemic flightless birds by hunting (See chapter 2).

Where man settles, destruction of the native vegetation follows. On islands, this can quickly lead to disappearance of distinctive vegetation types that evolved in isolation. Clearing and burning of island vegetation is an obvious cause of damage to native vegetation. Overall, however, introduced plants and mammalian herbivores have probably been responsible for the greatest damage to island vegetation.

Many plants, brought to islands as crop, forages, or ornamentals, have become problem weeds (Smith 1989). Many species invade natural communities, especially after disturbance, and crowd out or overgrow native species. Others, particularly perennial grasses that become dormant seasonally, increase the incidence of fire. Some exotics also cause major changes in soil water and nutrient status, which may be detrimental to native communities. Examples of these effects are given in the case studies later in the chapter.

Many kinds of herbivorous mammals have been introduced to islands. Early ships often introduced goats or other mammals to islands to create a meat supply for future use. Other large herbivores, such as pigs and donkeys, brought to the islands as domestic species, gave rise to feral populations. Feral populations of these species, lacking control by predators and exploiting vegetation unadapted to grazing or browsing pressures, often virtually strip islands of their plant cover. The tussock grasslands of the islands of the southern oceans were extremely vulnerable to overgrazing (Holgate and Wace 1961).

Damage to island vegetation by grazing and browsing mammals has been responsible for many extinctions of native plants and animals, as well (Holgate and Wace 1961, Brockie et al. 1988). Tristan da Cunha, in the South Atlantic, was originally covered by a rich forest and tussock grassland. Settlers in the early 1800s brought many kinds of domestic animals, which decimated the native vegetation. The tussock grass itself is now extinct on the island, together with several endemic birds (Holgate and Wace 1961).

Predators such as dogs, cats, mongooses, rats, and mice have been introduced deliberately or accidentally to many islands. Introduction of mongooses to several islands in the West Indies has led to extinctions of populations of native lizards. The introduction of the brown rat snake to Guam has resulted in the loss of almost all species of native birds (See chapter 2). Rats and mice, ubiquitous followers of man, reached many islands in shipments of cargo, or by shipwrecks. One or more species of *Rattus* have been introduced to about 80 percent of oceanic islands, where they have contributed to extinction of at least 31 species of birds. Together with cats they have also caused enormous reductions in the populations of seabirds on many islands (Moors and Atkinson 1985).

Introduced competitors have also been a cause of extinctions on islands. Introduced plants are often competitors for native species, as are introduced rodents for native species. Introduction of land birds to Hawaii, New Zealand and other island regions has also contributed to the decline of native bird faunas.

The Galápagos Islands (Fig. 9.1) are an oceanic archipelago lying on the equator about 1,000 km west of South America. The islands, part of Ecuador, are volcanic, and range in age from about 1 to 5 million years. The Galápagos Archipelago is one of the few oceanic island groups that was not colonized by humans in prehistoric time. Thus, the biota present when the islands were first visited by European voyagers represented the result of evolution unaffected by human activity. Island environments range from arid lowlands to humid highland forests. Recent volcanic activity has created landscapes of lava and ash in many places.

The biota of the Galápagos has been derived largely from South America, and is strongly endemic. Over 42 percent of the 541 native plants, many of the native birds, and all of the native mammals and reptiles are endemic (Fig. 9.4). Several genera of plants and animals have shown elaborate patterns of adaptive radiation in the archipelago, making it an important natural laboratory of evolution (Grant 1986, Cox 1990).

Human impacts date from the 16th century. The islands were discovered in 1535 by Spanish voyagers from Panama (Jackson 1985). From the 1500s through the 1700s they were visited by whalers and buccaneers. Settlement began in the early 1800s. Only four islands—Isabela, Santa Cruz, San Cristobal, and Floreana—have been settled, and the total human population is now about 9,000. In 1959, the uninhabited regions of the archipelago, about 90 percent of their total area, were declared a national park. In 1964, the Charles Darwin Research Station, located on Santa Cruz Island, was established to coordinate research and conservation activities in the archipelago. Tourism is now the major economic activity in the islands, with more than 20,000 visitors a year being attracted by the natural history of the archipelago (Fig. 9.5).

Hunting and killing have reduced or eliminated populations of several Galápagos species. Populations of giant tortoises (Fig. 9.6) were hunted to extinction, or severely reduced in numbers. These large tortoises were also taken primarily by crews of ships, because the animals could be kept alive and in good condition (as sources of meat) for months. It was common for a single ship to take aboard 100 to 400 tortoises at one time. The total number of animals removed from the Galápagos probably numbers in the hundreds of thousands (Steadman 1986). At least three races of the Galápagos tortoises were hunted to extinction, and a fourth survives as only a single individual. Recovery and protection of populations on six other islands is one of the major conservation goals of the Charles Darwin Research Station. More recently, on Floreana, killing of the native Galápagos hawk and barn owl by settlers was probably the final cause of local extinction of these species, since both were very tame (Steadman 1986).

Settlement, concentrated in the higher elevations where moisture permitted farming, resulted in the direct destruction of unique vegetation types. Forests of *Scalesia* trees, an endemic genus of the sunflower family, were severely affected by land clearing, as well as by other impacts. Only fragments remain of these original highland forests (Fig. 9.7).

The Galápagos Islands provide a good case study of the effects of both animal and plant introductions. Most of the larger

FIGURE 9.5

■ Tourist visitors to the Galápagos Islands, now numbering about 20,000 annually, must be accompanied by a licensed guide and must remain on designated trails.

FIGURE 9.4

■ The marine iguana is one of the ecologically unique, endemic reptiles of the Galápagos Islands.

FIGURE 9.6

■ Giant tortoises similar to this individual on Santa Cruz Island in the Galápagos originally occurred on oceanic islands in several parts of the world.

FIGURE 9.7

■ Forests of *Scalesia* trees are now restricted to small areas of the highlands of Santa Cruz in the Galápagos.

islands, especially those with human settlements, have 6 to 10 species of introduced mammals (Fig 9.8). Goats have been introduced to several of the smaller islands. In many cases, these introductions led to population explosions (Eckhardt 1972). On the island of Pinta, for example, one male and two female goats were introduced in 1959. By 1970 the population on this island of only 60 square kilometers was estimated to be between 5,000 and 10,000 goats. On Santiago Island, goats introduced in the early 1800s increased to 80,000 to 100,000 animals in the 1970s and 1980s (Schofield 1989). Goats consume a wide variety of plants, including ferns, herbaceous plants, shrubs, trees, and cacti (Schofield 1989). On several islands goat browsing decimated the lowland vegetation and severely damaged the forest understory at higher elevations. On Floreana, Eckhardt (1972) estimated that over 70 percent of the native plant species were reduced or eliminated by goat browsing.

Cattle, pigs, horses and donkeys have also become feral on some of the inhabited islands. Cattle grazing in highland areas of Santa Cruz has severely disturbed the native shrublands that occur above the *Scalesia* forests. In overgrazed areas of these shrublands exotic grasses tend to invade.

Many introduced plants have also become problems (Schofield 1989). Guava, *Psidium guajava,* has become the dominant shrub over large areas of the highlands in the wake of overbrowsing by goats and other animals. Recovery of the original highland vegetation is now inhibited in many places by this weedy shrub. The quinine tree, *Cinchona succirubra,* a tree that spreads by wind-blown seeds, has invaded much of the high-elevation forest and shrubland of Santa Cruz. Lantana, *Lantana camara,* a weedy tropical shrub, is another highly invasive exotic on Santa Cruz and San Cristobal.

Vegetation destruction has secondarily affected many animals (Steadman 1986). Damage and destruction of the *Scalesia* forests has probably caused the extinction of populations of the sharp-beaked ground finch, which has disappeared from all four islands that are inhabited by humans. Browsing by goats, donkeys, and other large mammals has probably been responsible for several extinctions of birds. On Floreana Island, for example, destruction of the forests of tree prickly pears by these animals probably led to the extinction of the native mockingbird and the large-billed ground finch, both of which depended on this plant.

In the Galápagos Islands, introduced black and Norway rats have probably led to the extinction of native rice rats on seven islands (Eckhardt 1972). Rice rats survive only on islands free of these species.

Major efforts have been made to control the spread of invasive plants and to eradicate introduced animals. Goats have been eliminated from five smaller islands, and have been greatly reduced or perhaps eliminated on Pinta, a medium-sized island. Major recovery of the vegetation has resulted in these locations.

FIGURE 9.8

■ Introduced mammals of many kinds occur on most of the Galápagos Islands, but populations of some species have recently been eradicated as part of efforts to restore the habitats of the islands.

The Hawaiian archipelago, a set of volcanic islands, atolls, and reefs (Fig. 9.1), is spread along almost 2,450 km of tropical ocean, the youngest islands at the eastern end of the chain (Wagner et al. 1985). Hawaii, the youngest of the main islands, is probably less than a million years in age; Kauai, the westernmost large island, is 3.8 to 5.6 million years old (Carlquist 1980). The rocks and atolls farther west are eroded remnants of still older islands, so the archipelago as a whole has been in existence for a much longer time, perhaps as long as 70 million years (Wagner et al. 1985).

The Hawaiian Islands are even richer than the Galápagos in biological diversity. Much of the islands has a humid tropical environment favorable to plant and animal life. About 95 percent of the plants and animals native to the islands are endemic (Fig. 9.9), perhaps the highest level of endemicity for any area of comparable size (Carlquist 1980). Of the 1,200 to 1,300 species of native flowering plants, 95 percent are endemic (Table 9.1), the result of colonization of the islands by about 270 ancestral forms (Fosberg 1948). Among invertebrates, 22 to 24 colonizations by land snails, followed by speciation, have resulted in an extraordinary fauna of about

1,000 species (Gagne and Christensen 1985). Similarly, 300 to 400 arthropod colonists have undergone speciation to produce more than 6,000 descendent species. Among land birds, a single ancestral finch radiated to produce 47 species and subspecies of songbirds specialized for diverse feeding niches. Many of these disappeared during the Polynesian period, and most of the survivors are now endangered (Scott et al. 1988).

When visited by Captain Cook in 1778, these islands had already sustained major biotic losses due to changes resulting from Polynesian settlement (See chapter 2). The Polynesian human population grew to about 200,000 to 250,000. Much of the destruction of the native lowland forests was accomplished during the Polynesian period, when perhaps 80 percent of the lowland forest was cleared for farming (Kirch 1982, Stone 1985). The Polynesians also hunted many of the native birds for food and feathers and introduced several predatory mammals and a number of plants. These effects evidently led to the extinction of many birds, and probably a variety of other plants and animals.

The rates of environmental disturbance and extinction of species have increased following European settlement, however. Large areas of native highland forests have been logged and replaced by cattle ranches and plantations of pines and eucalyptus. Koa, *Acacia koa,* one of the dominant native trees, is a valuable wood for furniture and cabinet making. 'Ohi'a, *Metrosideros polymorpha,* on the other hand, is less valuable for its wood, and has been harvested for fuel for electric power generation, even after native forests were reduced to less than a quarter of their original extent (Holden 1985). The greatest disturbance of native vegetation has resulted from the introduction of exotic plants and animals, however. Over 600 exotic plants have become established in the wild, and about 86 of these have become serious pests (Smith 1985). Some, such as guava and lantana, are the same species that are serious pests on the Galápagos. Several other trees, shrubs, vines, and grasses are serious invaders of native vegetation, however.

Firetree, *Myrica faya,* native to the Canary Islands, has invaded wet and moist sites, including open volcanic ash areas, on all of the main islands. This tree, which has nitrogen-fixing root nodules, forms pure stands that outcompete native plants (Vitousek and Walker 1989). This exotic thus modifies basic nutrient cycling processes of ash-soil ecosystems. Koa haole, *Leucaena*

leucocephala, a nitrogen-fixing leguminous shrub, is ubiquitous in dry to moist lowland areas throughout the islands. It also forms dense stands that crowd out all other plants.

The banana poka, *Passiflora mollissima,* a South American vine of the passion flower family, has caused serious damage to native upland forests. Banana poka (Fig. 9.10) its seeds spread by pigs that feed on its fruits, grows into the crowns of the native forest trees and smothers their foliage (Fig. 9.11). About 500 square kilometers of forest are being attacked by this vine (Mueller-Dombois and Loope 1990).

Broomsedge, *Andropogon virginicus,* a perennial grass from the southeastern United States has spread throughout large areas of former wet mountain forest. Although it is a perennial, broomsedge becomes dormant during the temperate zone winter. At this time broomsedge stands transpire little or no water, so that the soil can become saturated with water for long periods, leading to a serious landslide problem (Mueller-Dombois 1973). During the dormant season, broomsedge also becomes prone to fire, and is responsible for a major increase in fire frequency in areas where it occurs (Smith 1985).

FIGURE 9.9

■ The Haleakala silversword, a member of an endemic genus of the Sunflower Family, was almost driven to extinction by goat browsing before establishment of Haleakala National Park on Maui, Hawaii.

FIGURE 9.10

■ The banana poka, a passion flower native to South America, has been introduced to the Hawaiian Islands.

FIGURE 9.11

■ In the Hawaiian Islands, banana poka vines climb into the crowns of native forest and smother the trees.

Efforts have been made to find means of biological control of the most serious plant invaders. Partial control of lantana has been achieved by introduced insect enemies. Control of problem grasses in this fashion is difficult because the importation of their possible enemies is often opposed by the sugar cane industry (because of the possibility of sugar cane, a grass, also being attacked).

Introduced animals add to the destruction of the native vegetation. Pigs are a major cause of damage to the understory vegetation of native forests, reducing, as well, the regeneration of the native canopy trees. Dispersal of seeds of both the banana poka and guava is aided by pigs, which feed heavily on their fruits. Goats are a serious problem in dry to moist, more open areas. Goat browsing has seriously reduced a number of native species, as shown by the recovery of these forms in areas fenced to exclude them (Loope and Scowcroft 1985). Eradication of these animals is virtually impossible over large areas of rugged mountain topography. Many resident

people also hunt pigs and goats, and regard them as a desirable resource. Efforts to control these animals have therefore centered on trying to reduce their numbers to less destructive levels by hunting, and by excluding them from specific areas by fencing.

Rats, mice, mongooses, and feral cats are other problem exotics. Black rats damage the flowers, fruit, and bark of many shrubs and trees in the native forests. These smaller animals are also predators on native invertebrates and birds. The black rat led to the extinction of the Laysan rail on Midway Island, after the species had been translocated there from Laysan Island, where its habitat had been decimated by the grazing of introduced rabbits.

Many kinds of land birds have been introduced to the Hawaiian Islands, and few native birds survive in the lowland areas of the islands. Most of these species are confined to cities, farmland, and vegetation dominated by exotic plants. Nevertheless, competition from some of these introductions may be a factor in decline of native land birds (Mountainspring and Scott 1985). The introduction of avian pox and avian malaria to Hawaii, the latter resulting from the combined introduction of

mosquito vectors and exotic birds carrying the malaria parasites, has contributed to the extinction of native birds. These diseases are now one of the major threats to the remaining endemic birds (Scott and Sincock 1985, van Riper et al. 1986, Warner 1968).

As a result, the Hawaiian Islands contain more than a quarter of the threatened and endangered species in the United States. About 177 native plants are known or believed to have become extinct (Wagner et al. 1985), and 18 were listed as endangered or threatened in 1989. Over 30 percent of the species in the United States being considered for federal listing as endangered or threatened were from Hawaii. Since European settlement, 13 species of Hawaiian land birds have become extinct and several others have been reduced to less than 100 individuals (Scott et al. 1988). Most recently, in late 1989, the last Kauai o'o', a male that had responded to tape recordings of its species' call, could no longer be found. About 30 Hawaiian birds are now listed as threatened or endangered.

KEYSTONE EXOTICS AND THE STABILITY OF ISLAND ECOSYSTEMS

The examples described earlier suggest that certain types of species have a profound impact on island ecosystems. In general, these are species that differ substantially in way of life from those of the islands in question. Such species can be considered as **keystone exotics**—species capable of restructuring almost completely the ecosystems they invade. Good examples of keystone exotics include goats and donkeys (browsers) on the Galápagos Islands, the brown rat snake (an arboreal predator) in Guam, and pigs and the banana poka in Hawaii. These species all differ strikingly in their way of life from the animals or plants native to the islands, and native forms are simply not adapted to resisting their detrimental impacts.

This suggests that not all introductions to islands are likely to be deleterious. One illustration of this comes from New Zealand, to which hundreds of exotic animals and plants have been introduced. Some 15 species of ungulates have been introduced to, or spread as feral animals into, forested areas of New Zealand. In addition, the arboreal brush-tailed opossum, also a herbivore, has been introduced from Australia. These animals have caused severe, widespread damage to the native forests. In addition, many exotic birds have also been introduced to New Zealand, and these have replaced native forest species in many areas.

However, in the Hauraki Gulf, near Auckland on North Island, there are several forested islands that have remained free of introduced mammalian herbivores (Diamond and Veitch 1981). Although many of the introduced birds have reached these islands, they do not enter the forests and displace native forest birds. This suggests that birds from continental areas are not inherently superior to native island species. Thus, the introduced herbivores are keystone species. If the effects of these exotics can be controlled, the various introduced birds are not themselves likely to cause massive disruption of the New Zealand avifauna. We shall examine strategies for control of exotic species in detail in chapter 23.

KEY CONSERVATION STRATEGIES FOR ISLAND ECOSYSTEMS

1. Protection of native island vegetation against deliberate destruction and conversion by human activities.

2. Exclusion and removal of keystone exotics that trigger the restructuring of island plant and animal communities.

LITERATURE CITED

Brockie, R. E., L. L. Loope, M. B. Usher, and O. Hamann. 1988. Biological invasions of island nature reserves. *Biol. Cons.* **44:** 9–36.

Carlquist, S. 1965. *Island life.* Natural History Press, New York, NY.

Carlquist, S. 1980. *Hawaii: A natural history.* Pacific Tropical Botanic Garden, Honolulu, HI.

Cox, G. W. 1990. Centers of speciation in the Galápagos land bird fauna. *Evol. Ecol.* **4:**130–142.

Cox, G. W. and R. E. Ricklefs. 1977. Species diversity, ecological release, and community structuring in Caribbean land bird faunas. *Oikos* **28:**113–122.

Diamond, J. M. and C. R. Veitch. 1981. Extinctions and introductions in the New Zealand avifauna: Cause and effect? *Science* **211:**499–501.

Eckhardt, R. C. 1972. Introduced plants and animals in the Galápagos Islands. *BioScience* **22:**585–590.

Fosberg, F. R. 1948. Derivation of the flora of the Hawaiian Islands. Pp. 107–119 in E. C. Zimmerman (Ed.), *Insects of Hawaii.* Univ. of Hawaii Press, Honolulu.

Gagne, W. C. and C. C. Christensen. 1985. Conservation status of native terrestrial invertebrates in Hawai'i. Pp. 105–141 in C. P. Stone and J. M. Scott (Eds.), *Hawai'i's terrestrial ecosystems: Preservation and management.* Univ. of Hawaii, Honolulu.

Grant, P. R. 1986. *Ecology and evolution of Darwin's finches.* Princeton Univ. Press, Princeton, NJ.

Holden, C. 1985. Hawaiian rainforest being felled. *Science* **228:** 1073–1074.

Holgate, M. W. and N. M. Wace. 1961. The influence of man on the floras and faunas of southern islands. *Polar Record* **10:** 473–493.

Jackson, M. H. 1985. *Galápagos: A natural history guide.* Univ. Calgary Press, Calgary, Alberta, Canada.

Karr, J. R. 1982. Avian extinction on Barro Colorado Island, Panama: A reassessment. *Amer. Nat.* **119:**220–239.

Kirch, P. 1982. The impact of the prehistoric Polynesians on the Hawaiian ecosystem. *Pacific Sci.* **36:**1–14.

Loope, L. L. and P. G. Scowcroft. 1985. Vegetation response within exclosures in Hawaii: A review. Pp. 377–402 in C. P. Stone and J. M. Scott (Eds.), *Hawai'i's terrestrial ecosystems: Preservation and management.* Univ. of Hawaii, Honolulu.

Moors, P. J. and I. A. E. Atkinson. 1985. *Conservation of island birds: Case studies for the management of threatened island species.* Int. Council Bird Preservation, Cambridge, England.

Mountainspring, S. and J. M. Scott. 1985. Interspecific competition among Hawaiian forest birds. *Ecol. Monogr.* **55:**219–239.

Mueller-Dombois, D. 1973. A non-adapted vegetation interferes with water removal in a tropical rain forest area in Hawaii. *Trop. Ecol.* **14:**1–18.

Mueller-Dombois, D., K. W. Bridges, and H. L. Carson (Eds.). 1981. *Island ecosystems.* Academic Press, New York.

Mueller-Dombois, D. and L. L. Loope. 1990. Some unique ecological aspects of oceanic island ecosystems. *Monogr. Syst. Bot. Missouri Bot. Gard.* **32**:21–27.

Reid, W. V. and K. R. Miller. 1989. *Keeping options alive: The scientific basis for conserving biodiversity*. World Resources Institute, Washington, D.C.

Schofield, E. K. 1989. Effects of introduced plants and animals on island vegetation: Examples from the Galápagos Archipelago. *Cons. Biol.* **3**:227–238.

Scott, J. M. and J. L. Sincock. 1985. Hawaiian birds. Pp. 549–562 *in* R. L. Di Silvestro (Ed.), *Audubon wildlife report 1985*. National Audubon Society, New York.

Scott, J. M., C. B. Kepler, C. van Riper III, and S. I. Fefer. 1988. Conservation of Hawaii's vanishing avifauna. *BioScience* **38**:238–253.

Smith, C. W. 1985. Impact of alien plants on Hawai'i's native biota. Pp. 180–250 *in* C. P. Stone and J. M. Scott (Eds.), *Hawai'i's terrestrial ecosystems: Preservation and management*. Univ. of Hawaii, Honolulu.

Smith, C. W. 1989. Non-native plants. Pp. 60–69 *in* C. P. Stone and D. B. Stone (Eds.), *Conservation biology in Hawai'i*. Univ. of Hawaii Press, Honolulu.

Steadman, D. W. 1986. *Holocene vertebrate fossils from Isla Floreana, Galápagos*. Smithsonian Contrib. Zool. No. 413.

Stone, C. P. 1985. Alien animals in Hawai'i's native ecosystems: Toward controlling the adverse effects of introduced vertebrates. Pp. 251–297 *in* C. P. Stone and J. M. Scott (Eds.), *Hawai'i's terrestrial ecosystems: Preservation and management*. Univ. of Hawaii, Honolulu.

van Riper III, C., S. G. van Riper, M. L. Goff, and M. Laird. 1986. The epizootiology and ecological significance of malaria in Hawaiian land birds. *Ecol. Monogr.* **56**:327–344.

Vitousek, P. M. and L. R. Walker. 1989. Biological invasion by *Myrica faya* in Hawai'i: Plant demography, nitrogen fixation, ecosystem effects. *Ecol. Monogr.* **59**:247–265.

Wagner, W. L., D. R. Herbst, and R. S. N. Yee. 1985. Status of the native flowering plants of the Hawaiian Islands. Pp. 23–74 *in* C. P. Stone and J. M. Scott (Eds.), *Hawai'i's terrestrial ecosystems: Preservation and management*. Univ. of Hawaii, Honolulu.

Warner, R. E. 1968. The role of introduced diseases in the extinction of the endemic Hawaiian avifauna. *Condor* **70**:101–120.

Wilcox, B. A. 1978. Supersaturated island faunas: A species-age relationship for lizards on post-Pleistocene land-bridge islands. *Science* **199**:996–998.

FIGURE 10.6

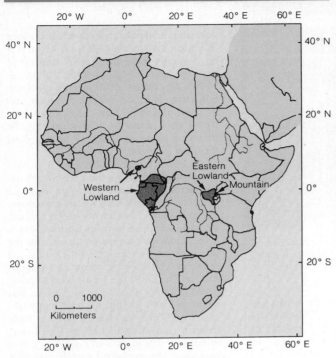

■ Ranges of the three subspecies of gorilla in Africa.

FIGURE 10.7

■ Mountain gorillas are almost entirely restricted to high elevations of the Virunga Mountains on the border of Zaire, Uganda, and Rwanda in East Africa.

form a continuous sanctuary for the gorillas. Poaching of adults, the capture of infants for sale as pets, and the accidental capture of gorillas in snares set for other animals have been causes of mortality in the Virunga population. Illegal clearing of the forest within preserve boundaries by squatters has also been a problem. The Karisoke Research Centre in Rwanda, established by Dian Fossey for studies of the Virunga gorillas, continues to play a major role in the protection of this population. World attention was focussed on this population by Fossey's studies, and tourist visitation of the preserves has become a significant aspect of the regional economy. This has led to increased protection of the gorilla population, which has increased to about 293 individuals (Aveling and Aveling 1989). The available habitat can probably support 400 to 480 individuals.

With the increased commercialization of many zoos and animal parks, the display value of certain animals has become enormous. The giant panda, now restricted to fewer than 1,000 individuals in a few mountain areas in China, and listed as endangered, is perhaps the best example. Loans of animals to western zoos by the government of China have been made for a price of about $500,000 for a three-month period, although much of this income has not been used for wildlife conservation in China (Roberts 1988). As a visitor attraction, however, pandas have been a great financial success. Many conservationists have argued that such exploitation is at the

expense of the wild population. These animals occur in six isolated locations, which are themselves divided into local populations of 50 or fewer individuals. Only about 90 to 100 animals exist in captivity, but the reproduction of captive populations is low. Thus, a strong impetus exists for the capture of wild animals to maintain zoo numbers and captive breeding programs. As a result of this controversy over commercial exploitation, in 1988 China curtailed the loan of animals to western zoos.

DESTRUCTIVE ANIMALS

Certain kinds of large animals are destructive to human property, especially farm crops and livestock. In Africa and Asia, elephants are notorious for invading small farms and destroying crops, and their propensity for this is one of the major conservation problems of the borders of parks containing elephants. In Malaysia, for example, elephant damage to oil palm and rubber tree plantations is estimated to be about $20 million annually (Sukumar 1989). In North America, white-tailed deer and mule deer cause damage to crops, orchards, and young conifer plantations on a smaller scale. Other large wildlife, such as elk, sometimes compete with livestock for forage on grazing lands.

In North America, black bears have been targeted as a destructive species by commercial lumber companies in the Pacific Northwest. In northern California, black bears occasionally strip bark from young redwood trees in the spring to reach the sweet sapwood beneath. The Louisiana-Pacific Corporation has estimated that this causes a loss of about one percent of the trees on about 13,000 acres of forest land, corresponding to about $300,000 annually. Based on these estimates of damage, black bears have been killed on corporation land between 1981 and 1987, some 45 being removed in 1987.

Five species of rhinoceroses exist. Two species occur in Africa: the black and the white rhinoceroses (Table 10.4). Three species are found in Asia: the Indian, Sumatran, and Javan rhinoceroses. The African black rhinoceros (Fig 10.8) was originally distributed throughout most of the savanna region of the continent. The African white rhinoceros was formerly widespread in the savannas of southern Africa and in small areas of savannas west of the Nile River from Uganda westward into the Central African Republic (Owen-Smith 1988). The range of the black rhinoceros has been reduced to isolated pockets in central, eastern, and southern Africa. Natural populations of the white rhinoceros (Fig. 10.9) survive only in Garamba National Park, Zaire, and in game reserves in Natal Province, South Africa (Owen-Smith 1988). The Indian rhinoceros formerly occurred along the foothills of the Himalayas from Pakistan to easternmost India; it now survives in isolated preserves in India, Bhutan, and Nepal. The Sumatran and Javan rhinoceroses formerly

FIGURE 10.8

■ During the 1980s, the black rhinoceros suffered heavy poaching for its horn in most of central and eastern Africa.

occurred from eastern India throughout most of southeastern Asia. The Sumatran rhino also occurred on Sumatra and Borneo, the Javan on Sumatra and Java. The Sumatran rhino now occurs only in scattered small pockets in its original range (Flynn and Abdullah 1984), the Javan rhinoceros only in western Java (Cohn 1988) and southern Vietnam (Schaller et al. 1990). Populations of all species have been reduced to less than 5,000 individuals, and in 1990 one, the Javan rhinoceros, comprised only about 54 individuals on the island of Java and 10 to 15 in mainland Vietnam (Schaller et al. 1990).

Poaching of rhinoceroses is carried out primarily to obtain the horn, which has purported medicinal and aphrodisiacal value in Asia, and is carved into elaborate dagger handles in countries of the Arabian Peninsula. In Asia, powdered rhino horn is a common component of medicines for skin diseases, muscle aches and sprains, and colds, as well as being considered an aphrodisiac. In 1989, the retail value of powdered horn for these uses in Taiwan was about $42,880 per kg, for an Asian horn, and $3,347 per kg, for an African horn (Martin and Martin 1989). The urine, dung, blood, and skin of Asian rhinoceroses also have medicinal or religious uses

TABLE 10.4

Populations of Rhinoceroses in the Wild.

SPECIES AND COUNTRY	1979	1984	1990
African Black	14,875	6,420	3,392
Ctr. Afr. Republic	3,000	170	0
Kenya	1,500	550	400
Tanzania	3,795	750	185
Zambia	2,750	1,650	40
Zimbabwe	1,400	1,680	1,700
South Africa	630	640	626
Other	1,710	980	441
African White	3,841	3,947	4,745
Northern			
Zaire	400	15	26
Sudan	400	10	0
Other	21	2	0
Southern			
Zimbabwe	180	200	200
South Africa	2,500	3,330	4,225
Other	340	390	295
Indian			1,200–1,500
Sumatran			500–900
Javan			65–70

Source: Data compiled from Gakahu (1991) and various other sources.

FIGURE 10.9

■ Healthy wild populations of white rhinoceroses occur only in Umfolozi and Hluhluwe National Parks in Natal Province, South Africa.

mainly in territorial fighting with other animals, rather than in feeding, and that loss of the horns will not affect the populations detrimentally. How fast the cut horns will regenerate is uncertain. Preliminary observations in Namibia suggest that after several months dehorned animals had suffered no ill effects, and that none had been killed by poachers.

Stable populations of both black and white rhinoceroses exist in several countries of southern Africa, particularly South Africa and Zimbabwe. In Natal Province of South Africa, Hluhluwe and Umfolozi Game Reserves exist largely for the conservation of these species. These populations are being used as a source of animals for translocation to places from which the species have been eliminated. A coordinated plan has been developed for the conservation of the black rhinoceros in South Africa and Namibia (Brooks 1989). Under this plan, the population, representing three subspecies, could be increased from about 1,000 animals to nearly 5,000. Black rhinos have also been increasing in Zimbabwe under effective protection (Tatham and Taylor 1989).

In Kenya, efforts are being made to aggregate black rhinoceroses from scattered small populations into larger groups in relatively secure areas. Lake Nakuru National Park, which is now completely fenced, has been chosen as one rhinoceros sanctuary (Lever 1990). Seventeen rhinos from a private ranch have been moved to the park, where only two animals had existed. Other fenced or carefully protected sanctuaries are being established in Tsavo, Aberdare, and Nairobi National Parks, as well as on certain private ranches (Cohn 1988).

in parts of Asia (Martin 1985). In various countries of the Arabian Peninsula, elaborately carved daggers are traditionally given by fathers to their sons at puberty, and the most prestigious of these have handles made from rhinoceros horn. An African rhino horn entering the dagger trade has a wholesale value of about $500 per kilogram, with the finished carved dagger handles retailing for up to $12,000 each.

The various uses of the African and Asian rhino horn resulted in the killing of about 2,580 animals per year in the late 1970s and early 1980s. During this period, populations of the black rhinoceros and the northern white rhinoceros were decimated

in several countries of central and eastern Africa (Table 10.4). The total population of the black rhinoceros declined from almost 15,000 in 1979 to little more than 3,500 in 1988 (Cohn 1988), and the northern white rhinoceros was almost exterminated in Zaire, Sudan, and the Central African Republic. A group of six white rhinoceroses that had been introduced to Meru National Park in Kenya from South Africa were massacred by poachers in 1988. In recent years, heavy poaching has also spread into more southern countries of Africa.

In Zimbabwe and Namibia, experimental programs have been initiated to capture and remove the horns of wild black rhinoceroses, in a drastic effort to discourage poaching. In these regions, the rhinoceroses are believed to use their horns

In several parts of western North America, conflicts between elk and livestock have arisen. The tule elk on the central valley of California was nearly driven to extinction because of competition with livestock. In the Jackson Hole area of northwestern Wyoming, where cattle ranching operations developed in the 1880s on areas used by elk in winter, conflicts between elk and cattle soon became serious (Boyce 1989). Originally, as many as 25,000 elk may have wintered in Jackson Valley. Farming and ranching usurped much of the winter range, leading to heavy mortality of elk, particularly during the severe winters of 1909–1911. Eventually, in 1912, this led to the establishment of the National Elk Refuge and to a program of supplementary winter feeding of elk (Fig.

10.10). The refuge, about 10,100 ha in area, now encompasses about a quarter of the original winter range in the valley. The Jackson elk herd now contains about 11,000 to 14,000 animals, and is one of the largest herds in North America. The animals occupy summer ranges in Yellowstone National Park, Grand Teton National Park, and the Bridger-Teton National Forest. In autumn they migrate up to 100 km to wintering locations in valleys in and near Jackson Hole. About 7,500 of these elk depend on supplementary feeding on the refuge and about 2,200 more at three locations in the adjacent Bridger-Teton National Forest (Boyce 1989), a program that has aroused much controversy. In 1988–1989, for example, the cost of supplementary feeding at the National Elk Refuge

FIGURE 10.10

■ The National Elk Refuge was established to provide winter habitat and food for elk that wintered in the valleys near Jackson, Wyoming.

was $750,000, a portion of which is repaid by the auction of antlers collected on the refuge. Without supplementary feeding, however, the herd would probably decline greatly, because its natural winter range is mostly occupied by private cattle ranching operations. Under the present management regime, a healthy herd is maintained, and an annual hunting harvest of about 3,000 animals is taken.

In western New Mexico, Rocky Mountain elk were introduced to the Gila National Forest to restore the species to an area from which the original Merriam elk had been eliminated in the early 1900s. Since their introduction in the 1930s, the elk have increased to more than 1,000 animals, and conflicts have developed between elk and cattle on national forest range leased by ranchers. The elk herd also supports a substantial amount of hunting, and the issue of whether the grazing resource on public land should be reserved exclusively for wildlife or should be shared by private ranchers has arisen.

Another aspect of the conflict of interest between wildlife management and ranching is the potential for disease transmission from wildlife to domestic animals. In the Greater Yellowstone Ecosystem, for example, both elk and bison carry **brucellosis.** This bacterial disease, which can also be transmitted to humans, can be very destructive to herds of domestic ungulates, and its occurrence in domestic animals requires quarantine and often destruction of entire herds in which infected animals are found. Since cattle are grazed on some of the national forest lands used by elk during their calving season, when the disease is normally transmitted between animals, a small risk of introduction of brucellosis to livestock exists. Although a brucellosis vaccine is available, it is only partially effective in protecting livestock.

KEY MANAGEMENT STRATEGIES FOR LARGE ANIMALS

1. Educational programs and management plans that minimize contacts between dangerous or destructive animals and humans or their property.

2. Effective international regulation of trade in valuable animal commodities and living animals of endangered species.

LITERATURE CITED

Armstrong, S. and F. Bridgeland. 1989. Elephants and the ivory tower. *New Scientist* **123(1679)**:37–41.

Aveling, C. and R. Aveling. 1989. Gorilla conservation in Zaire. *Oryx* **23**:64–70.

Boyce, M. S. 1989. *The Jackson elk herd.* Cambridge Univ. Press, Cambridge, England.

Brooks, P. M. 1989. Proposed conservation plan for the black rhinoceros *Diceros bicornis* in South Africa, the TBVC states, and Namibia. *Koedoe* **32(2)**:1–30.

Cauble, C. 1977. The great grizzly grapple. *Nat. Hist.* **86(7)**:74–81.

Cohn, J. P. 1988. Halting the rhino's demise. *BioScience* **38**:740–744.

Cohn, J. P. 1990. Elephants: Remarkable and endangered. *BioScience* **40**:10–14.

Feazel, C. T. 1990. *White bear.* Henry Holt and Co., New York.

Flynn, R. W. and M. T. Abdullah. 1984. Distribution and status of the Sumatran rhinoceros in peninsular Malaysia. *Biol. Cons.* **28**:253–273.

Gakahu, C. G. 1991. African rhinoceroses: Challenges continue in the 1990s. *Pachyderm.* **No. 14**:42–45.

Harcourt, A. H., K. J. Stewart, and I. M. Inahoro. 1989. Gorilla quest in Nigeria. *Oryx* **23**:7–13.

Herrero, S. 1970. Human injury inflicted by grizzly bears. *Science* **170**:593–598.

Herrero, S. 1985. *Bear attacks: Their causes and avoidance.* Winchester Press, Piscataway, NJ.

Ivory Trade Review Group. 1989. The ivory trade and the future of the African elephant. *Pachyderm* **No. 12**, pp. 32–37.

Lever, C. 1990. Lake Nakuru rhinoceros sanctuary. *Oryx* **24**:90–94.

Martin, E. B. 1985. Religion, royalty and rhino conservation in Nepal. *Oryx* **19**:11–16.

Martin, E. B. and C. B. Martin. 1989. The Taiwanese connection— a new peril for rhinos. *Oryx* **23**:76–81.

Martinka, C. J. 1988. *An experiment in grizzly bear conservation: Glacier National Park.* Ecological Society of America Annual Meeting (Oral presentation), 15 August, Davis, CA.

Moment, G. B. 1968. Bears: The need for a new sanity in wildlife conservation. *BioScience* **18**:1105–1108.

Moment, G. B. 1969. Bears and conservation: Realities and recommendations. *BioScience* **19**:1019–1020.

Morrell, V. 1990. Running for their lives. *Int. Wildlife.* **20(3)**:4–13.

Owen-Smith, R. N. 1988. *Megaherbivores: The influence of very large body size on ecology.* Cambridge Univ. Press, Cambridge, England.

Roberts, L. 1988. Conservationists in panda-monium. *Science* **241**:529–531.

Schaller, G. B., N. X. Dang, L. D. Thuy, and V. T. Son. 1990. Javan rhinoceros in Vietnam. *Oryx* **24**:77–80.

Sukumar, R. 1989. *The Asian elephant: Ecology and management.* Cambridge Univ. Press, Cambridge, England.

Tatham G. H. and R. D. Taylor. 1989. The conservation and protection of the black rhinoceros *Diceros bicornis* in Zimbabwe. *Koedoe* **32(2)**:31–42.

Western, D. and L. Vigne. 1985. The status of rhinos in Africa. *Swara* **8(2)**:10–12.

Yellowstone National Park. 1982. *Final environmental impact statement—grizzly bear management program.* U. S. National Park Service, Denver, CO.

CHAPTER 11

Migratory Birds

Seasonal migration by birds is one of the major phenomena of vertebrate biology. Billions of individuals of more than a thousand species of birds seasonally move distances varying from a few kilometers to more than 10,000 kilometers between breeding and non-breeding ranges. **Migratory birds** are at special risk from effects of environmental change because they require different habitats in different geographical areas during their annual cycle. Although the high mobility of migrants might seem to give such species great flexibility in the face of change, in reality it makes them vulnerable to impacts on any of the habitats on which they depend. Other migratory land animals, such as the African wildebeest and the monarch butterfly, and aquatic animals, such as whales, sea turtles, and salmon, face similar problems (Brower and Malcolm 1991).

PATTERNS OF MIGRATION

To emphasize the diversity of migration patterns of birds, we can group them into three categories: weather migrants, short-distance migrants and long-distance migrants. **Weather migrants** are species that move varying distances in direct response to severe weather conditions. In the temperate zone, for example, many species move southward, or to lower elevations in mountainous areas, in response to severe winter weather. The timing, distance of movement, and length of residence in non-breeding areas are all highly variable from year to year. **Short-distance migrants,** in contrast, show more regularly timed movements between breeding and non-breeding ranges separated by distances of a few hundred kilometers. Many such species breed in the higher temperate latitudes or in the arctic, and winter at lower latitudes within the temperate zone. Often the breeding and non-breeding ranges of these species overlap latitudinally. **Long-distance migrants** are those that annually move between breeding and non-breeding ranges thousands of kilometers apart. Many long-distance migrants breed at high temperate or arctic latitudes and winter in the tropics or in the opposite hemisphere. The timing and pattern of their movements are even more regular than those of short-distance migrants.

Adjustments of long-distance migrants to changes in their seasonal habitats are probably limited. For at least some species, the timing, route and distance of migration are genetically programmed. Furthermore, strong site fidelity is shown by many species to specific sites that were utilized during their first year, including areas visited during migration (Rappole and Warner 1978). The demography of long-distance migrants is also specialized (Cox 1985). Many of these species have extended their breeding ranges into high latitudes where total annual fecundity is severely limited by the short summer period. At higher latitudes, migrants can usually rear only a single brood. Often, poor summer weather conditions cause many nesting efforts to fail. Thus, species migrating to high latitudes must compensate by spending the non-breeding season in areas where survival is high. The migratory flights themselves are also risky, and long-distance migrants are adapted to carry these out as quickly as possible. Thus, environmental changes that reduce breeding productivity, increase time along the migration route, or reduce non-breeding survival can all contribute to the decline of migrant populations.

Three major systems of long-distance migration exist. In the Old World, the **Palearctic-African Migration System** involves about 183 of the 589 species of land and freshwater birds that breed in Europe and western Asia. These species move southward to winter in Africa south of the Sahara Desert (Moreau 1972). An estimated five billion birds undertake this southward migration, with about half of these surviving to return to their breeding grounds the following summer. The **Palearctic-Pacific Migration System** includes birds that breed in eastern Asia and parts of western Alaska, and winter in southeast Asia, Australia, and Pacific island regions. Finally, the **Nearctic-Neotropical Migration System** comprises species that breed in North America and winter in the New World tropics or in temperate South America. Of the 565 species of land and freshwater birds that breed in North America, 98 species winter exclusively in the New World tropics or in temperate South America, and an additional 151 species have winter ranges extending partly into the New World tropics. In all probability, this migration system involves as many individuals as the Palearctic-African system.

STAGING AND STOPOVER AREAS

Species that make flights of hundreds or thousands of kilometers must store large amounts of body fat just prior to these flights to provide the needed energy. In some cases, as much as 40 percent of the weight of a bird departing on a migratory flight consists of fat. Species that carry out their migrations in one or a few long flights are particularly dependent on habitats in which abundant foods suitable for fat deposition are available. Areas that provide these needs are known as staging and stopover areas.

Staging areas are locations at which birds gather prior to the first major leg of migration, whereas **stopover areas** are those at which birds stop to replenish their fat stores partway through migration. Shorebirds, in particular, are highly dependent on staging and stopover areas (Myers 1983, Myers et al. 1987). The Hudsonian godwit, for example, breeds in arctic Canada and then gathers at a staging area along the western shores of Hudson and James Bays. Most of the individuals then fly non-stop about 4,500 kilometers to the northern coast of South America (Morrison and Harrington 1979). Other species of shorebirds move shorter distances at a time, and use several stopover areas, where they remain for a few days and store fat (Fig. 11.1). Staging and stopover areas are used both during autumn and spring migration periods. In the spring, the last stopover before arrival on the breeding ground must provide for the last leg of the migration and often for the period of territory establishment and mating, when weather may be poor and food supplies low. In eastern North America, Delaware Bay (Fig. 11.2) is an especially important stopover area for shorebirds on northward migration (Myers et al. 1987). In southern Alaska, the Copper River delta is the stopover site for about 20 million shorebirds prior to the last leg of migration that takes them to arctic tundra breeding areas (Senner 1979). The fat stored at this last stopover site also serves to support the birds during early phases of the nesting cycle, when weather conditions on the tundra can be unfavorable to feeding.

Development, pollution and flooding threaten coastal staging and stopover areas (Smit et al. 1987). Global warming, and the associated sea level rise, will severely reduce the area of many coastal wetlands, particularly in temperate and tropical regions. These wetlands, already pressured by development and pollution, will be constricted between the rising water level and developed portions of the landscape that

FIGURE 11.1

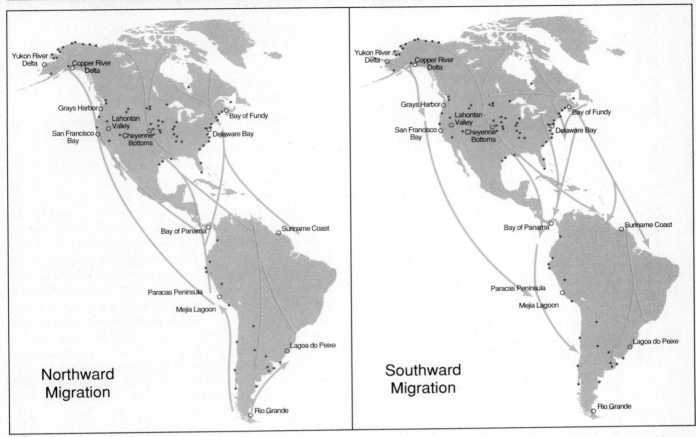

Northward Migration

Southward Migration

■ Fall and spring staging and stopover areas for migrating shorebirds in North and South America. The localities designated are part of the developing Western Hemisphere Shorebird Reserve Network. Circles indicate areas visited by 250,000 or more birds annually (or over 30 percent of a particular species' population), and dots areas visited by at least 20,000 birds (or at least 5 percent of a particular species' population) (Modified from Myers et al. 1987).

FIGURE 11.2

■ Delaware Bay is a major spring stopover area for migrating shorebirds, such as these ruddy turnstones and sanderlings, which feed heavily on the eggs of horseshoe crabs before moving on north to their arctic breeding grounds.

humans will try to protect against flooding (Lester and Myers 1989). The breeding areas of long-distance migrant shorebirds also lie at high latitudes, where climatic change is expected to be several times as great as in the tropics. The consequences of such change for shorebird breeding habitats are difficult to predict.

Although staging and stopover areas are clearly important for water birds, little is yet known about their importance for migrant land birds. Some coastal areas are known to be of major importance, however. Coastal chenier woodlands of Louisiana are important stopover sites for birds that have crossed the Gulf of Mexico (Moore and Kerlinger 1987). Many individuals spend one to several days in these woodlands and rebuild their fat stores (Fig. 11.3). In the autumn, Atlantic coastal areas from New England south to North Carolina are staging areas for a number of songbirds that fly long distances over water to wintering areas in the Bahamas, West Indies, or South America (Terborgh 1989). Inland, where forest migrants must rest and feed in whatever wooded areas remain, small woodlots may not serve as adequate stopover

FIGURE 11.3

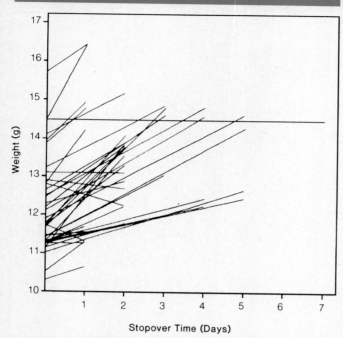

■ Changes in weights of migrant Kentucky warblers after arrival at a stopover site on the coast of Louisiana, following a spring migration crossing of the Gulf of Mexico (Modified from Moore and Kerlinger 1987).

areas. In eastern Illinois, for example, Graber and Graber (1983) found that migrating warblers were unable to gain weight when foraging in small woodlots, whereas they were able to store fat when foraging in the more extensive forest areas of southern Illinois.

LONG-TERM TRENDS OF MIGRATORY BIRDS

Strong evidence has now accumulated that migratory songbirds, particularly long-distance migrants breeding in forest habitats, are declining in much of North America (Aldrich and Robbins 1970, Hall 1984, Leck et al. 1988, Terborgh 1989) and probably also Europe (Grimmett 1987). In the Hutcheson Memorial Forest, a 24–ha woodlot in New Jersey, for example, a substantial decline occurred in long-distance forest migrants between 1960 and 1984 (Fig. 11.4). Even in virgin mountain forests in West Virginia, Hall (1984) found that 6 of 14 neotropical migrants had disappeared between 1947 and 1983, with the total density of the remaining species being only 63 percent of that in 1947. In North America, one of the strongest sources of evidence of the decline of migrants comes from the **North American Breeding Bird Survey (BBS).** The BBS is a standardized system of roadside counts taken annually at about 2,000 locations in Canada and the United States. Each count consists of observations at 50 points located at 0.8–km intervals along a 40–km route. BBS counts have been carried out since 1966. Analysis of data from these surveys show that most long-distance, forest migrants

FIGURE 11.4

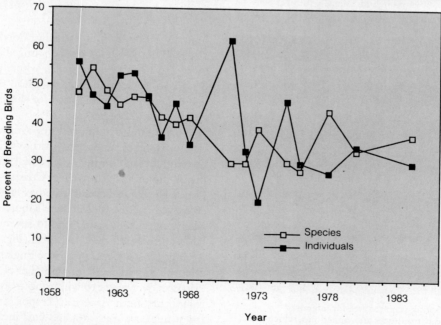

■ Changes in the percentage of long-distance migrants in the avifauna of the Hutcheson Memorial Forest in New Jersey between 1960 and 1984. Source: Data from C. F. Leck, et al., "Long-Term Changes in the Breeding Bird Populations of a New Jersey Forest" in *Biological Conservation,* 46:145–157, 1988, Elsevier Applied Science Publishers Ltd., Essex, England.

declined between 1978 and 1987, after having shown stable or increasing abundances from 1966 through 1977 (Robbins et al. 1989).

Independent evidence suggestive of a substantial decline of migrants wintering in the neotropics comes from radar observations of arriving spring migrants along the northern coast of the Gulf of Mexico (Gauthreaux 1992). Flocks of migrants that depart from the Yucatan Peninsula early in the night on one day reach the northern coastline late in the afternoon of the next day. Arriving flights of migrants are evident as masses of "blips" on weather surveillance radar screens. In the early 1960s, major flights were noted on 95 percent of days between early April and mid-May. From 1987 to 1989, however, major flights were noted on only 44 percent of all days, suggesting an overall trans-gulf migration only half as extensive.

In addition to declines of neotropical migrants, Whitham and Hunter (1992) found that significant declines of short-distance migrants may be occurring in eastern North America. Using BBS data, these observers found that many species breeding in early successional habitats in New England and wintering in similar habitats in the southeastern states had declined in numbers from 1966 to 1988. The decline of these species may relate to the disappearance of early successional habitats in the Southeast as biotic succession and commercial forestry convert more of the landscape to forests.

MIGRANT BIRDS ON THEIR BREEDING GROUNDS

Several studies have shown that forest fragmentation in eastern North America leads to a major change of the kinds of breeding birds (See chapter 3). Long-distance migrants make up 80 to 90 percent of the breeding birds in extensive tracts of eastern deciduous forest, but usually less than half of the breeders in small woodlots (Whitcomb et al. 1981). In eastern Illinois, for example, Blake and Karr (1984) found that forest interior, long-distance migrants were completely absent from woodlots less than 24 ha in area. In Maryland, Whitcomb et al. (1981) found that several long-distance migrants were absent from portions of the state 25 square kilometers in area, where forests had been fragmented, even though these areas had some woodlots 50 ha or more in size, indicating that fragmentation was leading to general regional extinction.

Much of the loss of forest interior species in eastern North America appears to result from increased predation and nest parasitism. Most neotropical migrants have open, cup-like nests (Fig. 11.5), and several of the species most sensitive to forest fragmentation nest on the ground or low in the vegetation. In addition, neotropical migrants tend to have small clutches and only one brood per season. Wilcove (1985) showed that predation on open-cup bird nests was much heavier in small woodlots than in large areas of forest, apparently because many kinds of avian and mammalian predators associated with open country or forest edges do extend their hunting some distance into the forest. As woodlots

FIGURE 11.5

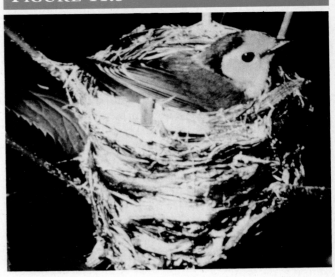

■ The hooded warbler is a neotropical migrant that breeds in forest areas of eastern North America, building an open-cup nest, and winters in mature tropical forest in southern Mexico and Central America.

become smaller, this penetration zone constitutes a larger and larger fraction of the total area. Nest parasites, such as the brown-headed cowbird in North America, also contribute to the decline of forest interior species in small woodlots. In eastern Illinois, Robinson (1992) found, for example, that nests of neotropical migrants in small woodlots often contained three or more cowbird eggs. Together with predation, cowbird parasitism reduced reproductive success to a level far below that necessary to maintain populations. Thus, populations of these migrants must be maintained by immigration from other areas where breeding success is higher.

Whitcomb et al. (1981) and May (1981) have suggested that maintenance of populations of migrants in individual forest units depends both on survival of adults that have previously bred in that unit and by settlement of first-time breeders, many of which are birds reared in other locations. With increased isolation of a woodlot, the young birds that leave to settle elsewhere are not balanced by those that come to the woodlot from other locations. Most migratory species thus consist of many small populations, occupying individual forest stands, that interact to form an overall regional population, or **metapopulation,** within which dispersal plays a major role in maintaining local populations.

MIGRANTS ON THEIR NON-BREEDING GROUNDS

Wintering migrants in the tropics often constitute a major component of the bird communities of their non-breeding areas. In the New World, for example, about half the birds that come from an area of 16.2 million square kilometers in North America crowd into an area of about 2.2 million square kilometers in Mexico, the Bahamas, and the West Indies. In these areas, therefore, their densities must often be greater than in their breeding ranges. Wintering migrants also form a major portion of the total bird population in such areas (Lynch 1989, Terborgh 1989).

During the non-breeding season, migrants use many different kinds of habitats, which may or may not be similar to those used during the breeding season. In some parts of the tropics, such as western Mexico, wintering land bird migrants seem to concentrate in disturbed habitats, in successional communities, or near superabundant food sources such as fruiting trees (Hutto 1989). Resident tropical birds are probably unable to exploit any of these situations fully. In other places, such as the Yucatan Peninsula, wintering migrants appear to use the full spectrum of habitats available, including mature forest (Lynch 1989). For the Neotropics, Terborgh (1980) estimated that at least 55 species of North American migrants winter partly or entirely in mature tropical forests. These include many flycatchers, thrushes, vireos, warblers, and tanagers (Fig. 11.5). Many of these birds, particularly those that are primarily insectivorous in winter, establish individual territories. Other insectivorous species may become members of mixed foraging flocks of resident and migrant species, the flock often acting as a territorial unit. On the other hand, species that are highly frugivorous in winter may form flocks that move from place to place and exploit a changing spectrum of fruit resources. In any case, wintering migrants become intimately integrated into the tropical bird community (Terborgh 1989).

Tropical deforestation has major implications for many migratory land birds, particularly in southeast Asia and the Neotropics. In analyses of BBS data for North America as a whole, Robbins et al. (1989) found that the declines of neotropical migrants between 1978 and 1987 were significantly greater for species wintering in forests than for those wintering in more open vegetation types. Continued destruction of tropical forests will reduce the habitat of certain species (Lynch 1989), and sooner or later, tropical deforestation will become a substantial contributor to the decline of long-distance migrants. Some of the decline in North America seems to be due to changes on the breeding grounds, especially forest fragmentation, with its diverse effects. This does not seem adequate to account for the recent decline at Hutcheson Forest, which has been isolated since the 1700s. The extent to which tropical deforestation has contributed to decline of neotropical migrants remains controversial (Holmes and Sherry 1988, Hutto 1988). In the southeastern United States, however, the Bachman's warbler has probably become extinct coincident with conversion of almost all of its tropical forest winter habitat in Cuba to sugar cane fields (Powell and Rappole 1986).

Other changes in the landscapes of wintering areas also pose conservation problems for northern hemisphere migrants. In the Palearctic-African region, the major wintering areas of many land birds lie in the sub-Saharan savanna zone. Here, desertification appears to have caused decline of several long-distance migrants (Grimmett 1987). Throughout the world, waterfowl, shorebirds and other water birds utilize coastal or freshwater shores and marshes as wintering sites. Some shorebirds, storks, and other water birds also winter in open grasslands and savannas. The destruction of coastal marshlands, especially in North America and Europe, and the conversion of native grasslands to cultivated cropland, particularly in South America, also threaten wintering populations of many species. Extensive changes are also occurring in African wetlands, where large numbers of water birds from western Eurasia winter (Grimmett 1987).

HUNTING OF MIGRATORY BIRDS

Hunters take large numbers of waterfowl and many other migratory birds, both legally and illegally. In Canada and the United States, hunting of migratory birds is carefully regulated, although many difficulties still exist in management of the populations involved.

In North America, waterfowl migrations follow a set of major pathways that have been grouped into four **flyways** (Fig. 11.6): **Atlantic, Mississippi, Central,** and **Pacific** (Hawkins et al. 1984). These flyways are not as sharply distinct as one might imagine, but birds breeding in the northern portions of a particular flyway do tend to stay within the overall flyway system in their movements during the non-breeding season. The flyways are perhaps as much administrative units as they are biological units. Surveys of reproductive success in the nesting regions of the different flyways have traditionally been used to determine bag limits and the length of the hunting season throughout the respective flyways.

During the 1970s, the total North American breeding population of ducks, including sea ducks and mergansers, averaged about 62 million birds, and yielded a fall flight of more than 100 million birds (U. S. Fish and Wildlife Service 1986). Continental goose populations range between four and six million, and those of tundra swans about 129,000. In addition, about 10,000 trumpeter swans occur in western North America. These populations, particularly those of ducks, are subject to great fluctuations in reproductive success. Many ducks lay clutches of eight to ten eggs, so that in good years reproduction can increase the total population three-fold or more. Because many ducks nest in marshlands or on small prairie ponds, however, drought years can be equally disastrous.

Recent hunting harvests of waterfowl in Canada and the United States have ranged from 10.8 million ducks in 1968 to 20.2 million in 1970, and from 1.9 million geese in 1974 to 2.5 million in 1980. About 80 percent of the total North American harvest occurs in the United States. The greatest harvests of ducks are in the Mississippi (37.5%) and Pacific (31.2%) flyways, and of geese in the Pacific flyway (30.4%). Subsistence harvests by native Americans in Alaska and Canada constitute about 5 percent of the total duck harvest and seven percent of the total goose harvest. Several thousand tundra swans are harvested on a subsistence basis, as

FIGURE 11.6

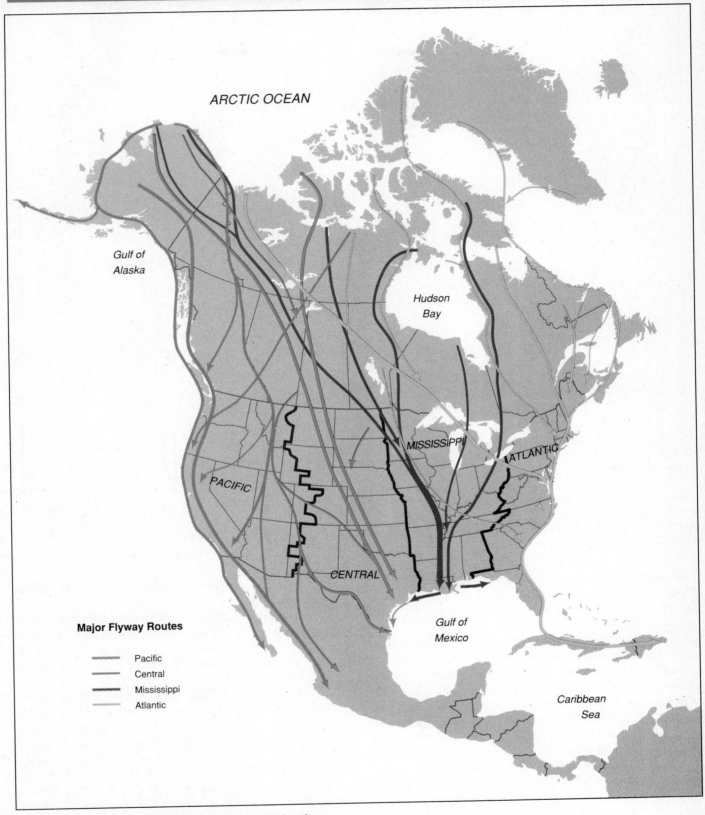

ARCTIC OCEAN

Gulf of
Alaska

Hudson
Bay

MISSISSIPPI

ATLANTIC

PACIFIC

CENTRAL

Gulf of
Mexico

Caribbean
Sea

Major Flyway Routes

Pacific
Central
Mississippi
Atlantic

■ Waterfowl flyways in North America. Sources: U.S. Fish and
Wildlife Service, *North American Waterfowl Management Plan,* 1986, U.S.
Department of the Interior, Washington, D.C.; and *Environment
Canada,* Ottawa, Canada.

FIGURE 11.7

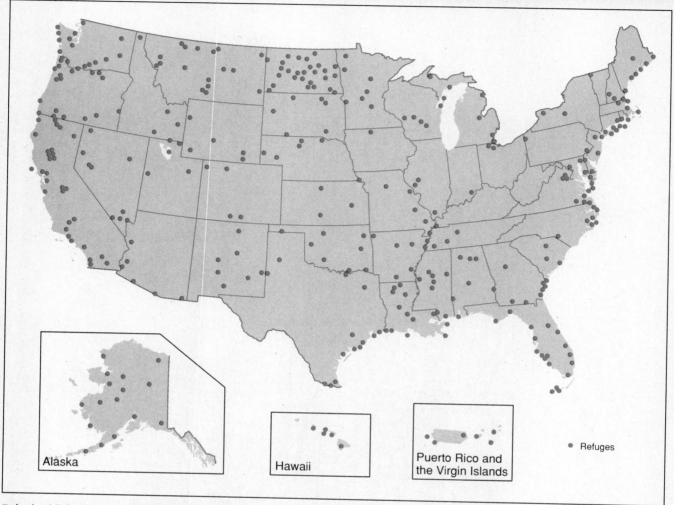

■ Federal wildlife refuges in the United States and Canada. Source: The National Wildlife Refuge System, U.S. Fish and Wildlife Service, Department of the Interior.

well, and a very small recreational harvest occurs in the United States. For ducks, hunting mortality is estimated to be roughly 50 percent of total over-winter mortality, and that for geese somewhat higher.

In addition to direct hunting harvest of waterfowl, two other causes of mortality associated with hunting must be recognized. Many birds are crippled by hunters, but not recovered, and most of these ultimately die. Crippling losses for waterfowl are about 18 percent of the number harvested, or 2.5 to 2.7 million ducks and 270,000 geese annually. Waterfowl also ingest lead pellets while feeding, taking them in as they would grit to promote the action of the gizzard. Excessive lead intake causes lead poisoning, which is estimated to kill about 1.6 to 2.4 million ducks annually. The problem of lead poisoning has stimulated efforts to replace lead shot with steel shot in areas where lead poisoning is serious. Steel shot yields a more condensed pellet distribution, and is less effective at greater distance, and many hunters argue

that switching to steel shot may increase the problem of bird loss by crippling.

Several other migratory species are legal gamebirds in North America. These include sandhill cranes, shorebirds such as Wilson's snipe and woodcock, and various doves and pigeons. The annual harvest of mourning doves in the United States is about 50 million birds.

The **National Wildlife Refuge System** of the United States, and the comparable system in Canada, function largely for protection of migratory birds, both game and nongame species. National Wildlife Refuges exist in 49 of the 50 states, and include over 367 areas owned and 175 additional areas administered by the U.S. Fish and Wildlife Service, with a combined area exceeding 13.7 million hectares (Fig. 11.7). Most of these refuges have the primary purpose of protecting breeding, migration, or wintering areas of waterfowl and other migratory birds (Fig 11.8). Many others combine preservation of migratory birds and protection of other forms of wildlife.

FIGURE 11.8

■ The Sacramento National Wildlife Refuge is a major wintering area for snow geese that have moved south along the Pacific Flyway.

FIGURE 11.9

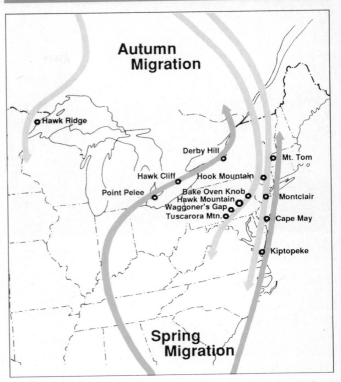

■ Major spring and fall migration routes of raptors in eastern North America, with important observation stations (Modified from Brett 1986).

FIGURE 11.10

■ Hawk Mountain, Pennsylvania, was once a major shooting gallery for hawks during autumn migration. These are part of more than 230 hawks killed near the main lookout on one day in October, 1932.

Formerly, when most raptors were considered to be enemies of both wildlife and domestic fowl, hawk shooting was a popular autumn practice in some parts of North America. Shooting was concentrated in areas where southward migrations of these birds were concentrated. Major southward flyways of raptors follow the Atlantic coastline and the shorelines of other large water bodies such as the Great Lakes (Fig. 11.9). Large numbers thus become concentrated in locations such as Cape May, at the southern tip of New Jersey, and Point Pelee, a peninsula on the northern shore of Lake Erie. Other major flyways follow the ridges of the Appalachian Mountains. Here, the birds glide southward in the updrafts created as winds strike the western face of the major ridges. Since the major Appalachian ridges extend unbroken for hundreds or thousands of kilometers, they provide a "free ride" for raptors from New England to Georgia. In the early 1900s, hawks were shot at many ridge peaks along the Appalachians.

In 1934, one of the most notorious of these shooting locations (Fig. 11.10) was converted into **Hawk Mountain Sanctuary** (Brett 1986). Hawk Mountain became both a center for study of raptor migrations, and a center of conservation effort for protection of raptors. Now that these birds are protected, many of the former shooting points have become observation sites for ornithologists and bird watchers interested in raptors (Fig. 11.11).

In other parts of the world, hunting of migratory birds is practiced in manners unfamiliar to most North Americans. In Europe, North Africa, and the Middle East, enormous numbers of migratory songbirds are hunted for food. In Italy,

FIGURE 11.11

■ Today, Hawk Mountain is a sanctuary dedicated to observation and study of raptor migration.

for example, it is estimated that about 240 million songbirds are harvested annually. Birds are shot, captured with nooses, taken in nets made of fine nylon mesh and trapped with birdlime. Numbers in excess of 15 million are also taken in Spain, France, and Lebanon. Large numbers are also killed on the islands of Mallorca, Cyprus, and Malta, and some harvest occurs in all countries bordering the Mediterranean. All told, about 500 million small birds are harvested annually in this manner.

OTHER MIGRATION HAZARDS

Tall lighted structures create a hazard to songbirds migrating at night, especially during their fall migration. The structures involved include television and radio towers (Crawford 1981), lighthouses, refinery flare stacks (Bjorge 1987), and even structures such as the Washington Monument. Migrating birds evidently become disoriented in the vicinity of these structures, and are killed by colliding with them. Formerly, when most airports used fixed-beam ceilometers, birds often became disoriented and were killed by collision with the ground. Most mortality occurs on overcast nights when concentrated southward flights have been triggered by the passage of a cold front. The precise cause of disorientation is not understood, but may involve temporary blinding or orientation of birds to the light as if it were the moon.

At the WCTV tower near Tallahassee, Florida, scientists from the Tall Timbers Research Station and Florida State University have monitored mortality since 1956. Over this period, the tower, 308 m tall, has claimed over 42,000 birds of 189 species. This equals about 1,600 birds annually. Occasional massive kills have been recorded at other locations. On the nights of September 18 and 19, 1963, for example, about 30,000 birds were killed at one tower near Eau Claire,

Wisconsin. Overall, perhaps half a million birds may be killed annually in this manner in North America.

CONSERVATION OF MIGRATORY BIRDS

In North America, migratory birds have been the subject of agreements among several countries, particularly Canada, the United States and Mexico. The earliest of these agreements was the **Migratory Bird Treaty Act** (1918), in which Canada and the United States protected non-game species and specified coordinated management policies for game species. In 1980, the United States passed the **Fish and Wildlife Conservation Act,** which requires the Fish and Wildlife Service to monitor populations of all non-game birds, and to identify the conservation needs of any that are approaching endangerment. Specific funding was not provided for this effort, although the United States Congress appropriated $1.75 million for non-game activities in 1988 and 1989. The Office of Migratory Bird Management in the Fish and Wildlife Service devoted about 10 percent of its $8 million basic budget to non-game species in 1989 (Gradwohl and Greenberg 1989). Specific objectives for coordination of research on non-game birds by the Fish and Wildlife Service and other private and state groups have been developed, however (Office of Migratory Bird Management 1990).

To promote the preservation of migratory waterfowl, the United States and Canada have developed the **North American Waterfowl Management Plan** (U.S. Fish and Wildlife Service 1986). This plan emphasizes habitat protection and improvement, and aims to restore duck populations at the average levels prevailing in the 1970s, and maintain goose populations at close to the high levels of the 1980s. Specific attention is given to several populations that have been declining, such as those of the black duck and the cackling and dusky races of the Canada goose. The plan identifies specific regions of major importance to waterfowl, in which habitat conservation activities would be concentrated (Fig. 11.12). It covers the period from 1986 through 2000, and is subject to review at five-year intervals. To implement this plan, Canadian and United States agencies must spend $1.5 billion, two-thirds of it in Canada. About 75 percent of this cost would be provided by the United States.

For migratory shorebirds, an international consortium of private and governmental organizations is working to develop a **Western Hemisphere Shorebird Reserve Network** (Myers et al. 1987). The goal of this effort is to promote a network of sites throughout North and South America that will guarantee adequate breeding, staging, stopover, and wintering habitats for shorebirds (Fig. 11.1). By 1987, 23 state, provincial, or national agencies in Canada, the United States and Peru had become involved in this effort, together with several private conservation groups. Over 90 sites of major conservation importance had been identified, and efforts begun to achieve their protection.

FIGURE 11.12

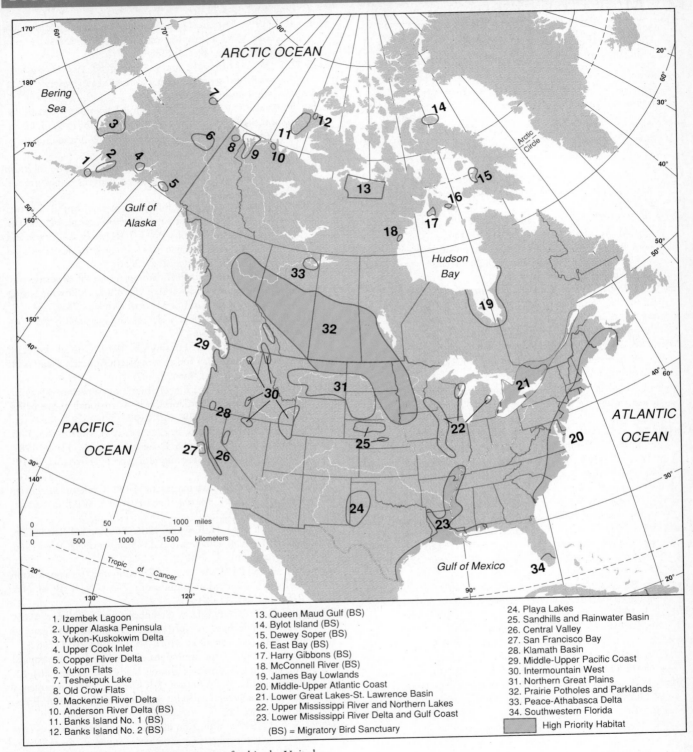

1. Izembek Lagoon
2. Upper Alaska Peninsula
3. Yukon-Kuskokwim Delta
4. Upper Cook Inlet
5. Copper River Delta
6. Yukon Flats
7. Teshekpuk Lake
8. Old Crow Flats
9. Mackenzie River Delta
10. Anderson River Delta (BS)
11. Banks Island No. 1 (BS)
12. Banks Island No. 2 (BS)

13. Queen Maud Gulf (BS)
14. Bylot Island (BS)
15. Dewey Soper (BS)
16. East Bay (BS)
17. Harry Gibbons (BS)
18. McConnell River (BS)
19. James Bay Lowlands
20. Middle-Upper Atlantic Coast
21. Lower Great Lakes-St. Lawrence Basin
22. Upper Mississippi River and Northern Lakes
23. Lower Mississippi River Delta and Gulf Coast

(BS) = Migratory Bird Sanctuary

24. Playa Lakes
25. Sandhills and Rainwater Basin
26. Central Valley
27. San Francisco Bay
28. Klamath Basin
29. Middle-Upper Pacific Coast
30. Intermountain West
31. Northern Great Plains
32. Prairie Potholes and Parklands
33. Peace-Athabasca Delta
34. Southwestern Florida

High Priority Habitat

■ Specific regions of major importance to waterfowl in the United States and Canada. Source: U.S. Fish and Wildlife Service, 1986.

Key Management Strategies for Migratory Birds

1. Development of international agreements relating to the exploitation of migratory birds and their habitats.

2. Recognition of the importance of staging and stopover areas, as well as breeding and wintering habitats, to migratory species.

Literature Cited

Aldrich, J. W. and C. S. Robbins. 1970. Changing abundance of migratory birds in North America. *Smithsonian Contrib. Zool.* **26:**17–26.

Bjorge, R. R. 1987. Bird kill at an oil industry flare stack in northwest Alberta. *Can. Field Nat.* **101:**346–350.

Blake, J. G. and J. R. Karr. 1984. Species composition of bird communities and the conservation benefit of large versus small forests. *Biol. Cons.* **30:**173–187.

Brett, J. J. 1986. *The mountain and the migration.* Kutztown Publishing Co., Kutztown, PA.

Brower, L. P. and S. B. Malcolm. 1991. Animal migrations: Endangered phenomena. *Amer. Zool.* **31:**265–276.

Cox, G. W. 1985. The evolution of avian migration systems between temperate and tropical regions of the New World. *Amer. Nat.* **126:**451–474.

Crawford, R. L. 1981. Bird casualties at a Leon County, Florida TV tower: A 25-year migration study. *Bull. Tall Timbers Res. Stn.* **22,** 30 pp.

Gauthreaux, S. A. 1991. The use of weather radar to monitor long-term patterns of trans-Gulf migration in spring. *In press in* J. M. Hagan and D. W. Johnston (Eds.), *Ecology and conservation of neotropical migrant landbirds.* Smithsonian Institution Press, Washington, DC.

Graber, J. W. and R. R. Graber. 1983. Feeding rates of warblers in spring. *Condor* **85:**139–150.

Gradwohl, J. and R. Greenberg. 1989. Conserving nongame migratory birds: A strategy for monitoring and research. Pp. 297–328 *in* W. J. Chandler (Ed.), *Audubon wildlife report 1989/1990.* Academic Press, San Diego, CA.

Grimmett, R. 1987. A review of the problems affecting Palearctic migratory birds in Africa. *Int. Council Bird Pres., Study Rep.* 22.

Hall, G. A. 1984. Population decline of neotropical migrants in an Appalachian forest. *American Birds* **38:**14–18.

Hawkins, A. S., R. C. Hanson, H. K. Nelson, and H. M. Reeves (Eds.). 1984. *Flyways: Pioneering waterfowl management in North America.* U.S. Fish and Wildlife Service, Washington, DC.

Holmes, R. T. and T. W. Sherry. 1988. Assessing population trends of New Hampshire forest birds: Local vs. regional patterns. *Auk* **105:**756–768.

Hutto, R. L. 1988. Is tropical deforestation responsible for the reported declines in neotropical migrant populations? *American Birds* **42:**375–379.

Hutto, R. L. 1989. The effect of habitat alteration on migratory birds in a west Mexican tropical deciduous forest: A conservation perspective. *Cons. Biol.* **3:**138–148.

Leck, C. F., B. G. Murray, Jr., and J. Swinebroad. 1988. Long-term changes in the breeding bird populations of a New Jersey forest. *Biological Conservation* **46:**145–157.

Lester, R. T. and J. P. Myers. 1989. Global warming, climate disruption, and biological diversity. Pp. 177–221 *in* W. J. Chandler (Ed.), *Audubon wildlife report 1989/1990.* Academic Press, San Diego, CA.

Lynch, J. F. 1989. Distribution of overwintering nearctic migrants in the Yucatan Peninsula, I. General patterns of occurrence. *Condor* **91:**515–544.

May, R. M. 1981. Modeling recolonization by neotropical migrants in habitats with changing patch structure, with notes on the age structure of populations. Pp. 207–213 *in* R. L. Burgess and D. M. Sharpe (Eds.), *Forest island dynamics in man-dominated landscapes.* Springer-Verlag, New York.

Moore, F. and P. Kerlinger. 1987. Stopover and fat deposition by North American wood-warblers (Parulinae) following spring migration over the Gulf of Mexico. *Oecologia* **74:**47–54.

Moreau, R. E. 1972. *The Palearctic-African bird migration systems.* Academic Press, London.

Morrison, R. I. G. and B. A. Harrington. 1979. Critical shorebird resources in James Bay and eastern North America. *Trans. N.A. Wildl. Nat. Res. Conf.* **44:**498–507.

Myers, J. P. 1983. Conservation of migrating shorebirds: Staging areas, geographic bottlenecks, and regional movements. *American Birds* **37:**23–25.

Myers, J. P., R. I. G. Morrison, P. Z. Antas, B. A. Harrington, T. E. Lovejoy, M. Sallaberry, S. E. Senner, and A. Tarak. 1987. Conservation strategy for migratory species. *Amer. Sci.* **75:**19–26.

Office of Migratory Bird Management. 1990. *Conservation of avian diversity in North America.* U.S. Fish and Wildlife Service, Washington, D.C.

Powell, G. V. N. and J. H. Rappole. 1986. The hooded warbler. Pp. 827–853 *in* R. L. DiSilvestro (Ed.), *Audubon wildlife report 1986.* Nat. Audubon Soc., N.Y.

Rappole, J. H. and D. W. Warner. 1978. Migratory bird population ecology: Conservation implications. *Trans. N. A. Wildl. Nat. Res. Conf.* **43:**235–240.

Robbins, C. S., J. R. Sauer, R. S. Greenberg, and S. Droege. 1989. Population declines in North American birds that migrate to the neotropics. *Proc. Natl. Acad. Sci. USA* **86:**7658–7662.

Robinson, S. K. 1992. Population dynamics of breeding neotropical migrants in a fragmented Illinois landscape. *In press in* J. M. Hagan and D. W. Johnston (Eds.), *Ecology and conservation of neotropical migrant landbirds.* Smithsonian Institution Press, Washington, DC.

Senner, S. E. 1979. An evaluation of the Copper River delta as critical habitat for migrating shorebirds. *Studies in Avian Biology* **2:**131–145.

Smit, C. J., R. H. D. Lambeck, and W. J. Wolff. 1987. *Threats to coastal wintering and staging areas of waders.* Wader Study Group 49 (Supplement):105–113.

Terborgh, J. H. 1989. *Where have all the birds gone?* Princeton Univ. Press, Princeton, NJ.

Terborgh, J. W. 1980. The conservation status of neotropic migrants: Present and future. Pp. 21–30 *in* A. Keast and E. S. Morton (Eds.), *Migrant birds in the Neotropics: Ecology, behavior, distribution, and conservation.* Smithsonian Institution Press, Washington, D.C.

U.S. Fish and Wildlife Service. 1986. *North American waterfowl management plan.* U.S. Department of Interior, Washington, D.C. and Environment Canada, Ottawa, Canada.

Whitcomb, R. F., C. S. Robbins, J. F. Lynch, B. L. Whitcomb, M. K. Klimkiewicz, and D. Bystrak. 1981. Effects of forest fragmentation on avifauna of the eastern deciduous forest. Pp. 125–205 *in* R. L. Burgess and D. M. Sharpe (Eds.), *Forest island dynamics in man-dominated landscapes.* Springer-Verlag, New York.

Whitham, J. W., Jr. and M. L. Hunter, Jr. 1992. Population trends of neotropical migrant landbirds in northern coastal New England. *In press in* J. M. Hagan and D. W. Johnston (Eds.), *Ecology and conservation of neotropical migrant landbirds.* Smithsonian Institution Press, Washington, DC.

Wilcove, D. S. 1985. Nest predation in forest tracts and the decline of migratory songbirds. *Ecology* **66**:1211–1214.

CHAPTER 12

Predators and Predator Management

The ecological role of predation and how predators should be managed in wildlife ecosystems are among the most controversial topics in conservation ecology. Predators are diverse in nature, comprising the members of the carnivore and top carnivore trophic levels of the ecosystem. Tiger beetles and insectivorous songbirds must thus be considered predators, along with killer whales and tigers. The activities of some predators may be the dominant factor in structuring the composition of some biotic communities. Predation by sea otters, for example, appears to control virtually the entire composition of the kelp bed ecosystem (See chapter 17).

Concern over the conservation and management of predators, however, has centered on the larger mammalian carnivores whose prey include wild ungulates and occasionally livestock. Survival of many of these large carnivores in the wild is one major concern. Tigers, leopards, lions, cheetahs, jaguars, wolves, African wild dogs and other similar species are now extinct or endangered over much of their original range, and present urgent problems of conservation management. The effects of large carnivores on their prey is another concern. For predators on game animals, a major question is whether predation does or does not limit prey populations at levels well below their carrying capacity, where predator control might increase substantially the numbers of animals available to hunters. For predators that sometimes kill domestic animals, basic questions include whether losses are great enough to justify the costs of control and, if so, how control should be exerted without disrupting other ecosystem relations involving the predator.

As we shall see, the predator-prey relationship is complex, even when we restrict our attention to large mammalian predators. Scientific studies of predation are continually revealing new aspects of the relationship, so that our understanding of the phenomenon is constantly changing. In any case, we should begin by noting that predation may range from being an unimportant factor in the dynamics of the prey species to the opposite extreme—a severe limitation of prey at a low density. The latter relationship is employed to human benefit in the biological control of agricultural pests by their predators.

ECOLOGY AND BEHAVIOR OF PREDATORS

If we limit our attention to vertebrate predators, we find that they show distinctive features of behavior and ecology. These include systems of territoriality, basic patterns of response to changes in prey abundance, and patterns of selectivity in the capture of prey individuals.

Territoriality is the defense of an area of habitat by an individual, a pair, or a group of individuals against the intrusion of others of the same species, or sometimes those of related or ecologically competing species. Individual mountain lions, for example, seem to defend at least the core areas of their overall ranges, although these ranges tend to overlap considerably, especially for males and females (Hornocker 1970, Neal et al. 1987). Pairs of many species of raptors, such as golden eagles and peregrine falcons, defend territories. Packs of gray wolves (Pimlott 1967) and prides of lions (Schaller 1972) defend group territories (Fig. 12.1). In all cases, direct aggressive challenge of invaders is the ultimate defense. In many vertebrates, however, vocalizations serve to define territorial boundaries. These include the howling of wolves, the roaring of lions, and the singing of birds. In mammals, scent-marking is a frequent way in which territorial boundaries are marked. Most animals use more than one territorial defense mechanism; wolves scent-mark territory boundaries, proclaim ownership by howling, and viciously attack trespassing animals.

Predators vary greatly in their response to changing prey abundance. Their responses are of two basic types: functional and numerical (Holling 1959). A **functional response** is a change in the effort exerted in hunting a particular type of prey. For example, as the numbers of lemmings increase in an arctic tundra area, arctic foxes and snowy owls may shift more of their hunting effort to these animals. When lemmings decline, these predators may switch back to alternate prey, such as arctic hares. A **numerical response** is a change in the density of predators, due to reproduction and mortality. Increased lemming numbers may allow arctic foxes and snowy owls to produce more young, thus increasing their population size. When lemmings crash, however, predator reproduction may fail, and many predators may die, reducing their numbers. In a given situation, the overall response of predators is thus a combination of their functional and numerical responses (Fig. 12.2). Given the widely differing characteristics of predators, this overall response can be quite variable. In general, however, the smaller the predator and the higher its reproductive potential, the greater is the importance of the numerical response. The larger the predator and the lower its reproductive potential, the more important its functional response becomes. Nevertheless, most predators show a combination of both numerical and functional responses to changes in prey abundance (Fig. 12.3).

A third important feature of predation is **predator selectivity,** the extent to which a predator concentrates on substandard prey—individuals that are very young or old, weak,

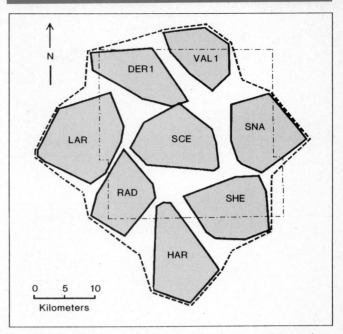

FIGURE 12.1

■ Winter territories of gray wolf packs in northeastern Minnesota in 1982–1983 as revealed by tracking radio-collared animals on the 839-square kilometer Bearville Study Area (dot-dash line) about 50 km north of Grand Rapids, Minnesota. The dashed line enclosed the annual census area for wolf packs (Modified from Fuller 1989).

diseased, or injured. High selectivity, in this sense, means that the predator is taking prey that are likely to contribute little to prey population growth, and that are likely to die soon in any case. Low selectivity means that many of the prey taken would probably contribute significantly to prey population growth.

Information on selectivity of predators is difficult to obtain. Temple (1987), however, studied selectivity in hunting by a tame red-tailed hawk for various kinds of prey. He compared the condition of prey captured by the hawk with that of a random sample of individuals from the prey populations. He found that selectivity was high when prey were hard to capture and the success rate of attempted captures was low. In effect, capture attempts constituted a "test" of the condition of the prey, in which any substandard quality of the prey increased the chance of capture. Temple found that gray squirrel prey, which were difficult to capture, were lower in body fat, more heavily parasitized, and had more physical defects than animals in the population at large. These differences were not seen for chipmunk prey, which were more easily captured. Reviewing the literature, Temple found that high selectivity for prey difficult to capture was shown by many other predators (Table 12.1). Pursuit predators such as lions, cheetahs, African wild dogs, and wolves tend to test the condition of their prey, and thus often show strong selectivity (Fig. 12.4).

FIGURE 12.2

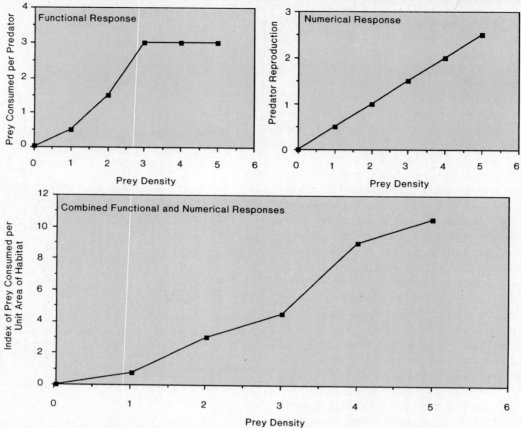

■ Functional, numerical, and total responses of predators to change in prey density.

FIGURE 12.3

■ Medium-sized predators such as this Eurasian eagle owl show both major functional and numerical responses to changes in abundance of their prey.

FIGURE 12.4

■ Pursuit predators such as African lions tend to show strong selectivity in their predation on animals such as wildebeest.

TABLE 12.1

Difficulty of Prey Capture in Relation to Degree of Selectivity for Substandard Individuals by Large Mammal Predators.

PREDATOR	PREY	DIFFICULTY OF CAPTURE	SELECTION FOR SUBSTANDARD INDIVIDUALS
Tiger	Deer	Difficult	Yes
	Pig	Easy	No
Lion	Wildebeest	Difficult	Yes
	Zebra	Difficult	Yes
	Cape Buffalo	Difficult	Yes
	Gazelle	Easy	No
Cheetah	Wildebeest	Difficult	Yes
	Gazelle	Easy	No
Puma	Elk	Difficult	Yes
	Deer	Easy	No
Wolf	Moose	Difficult	Yes
	Caribou	Difficult	Yes
	Deer	Easy	Sometimes
Wild Dog	Wildebeest	Difficult	Yes
	Zebra	Difficult	Yes
	Gazelle	Easy	No
Coyote	Deer	Difficult	Yes
	Pronghorn	Difficult	Yes
	Sheep	Easy	No
Hyena	Wildebeest	Difficult	Yes
	Gazelle	Easy	No

Source: Data from S. A. Temple, "Do Predators Always Capture Substandard Individuals Disproportionately from Prey Populations?" in *Ecology*, 68:669–674, 1987, Ecological Society of America, Tempe, AZ.

PREDATION IN AN ECOSYSTEM CONTEXT

Predation is, in addition, influenced strongly by many other physical and biotic factors of the environment, such as weather conditions, fire influences, structure of the vegetation, alternate prey availability, and the kinds of predators present. These conditions may allow predation to range from being of minor importance to having a severe depressing influence on prey populations. Management efforts may also compound the complexity of the relationship. Small areas of protected, high-quality habitat, for example, may become predation traps by attracting predators to locations where prey have become concentrated and are easily found (Bergerud 1987).

A good example of the complex influence of ecosystem conditions is provided by predation on mule deer in California. Mule deer herds in many parts of the state have declined in numbers in recent decades, and predation is commonly cited as being a major cause. Most wildlife biologists believe that the main cause of declines is habitat loss and deterioration (McCullough et al. 1990). The mechanisms are complex, however, and may involve basic changes in several aspects of wildlands ecology.

The North Kings herd, a migratory population in the Sierra Nevadas east of Fresno, declined from 17,000 animals in 1952 to 3,400 animals in 1977 (Salwasser et al. 1978) and 2,000 animals in 1986 (Neal et al. 1987), for example. This herd winters in the foothills bordering the San Joaquin Valley, where it escapes the deep snows and cold of the mountains. In summer, the herd moves to higher elevations in the Sierras, where milder, more moist conditions than those of the foothills favor summer forage production. The fawns are born in June and July at these higher elevations. Studies have revealed that the decline of the herd is due to low recruitment of young animals, yet fecundity is very high. Females produce an average of 1.5 fawns annually. However, many of these fawns do not survive to enter the adult population.

Studies of coyote predation in the summer range of the North Kings herd in the 1970s showed that for most of the year rodents and rabbits were the primary prey. Without these alternate prey, the coyote population could obviously not survive. From mid-June to the end of July, however, the coyote diet consisted mostly of fawns (Salwasser 1974). This finding did not indicate whether coyotes were mostly preying on healthy fawns, were culling inadequately nourished fawns that would have died soon in any case, or were simply scavenging carcasses of stillborn or dead fawns.

FIGURE 12.5

■ The mountain lion has been protected against hunting in California by a voter initiative proposition.

Further studies revealed that the summer range of the North Kings herd was deficient in food and cover for deer, due primarily to overprotection from fire. Analysis of fire scars on large trees indicated that between 1580 and 1920 A.D. fires occurred at an average interval of eight to nine years, but that by the 1970s no widespread fire had occurred in over 60 years. The result was that growth of high brush and dense forest stands provided little food for does late in pregnancy or for the fawns after they were born. Thus, coyotes found easy prey in weakened fawns. Coyote predation, in this instance, appeared to be largely an indication of poor range condition.

Efforts in the late 1970s and 1980s to improve range conditions of the North Kings herd, using techniques such as prescribed burning to open up overly mature brushlands, did not lead to the recovery of the herd, however, and it has continued to decline. Some investigators now suggest that the deer population may have fallen below a critical ratio with coyotes and mountain lions, the latter having been under protection for a decade (Bertram 1984, Neal et al. 1987). Moun-

tain lions in the range of the North Kings herd may have increased from about 12 individuals in the early 1970s to about 40 individuals in the late 1980s (Neal et al. 1987). Studies with radio-collared fawns and does suggest that predators may kill as much as 48 percent of fawns and 22 percent of does annually. Inasmuch as Smith (1990) in 1985–1986 found that coyotes were feeding very little on deer, even during the period of fawn drop, much of this predation appears to be due to mountain lions. Thus, recovery of this deer herd, even with improved range conditions, may now be inhibited by the high abundance of mountain lions relative to deer.

Dealing with this situation has now become complicated by California politics. In 1990, an initiative proposition giving complete protection to the mountain lion (Fig. 12.5) was passed by California voters. This action means that reduction of the mountain lion population, either by culling or controlled hunting, cannot be used to aid recovery of the deer population.

Perhaps because of its notoriety, the North American gray wolf (Fig. 12.6) has become one of the best-studied large predators (Pimlott 1967, Mech 1979, Allen 1979, Fuller 1989). Yet even in this case, a full understanding of the relationship of the wolf and its prey is far from being obtained. Continuing studies of wolf predation are under way in several locations in Michigan, Minnesota, Canada, and Alaska.

Isle Royale, an island of 570 square kilometers in area lying near the north shore of Lake Superior, is the site of one of the longest studies of wolf ecology (Fig. 12.7). Isle Royale was lumbered and depopulated of large game in the early 1900s. In 1912, however, moose reached the island by walking over the frozen lake surface in winter. Moose found an early successional forest habitat that was ideal in quality, and their population exploded to between 1,000 and 3,000 individuals in the early 1940s. At this time severe over-browsing was evident.

Wolves reached the island in 1949, crossing the ice in similar fashion. Soon, an apparent "steady state" relationship of about 20 to 22 wolves and about 600 moose developed (Fig. 12.8). These populations corresponded to a moose:wolf ratio of about 30:1. Under this regime, wolves harvested about 142 to 150 moose per year—essentially a quarter of the population—or about 6.8 to 7.5 moose per wolf. Under these conditions, wolf predation appeared to be limiting the moose population below the levels apparently set by food availability in the 1930s and 1940s (Pimlott 1967).

Other studies at about this same time suggested that the wolf density on Isle Royale was typical of the maximum density that wolf populations could achieve, about one wolf per 26 square kilometers. This limit seemed to be set by the behavior of the wolf. Pack territoriality, combined with a system of behavioral dominance that limits the size and structure of packs, appeared to set an upper limit to the number of wolves that could exist in a region. This was interpreted to be an intrinsic population control mechanism, and even more interestingly, one that appeared to be tuned to a fairly low density of large prey, such as that probably existing in mature forests of North America prior to European settlement (Pimlott 1967). Before European settlement, in fact, many biologists feel that

FIGURE 12.6

b.

■ Predation on moose (a) by gray wolves (b) has been studied on Isle Royale since wolves recolonized the island in 1949.

a.

FIGURE 12.7

■ A large pack of gray wolves travels through deep snow on Isle Royale, Michigan, the site of long-term studies of wolf predation.

large predators probably limited the densities of prey such as moose, deer, and caribou in North America (Ballard et al. 1987).

Since the late 1960s, however, the wolf and moose populations of Isle Royale have tended to fluctuate widely (Peterson and Page 1988). Moose numbers have increased to 1,000 to 1,500 animals and fallen to less than 500. Wolf numbers have oscillated between 11 and 50 (Fig. 12.4). These fluctuations seem to be related partly to changes in quality of food supply of moose, and thus the average health and vigor of individual moose, which are able to defend themselves effectively when they are in good condition. In 1990, however,

the population fell to four animals, apparently as the result of a disease, canine distemper, introduced to the island in some manner. Furthermore, the wolf population shows no genetic variability, suggesting that problems may also have resulted from inbreeding of the very small wolf population (See chapter 24). In any case, it now appears that even on Isle Royale, wolves may limit the size of the moose population at certain times, and not at others.

FIGURE 12.8

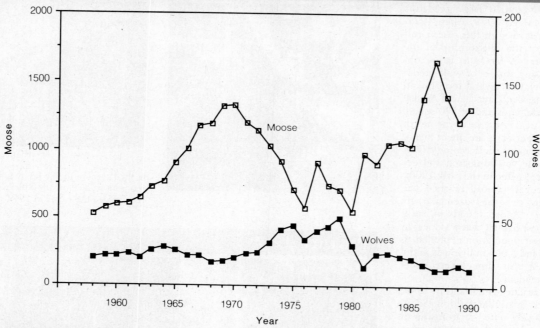

■ Estimated numbers of moose and wolves on Isle Royale since 1958 (Source: from Rolf O. Peterson).

Studies in Alaska give another perspective on this relationship (Gasaway et al. 1983, Bergerud 1988). On the Tanana Flats, an area of 15,500 square kilometers between Fairbanks and the Alaska Range to the south, moose and caribou populations declined catastrophically between 1965 and 1975 (Table 12.2). This decline was attributed to the combined effects of severe winters with deep snow, hunting harvest by humans, and wolf predation. In 1965, the combined ratio of these large ungulates to wolves was about 140:1, far above the 30:1 ratio at which wolves seemed to be controlling moose on Isle Royale. By 1975, however, this ratio had fallen to only 20:1, well below the level at which limitation of prey populations was likely. From 1976 to 1979, deliberate reduction of the wolf population was carried out, the intention being to increase the ratio of prey to wolves. By 1978, this experiment showed success, and prey had reached ratios of 83:1 and were increasing (Gas-

away et al. 1983). Thus, in Alaska, the pattern seemed to be that at a prey:wolf ratio of 20:1, the effect of wolf predation was to limit or further depress prey abundance. At ratios of 20–30:1, the effect varied depending on other mortality factors. At ratios above 30:1, the prey tended to increase unless other mortality factors were exceptional. Caribou populations seem to show a similar relationship, being depressed by wolf predation in forested areas where wolf populations are high, due to the availability of moose as alternate prey, but being independent of wolf predation in tundra areas where wolf numbers are low (Bergerud 1988).

In Minnesota, radio-tracking of wolves has been used to obtain detailed information on movements, territories, and predation on white-tailed deer, their primary prey (Fuller 1989). Wolf densities in the study area were near the maximum of one animal per 26 square kilometers. The ratio of white-tailed deer to wolves was about 158:1, and a small population of moose was also present. On average, individual wolves killed about 19 deer annually, 11 of these being fawns.

These studies suggest that critical ratios of prey to wolf exist for the main-

tenance of stable ungulate populations, both with and without hunting harvest of the ungulate prey (Keith 1983). In the absence of hunting where moose are the primary prey, a ratio of 35:1 appears to be minimal, and where the smaller white-tailed deer are the primary prey, the minimal ratio is about 90:1. Ratios necessary to sustain these ungulate populations with additional hunting harvest would be larger.

Thus, even for individual predator species, the effect of predation may vary from controlling to non-controlling. Favorable range conditions, however, can permit prey species such as moose, mule deer, and other early successional ungulates to exceed the critical ratios of prey to wolf below which predation is a controlling factor. Improvement of range condition for ungulates thus should be the primary aim of management, with direct predator control being necessary only to correct imbalances that have developed because of unusual events.

TABLE 12.2

Populations of Moose, Caribou, and Wolves on the Tanana Flats, Alaska Before and After Artificial Reduction of the Wolf Population.

				RATIOS	
YEAR	MOOSE	CARIBOU	WOLF	MOOSE/WOLF	CARIBOU/WOLF
Before					
1965	23,000	5,000	200	115:1	25:1
1975	2,800	1,800	240	12:1	8:1
After					
1978	3,500	3,100	80	44:1	39:1

Source: Data from W. C. Gasaway, et al., "Interrelationships of Wolves, Prey, and Man in Interior Alaska," in *Wildlife Monographs*, No. 84:1–50, 1983.

FIGURE 12.9

■ Predation of coyotes on sheep still remains as one of the most controversial issues in predator management in North America.

PREDATORS AND LIVESTOCK DEPREDATIONS

A second focus of public concern is predation on livestock by large carnivores such as wolves, mountain lions, and coyotes. In North America, this concern has centered on the coyote (Fig. 12.9), the predator that remains most abundant on rangeland, and on sheep, the domestic animal most often taken. Once again, this relationship is far from being fully understood.

Most quantitative information on the extent of coyote predation on sheep comes from questionnaires prepared by sheep ranchers. This, of course, is a somewhat suspect source of data, since sheep ranchers are also the primary group agitating for increased predator control efforts. Nevertheless, many ranchers are knowledgeable about the causes of livestock mortality and they are the only practical source of comprehensive information on predation losses.

TABLE 12.3

Mortality of Adult Sheep and Lambs from Various Causes in 15 Western States as Determined for 1974.

	PERCENT MORTALITY	
CAUSE OF DEATH	LAMBS	ADULTS
Coyote Predation	8.1	2.5
Other Predators	3.3	0.9
Subtotal	**11.4**	**3.4**
Other Causes	8.2	5.1
Unknown	3.6	1.9
Subtotal	**11.8**	**7.0**
TOTAL	23.2	10.4

Source: U.S. Department of Agriculture, *Econ. Report No. 408*, 1978.

TABLE 12.4

Mortality of Adult Sheep and Lambs from Various Causes in Central California from 1973 to 1983.

	PERCENT MORTALITY	
CAUSE OF DEATH	LAMBS	ADULTS
Coyote Predation	2.4	1.3
Other Predators	0.3	0.2
Subtotal	**2.7**	**1.5**
Other Causes	0.8	1.6
Unknown	2.5	2.6
Subtotal	**3.3**	**4.2**
TOTAL	6.0	5.7

Source: Data from J. H. Schrivener, et al., "Sheep Losses to Predators on a California Range, 1973–1983" in *Journal of Wildlife Management*, 38:418–421, 1985.

The most comprehensive evaluation of sheep losses to predators is a 1974 USDA study (Table 12.3) in which losses to coyotes, other predators, and other causes were evaluated (Gee et al. 1977). This study, carried out in 15 western states, covered 78 percent of the United States sheep ranching area. Results show that heaviest losses are for lambs, but that substantial losses of adult sheep also occur. What is also evident is that documented predation losses for both lambs and adult sheep are less than half of all mortality of animals, and that losses to coyotes amount to only about a third for lambs, or a quarter for adults, of total losses. A similar picture (Table 12.4) was presented by a more recent study of sheep losses in central California between 1973 and 1983. Thus, predation losses to coyotes are only one part of the problem of sheep ranching, an industry that has been declining in recent decades because of marginal production economics.

Considerable amounts of money have been spent on coyote control, with very little evaluation of the benefits resulting. In 1974, federal and local government, as well as private funds for coyote control totalled $6,883,000, spent largely to hire professional predator control personnel. These funds resulted in the killing of 88,092 coyotes, at an average cost of $78 per animal. Various methods were used for coyote killing. For methods such as snaring, trapping, aerial or ground gunning, or denning (locating dens and killing the adults and young), costs were similar, ranging from about $70 to $100 per kill. Use of the **M-44 device,** a metal tube that uses a shotgun shell to shoot cyanide into the mouth of any animal that pulls on the baited end, kills coyotes at a cost of only $37 per animal. Poisoning carcasses of dead livestock with **Compound 1080** (sodium fluoroacetate) leads to coyote kills at a cost of only $11.50 per animal. These latter methods are relatively unselective, and frequently kill other predator and scavenger species. Their cheapness, thus, is the advantage that is behind the support for their use.

No real evaluation has been attempted of the benefits of coyote control efforts on an industry-wide basis. In 1974, however, the industry estimated a total economic loss to coyotes of $27 million, representing the death of about 1.08 million lambs and adult sheep. This suggests that an individual sheep lost to predation had a value of about $25. Thus, at a cost of $78 per coyote kill, the control program would have to reduce sheep losses by slightly over three animals per coyote killed to break even in cost effectiveness. Whether this benefit is realized overall is highly uncertain, but at the very least it suggests that general coyote control efforts are probably marginal in benefit/cost relationships. Of course, the above analysis is simplistic, and does not consider market adjustments in value of individual animals relative to the number of animals marketed, or the carry-over of benefits of control in one year to the next.

As for predation on game animals, alternatives to continuing predator control programs exist to reduce the problem of predation on domestic animals (Robel et al. 1981). Specific techniques that have proven effective in reducing sheep losses include burial or removal of carcasses of dead animals, which reduces the food supply for coyotes on sheep range areas and lessens the tendency of coyotes to become conditioned to seeking sheep as food. The use of trained sheep dogs to protect herds from predators is a second effective technique. Confinement of sheep in corrals, especially if these are lighted, reduces sheep losses considerably. Management of the reproductive cycle of sheep so that the lambs are born in fall, rather than spring, reduces the predation on lambs, because coyotes are usually better fed and have more alternate prey available in fall than after a long winter period of low availability of alternate prey. Finally, the selective killing of individual coyotes that have become problem predators on sheep, as opposed to a general coyote suppression program, is an effective, and sometimes necessary, technique. These approaches are an example of integrated pest management, which we shall discuss in more detail in chapter 13.

FIGURE 12.10

■ Efforts are being made to introduce the red wolf, a species driven to extinction in the wild, to the Alligator River National Wildlife Refuge in North Carolina.

REINTRODUCTION OF PREDATORS TO PARKS AND PRESERVES

Large predators have been eliminated from many parts of their original ranges by human persecution, in some cases involving governmental predator eradication programs. Restoration of such species to major parks and preserves that once supported them is now being encouraged by many conservation groups. Black bears have been restored to national forest areas of Arkansas, for example (Smith et al. 1991). Another effort, begun in 1989, is reintroduction of the Canada lynx to Algonquin State Park in New York. In the winter of 1989–1990, 18 lynx were translocated to the park from northwestern Canada and Alaska, with an additional 40 animals reintroduced in 1990–1991.

In 1986, the United States Fish and Wildlife Service reintroduced red wolves (Fig. 12.10) to the Alligator River National Wildlife Refuge in North Carolina, the first step in the possible restoration of the species to coastal islands along the Carolinas, Florida, and Mississippi. By 1990, 21 captive-bred animals (See chapter 25) had been released, and two young had been reared successfully in the wild (Phillips 1990). Plans are also being made for reintroduction of the Mexican wolf, a subspecies of the gray wolf, to White Sands Missile Range, New Mexico, and possibly to Big Bend National Park, Texas.

The most controversial predator reintroduction proposal is for return of the gray wolf to the Greater Yellowstone Ecosystem in Wyoming and neighboring states. Wolves were originally present in Yellowstone, but were probably eliminated in about 1926, at the end of a 10–year period when 136 animals were killed by trapping, shooting, or poisoning. As a designated endangered species in the northern Rocky Mountains, recovery of the population is mandated by the Endangered Species Act. Reintroduction of the species would also be consistent with the objective of creating in Yellowstone Park a complete, naturally functioning ecosystem like that existing prior to European arrival in North America.

In 1989, a bill was introduced to the United States House of Representatives to expedite wolf reintroduction to the Yellowstone region. This action, however, requires an environmental impact assessment by the United States Fish and Wildlife Service, an effort not yet completed. Reintroduction would likely be coupled with a provision for allowing animals that wander outside park and wilderness areas to be killed. In addition, a fund for reimbursement of damages caused by wolves to livestock and other human property would probably be established.

It is estimated that Yellowstone Park area could support about 11 to 15 packs, or a total of about 110 to 150 animals (Singer 1990). Models of the interaction of the wolves and prey suggest that wolf predation might reduce the size of elk herds by about 15 to 25 percent, with lesser or very minor effects on other species (Boyce 1990). The total population of elk, the most abundant large ungulate in this ecosystem, is now about 40,000 to 50,000 animals.

Small numbers of gray wolves now occur in Glacier National Park, Montana, and in the Frank Church Wilderness, Idaho. Conceivably, animals from these populations might disperse into the Yellowstone region on their own, establishing a natural population with full protection under the Endangered Species Act. Under existing regulations, such animals would be protected wherever they established themselves—inside or outside the park.

Opposition to wolf reintroduction centers on possible depredations on livestock in areas outside the park and wilderness areas. Some livestock have been killed in the vicinity of Glacier National Park, by the small wolf population there. In northeastern Minnesota, where about 1,200 wolves occur, about 10 cattle and sheep are taken annually, but in some years losses have exceeded 100 livestock. In addition, hunting interests oppose reintroduction of wolves on the grounds that many elk and deer summering within the park migrate to areas outside the park, where they can be hunted in autumn. Reduction of herds with summer ranges in the park might therefore reduce hunting harvests.

KEY MANAGEMENT STRATEGIES FOR PREDATORS

1. Primary efforts should be made to manipulate range conditions or domestic herd management to favor game species or protect domestic animals rather than engaging in sustained programs of predator control.

2. Limited use of direct predator control may be justified when the predator/prey ratio remains high when range conditions are favorable to prey, or when specific animals become problem predators on domestic animals.

LITERATURE CITED

Allen, D. L. 1979. *The wolves of Minong.* Houghton-Mifflin, Boston.

Ballard, W. B., J. S. Whitman, and C. L. Gardner. 1987. Ecology of an exploited wolf population in south-central Alaska. *Wildl. Monogr.* **98:**1–54.

Bergerud, A. T. 1987. Increasing the numbers of grouse. Pp. 686–731 in A. T. Bergerud and M. W. Gratson (Eds.), *Adaptive strategies and population ecology of northern grouse.* Univ. of Minnesota Press, Minneapolis.

Bergerud, A. T. 1988. Caribou, wolves and man. *Trends in Ecol. Evol.* **3:**68–72.

Bertram, R. C. 1984. *North Kings deer herd study.* Calif. Dept. Fish and Game, Sacramento, CA.

Boyce, M. S. 1990. Wolf recovery for Yellowstone National Park: A simulation model. Pp. 3:3–58 in *Wolves for Yellowstone? A report to the United States Congress.* Volume II. U.S. Government Printing Office, Billings, MT.

Fuller, T. K. 1989. Population dynamics of wolves in north-central Minnesota. *Wildl. Monogr.* **105:**1–41.

Gasaway, W. C., R. O. Stevenson, J. L. Davis, P. E. K. Shepherd, and O. E. Burris. 1983. Interrelationships of wolves, prey, and man in interior Alaska. *Wildl. Monogr.,* **84:**1–50.

Gee, C. K., R. S. Magelby, W. R. Bailey, R. L. Gum, and L. M. Arthur. 1977. *Sheep and lamb losses to predators and other causes in the western United States.* USDA Agr. Econ. Rep. No. 369. 41 pp.

Gum, R. L., L. M. Arthur, and R. S. Magleby. 1978. *Coyote control: A simulation evaluation of alternative strategies.* USDA Agr. Econ. Rep. No. 408. 49 pp.

Holling, C. S. 1959. The components of predation as revealed by small mammal predation of the European pine sawfly. *Canad. Entomol.* **91:**290–320.

Hornocker, M. G. 1970. An analysis of mountain lion predation upon mule deer and elk in the Idaho Primitive Area. *Wildl. Monogr.* **21:**1–39.

Keith, L. B. 1983. Population dynamics of wolves. Pp. 66–77 in L. N. Carbyn (Ed.), *Wolves in Canada and Alaska.* Can. Wildl. Rep. Ser. 45.

McCullough, D. R., D. S. Pine, D. L. Whitmore, T. M. Mansfield, and R. H. Decker. 1990. Linked sex harvest strategy for big game management with a test case on black-tailed deer. *Wildlife Monogr.* No. 112.

Mech, L. D. 1979. *The wolf: Ecology and behaviour of an endangered species.* Nat. Hist. Press, NY.

Neal, D. L., G. N. Steger, and R. C. Bertram. 1987. *Mountain lions: Preliminary finding on home-range use and density in the central Sierra Nevada.* USDA Forest Service Res. Note PSW-392. 6 pp.

Peterson, R. O. and R. E. Page. 1988. The rise and fall of Isle Royale wolves, 1975–1986. *J. Mammal.* **69:**89–99.

Phillips, M. K. 1990. *Restoration of endangered red wolves in northeastern North Carolina.* National Meeting of the Society for Ecological Restoration, Chicago, IL.

Pimlott, D. H. 1967. Wolf predation and ungulate populations. *Amer. Zool.* **7:**267–278.

Robel, R. J., A. D. Dayton, F. R. Henderson, R. L. Meduna, and C. W. Spaeth. 1981. Relationships between husbandry methods and sheep losses to canine predators. *J. Wildl. Manag.* **45:**894–911.

Salwasser, H. 1974. Coyote scats as an indicator of the time of fawn mortality in the North Kings deer herd. *Calif. Fish and Game* **60:**84–87.

Salwasser, H., S. A. Holl, and G. A. Ashcraft. 1978. Fawn production and survival in the North Kings River deer herd. *Calif. Fish and Game* **60:**38–52.

Schaller, G. B. 1972. *The Serengeti lion.* Univ. of Chicago Press, Chicago.

Schrivener, J. H., W. E. Howard, A. H. Murphy, and J. R. Hays. 1985. Sheep losses to predators on a California range, 1973–1983. *J. Wildl. Manag.* **38:**418–421.

Singer, F. J. 1990. Some predicted effects concerning a wolf recovery into Yellowstone National Park. Pp. 4:3–34 in *Wolves for Yellowstone? A report to the United States Congress.* Volume II. U.S. Government Printing Office, Billings, MT.

Smith, J. R. 1990. Coyote diets associated with seasonal mule deer activities in California. *Calif. Fish and Game* **76:**78–82.

Smith, K. G., J. D. Clark, and P. S. Gipson. 1991. History of black bears in Arkansas: Over-exploitation, elimination, and successful reintroduction. *Eastern Workshop on Black Bear Research and Management* **10:**5–14.

Temple, S. A. 1987. Do predators always capture substandard individuals disproportionately from prey populations? *Ecology* **68:**669–674.

CHAPTER 13

Pesticides and Wildlife

The widespread use of synthetic organic pesticides following World War II has posed enormous problems for wildlife, both aquatic and terrestrial. Hundreds of basic pesticide chemicals are now available, and these are utilized in thousands of specific formulations and brand-name products. Several pesticides have become general contaminants of the biosphere, having been transported worldwide by systems of air and water circulation.

Pesticides are usually targeted against a single species or groups of ecologically similar species. Experience has shown, however, that many nontarget organisms are usually affected, and that the overall impacts of pesticides on ecosystem structure and function can be very complicated and difficult to predict. Furthermore, some long-life pesticide chemicals establish biogeochemical movement patterns that carry them to distant localities and concentrate them in unexpected ways. Repeated application of these very powerful chemicals has also made them major agents of evolutionary change in many animal and plant species. Pesticide technology is a rapidly changing field, in which new formulations are continually appearing. Pesticide impacts have become a continuing concern of conservation ecology.

We shall first consider the pesticides belonging to a particular chemical family, the chlorinated hydrocarbons, as an example of the far-ranging impacts that pesticides can have. Then we shall consider some of the conservation problems posed by pesticides of other types. Finally, we shall examine briefly some of the alternatives that must be sought to the massive use of pesticides in agriculture and in control of disease vectors.

CHLORINATED HYDROCARBON PESTICIDES AND THEIR EFFECTS

The pesticidal properties of DDT (dichloro, diphenyl, trichloroethane), an organic chemical first synthesized in the late 1800s, were first noted in 1939 by Paul Muller, a Swiss biochemist (Dunlap 1981). During World War II, DDT was first used to control the human body lice responsible for transmitting typhus in war-ravaged cities and to protect allied troops against malaria-carrying mosquitos. After the war, DDT rapidly came into use for the control of agricultural pests and vectors of human diseases such as malaria. Muller received the 1948 Nobel Prize in Chemistry for his discovery, then hailed as the means of ridding the world of insects detrimental to agriculture and human health (McEwen and Stephenson 1979).

DDT is a member of a group of petroleum-derivative chemicals known as **chlorinated hydrocarbons,** which have a structure consisting of carbon rings to which chlorine atoms are attached in varying fashion. Chlorinated hydrocarbons include such pesticide chemicals as dieldrin, aldrin, endrin, chlordane, heptachlor, toxaphene, mirex, kepone, and many others. The members of this chemical group share, to a greater or lesser degree, four characteristics that, in combination, make them serious environmental pollutants from the standpoint of danger to plant and animal life (Wurster 1969). These characteristics are:

1. Chemical stability. Chlorinated hydrocarbons are persistent, and are degraded very slowly in nature. DDT is converted to metabolites such as DDD and DDE, which still have many of the properties of the original compound. DDT and these major metabolites, however, have an environmental half-life of 10 to 15 years, meaning that after this period, half of the quantity that was applied in a particular location still survives somewhere, and that after two half-lives, a quarter of the original amount still survives. Since DDT has only been in use for a period equal to three to four half-lives, this means that much of the chemical that was ever produced still exists somewhere in nature.

2. High mobility. Chlorinated hydrocarbons are carried widely by movements or air and water. Winds transport them in vapor form, as fine droplets, and as material adsorbed on dust particles. Although their solubility in water is low, the large volume of water that moves through most ecosystems means that large quantities can ultimately be moved in solution. Water carries even larger quantities of these pesticides in particulate form or adsorbed on particles carried in suspension. Several of the chlorinated hydrocarbons have become biosphere-wide in distribution as a result, their residues being detectable in mid-ocean waters and in antarctic snow.

3. High solubility in lipids. Although chlorinated hydrocarbons have a very low solubility in water, their solubility in lipids (fats and oils) is very high. This means that wherever water comes in contact with lipids, chlorinated hydrocarbons tend to move into the lipids. Since lipids are a basic component of living organisms, this means that organisms tend to take up these pesticides from the water that surrounds them, or from the food or water they ingest.

4. Toxicity and biological activity. Pesticides, by definition, are toxic to their target organisms, but many are toxic to a much wider range of organisms. Even when they are not lethal, however, these chemicals are biologically active, and can disrupt physiological processes, especially those involved in reproduction and development.

It is the combination of these four characteristics that makes chlorinated hydrocarbons so dangerous, in fact. If any one of these characteristics is taken away, the result is a chemical with a much less serious threat of environmental pollution.

Direct Side Effects

Soon after DDT came into general use, a number of direct side effects were observed (Carson 1962, Rudd 1964). These involved the mortality of non-target organisms in areas of heavy applications. For example, DDT was used against the elm bark beetle, the vector for the Dutch elm disease fungus that had invaded North America and was killing native elms. High-pressure spraying rigs were used to spray individual trees with 0.4 to 0.6 kg of DDT twice a year. Such spraying resulted in the mortality of songbirds in areas of the midwestern and northeastern United States in the 1950s and 1960s. Heavy spraying in efforts to prevent the spread of the Japanese beetle likewise killed many non-target birds and mammals. Major spray programs were aimed at outbreaks of the spruce budworm in forests of Canada and the northern United States. In New Brunswick, for example, over 6,800 tons of DDT were used on millions of hectares of forest from 1952 through 1968 (McEwen and Stephenson 1979). This program led to heavy mortality of salmonid fish in streams exposed to the applications or receiving runoff carrying DDT.

Such direct side effects attracted the concern of biologists, foremost among whom was Rachel Carson, who documented these events dramatically in her 1962 book, *Silent Spring*. This book, in turn, stimulated concern in the public at large about pesticide use, as well as arousing considerable controversy among biologists themselves.

Indirect Side Effects

Rachel Carson also recognized some of the first indirect side effects of pesticide use—effects distant in time or far removed

FIGURE 13.1

■ Mortality of western grebes at Clear Lake, California, provided one of the first examples of bioaccumulation and biomagnification of chlorinated hydrocarbon pesticides.

FIGURE 13.2

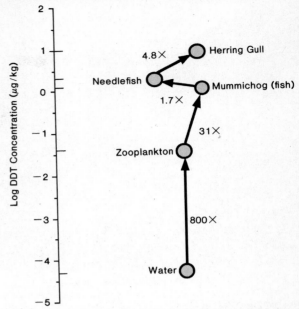

■ Bioaccumulation and biomagnification of DDT in the food chain of an estuary on Long Island, New York, based on studies in the 1960s. Source: Data from G. M. Woodwell, et. al., "DDT Residues in an East Coast Estuary: A Case of Biological Concentration of a Persistent Pesticide" in *Science,* 156:821–823, 1967, American Association for the Advancement of Science, Washington, D.C.

in space from the actual applications of pesticides. Such indirect effects have proved to be the greatest threat to wildlife.

At Clear Lake, in northern California, for example, the pesticide DDD (a close relative of DDT) was used in attempts to control a nuisance problem caused by a midge fly with an aquatic larva (Hunt and Bischoff 1960). The adult midges, non-biting insects about the size of a mosquito, emerged from the lake in enormous numbers and were attracted to lights at houses surrounding the lake. To try to eliminate the problems caused by these billions of insects, in September, 1949, DDD was introduced directly into the lake water at a concentration of 14 ppb (parts per billion). The result was an estimated 99 percent reduction of the midge problem. After several years, however, the problem reappeared, and the lake was treated again with DDD at 20 ppb in September, 1954. In December, 1954, three months after the treatment, more than 100 western grebes (Fig. 13.1), a fish-eating, diving bird, were found dead around the lake, and in March, 1955, six months after the treatment, still more dead grebes were found. In September, 1957 the lake was once again treated as in 1954. In December, 1957 another die-off of about 75 grebes was noted. On this occasion the birds were collected and their tissues analyzed for DDD. The body fat of these birds contained up to 1,600 ppm (parts per million) of DDD, about 80,000 times the maximum concentration introduced into the lake water.

Analyses of DDD in the tissues of plankton and fish showed that the entire food chain was contaminated, and that the concentration increased along the food chain. This was one of the first observations of pesticide **bioaccumulation,** the concentration from an organism's physical environment, and **biomagnification,** the concentration from one link to the next in a food chain. Phytoplankton organisms bioaccumulated DDD from the water because of the solubility relationships of DDD in water and fats. Zooplankton that fed on phytoplankton retained most of the DDD from the phytoplankton they consumed over their lives, and fish did the same for the DDD in the zooplankton they consumed. This biomagnification continued with the feeding of grebes on fish, leading to levels of DDD that were lethal.

In a salt marsh on the south shore of Long Island, Woodwell et al. (1967) examined DDT concentration in all components of the ecosystem. Within living organisms, DDT concentration varied through three orders of magnitude. In zooplankton, the concentration was about 40 parts per billion, while in various water birds values ranged from 3 to 75 parts per million. All biotic components of the marsh system, both aquatic and terrestrial, were contaminated, however (Fig. 13.2).

Bioaccumulation and biomagnification have subsequently been recorded in many other situations, and for several other chlorinated hydrocarbons (Matsumura 1985). Bioaccumulation alone can increase the concentrations of chlorinated hydrocarbons in aquatic organisms by a factor of 10^3 to 10^6. Biomagnification at one food chain transfer tends to be less, but may be as much as 30-fold (L. J. Blus, pers. comm.). Several other environmental pollutants (See chapter 19) also have been found to show bioaccumulation and biomagnification.

FIGURE 13.3

■ Ospreys nesting at the mouth of the Connecticut River, on Long Island Sound, were one of the first birds to experience reproductive failure from the effects of DDT.

FIGURE 13.4

■ The peregrine falcon disappeared from much of eastern North America due to egg-shell thinning induced by DDT contamination.

Avian Reproductive Failure and Shell-Thinning

It soon became apparent, in addition, that contamination of wildlife populations by chlorinated hydrocarbons was having many effects other than direct mortality of adult animals. At the mouth of the Connecticut River on Long Island Sound the numbers of pairs of nesting ospreys (Fig. 13.3) had declined from 200 in 1938 to only 22 in 1963 (Ames 1966). Studies of the remaining pairs showed that the decline was due to failure of most of the eggs to hatch, and that the eggs were heavily contaminated with DDT, as was the entire food chain leading up to the osprey.

The most widespread and insidious effect of chlorinated hydrocarbons was, and still is, shell-thinning in birds at the ends of long food chains. Shell-thinning was first recognized in 1967 for the peregrine falcon, golden eagle, and European sparrow hawk in Great Britain by Ratcliffe (1967). Ratcliffe noted that decline of populations of these species had been correlated with an unusually high frequency of breakage of eggs in nests. Comparing the thicknesses of the shells of eggs collected and placed in museum collections prior to 1945, with those of eggs taken from the wild, he found a substantial thinning. Ratcliffe also noted that these birds carried high body burdens of chlorinated hydrocarbons.

Subsequent studies showed that the shell-thinning phenomenon was extensive in four groups of birds: 1) bird-eating raptors such as the peregrine falcon (Fig. 13.4), prairie falcon, merlin, goshawk, Cooper's hawk, sharp-shinned hawk, and European sparrow hawk; 2) fish-eating raptors such as the osprey and bald eagle (Fig. 13.5); 3) fish-eating water birds including many loons, grebes, pelicans, cormorants, ibis,

FIGURE 13.5

■ The bald eagle was one of many fish-eating raptors and water birds to suffer population decline due to egg-shell thinning caused by DDT.

herons, egrets, gulls, and terns; and 4) carrion feeders such as the California condor, turkey vulture, carrion crow, and rook.

Shell-thinning was associated with declining populations of a number of species. From studies of populations of the species affected, the generalization emerged that no raptor that experienced an average shell-thinning of more than 18 percent was able to maintain a stable population (Lincer 1975). Counts of migrating hawks at Hawk Mountain Sanctuary,

Pennsylvania, show that bald eagles, golden eagles, sharp-shinned hawks, Cooper's hawks, and peregrine falcons declined significantly during the period from 1946 to 1972 (Bednarz et al. 1990). The birds tallied in these counts represent much of eastern Canada and the northeastern United States.

Experimental studies in which kestrels were fed diets containing different levels of DDT eventually showed that this pesticide was, in fact, the cause of shell-thinning (Lincer 1975). A diet containing 10 ppm DDT resulted in egg shells more than 29 percent thinner than normal. The precise mechanism of thinning is still uncertain, and may involve several hormonal and enzymatic mechanisms that mediate calcium removal from storage centers in medullary bone and its deposition in the shell gland of the bird oviduct (McEwen and Stephenson 1979).

With broad restrictions on the use of chlorinated hydrocarbons in the developed countries of North America and Europe in the 1970s and 1980s, many water bird species that experienced declines due to egg-shell thinning are recovering (Blus 1982, Anderson and Gress 1983, Chapdelaine et al. 1987). The various raptors that declined due to chlorinated hydrocarbon contamination have shown increasing numbers in annual counts at Hawk Mountain from 1973 through 1986 (Bednarz et al. 1990). The osprey populations of Long Island Sound, one of the first populations to show DDT-induced decline, have recovered rapidly (Spitzer and Poole 1980). In Ontario, the reproductive success of bald eagles, which had declined to less than one young fledged for every two nests in 1966, climbed to over one bird per nest in 1981 (Grier 1982).

Chlorinated hydrocarbon problems persist even in North America, however. Snowy egrets breeding in Idaho (Findholt 1984) and white-faced ibis breeding in the Great Basin (Henny and Herron 1989) still show thin-shelled eggs and reduced breeding success due to DDT and DDE pesticide loads picked up on their winter ranges in Mexico. In Florida, both adult and locally-reared juvenile fulvous whistling-ducks show substantial body burdens of several chlorinated hydrocarbon pesticides, including chemicals such as aldrin that are rapidly converted to distinct metabolites (Turnbull et al. 1989). This suggests that illegal use of banned pesticides occurs in the region.

In many developing countries, such as Kenya, however, use of chlorinated hydrocarbon pesticides remains high, or is increasing, and environmental pollution is serious (Lincer et al. 1981). Reports of declining reproduction by raptors in Zimbabwe, due to DDT and other chlorinated hydrocarbons (Tannock et al. 1983), indicate that pollution by these pesticides is far from eliminated on a world-wide basis.

Factory Area Contamination

The Southern California coastal marine environment experienced massive contamination by DDT in the 1950s and 1960s. During this period, the peregrine falcon and bald eagle

FIGURE 13.6

■ Brown pelicans off the coast of southern California suffered almost complete reproductive failure due to egg-shell thinning caused by industrial pollution of coastal waters by DDT.

disappeared as nesting birds on the Channel Islands off Southern California, and populations of brown pelicans (Fig. 13.6) declined rapidly as a result of almost complete nesting failure due to shell-thinning. Breeding populations of the western gull also declined severely.

The source of the DDT, once thought to be come as runoff from agricultural land, turned out to be the Montrose Chemical Corporation plant in Los Angeles, which began producing DDT in 1950 (MacGregor 1974). Chemical wastes from this plant were discharged directly into the Los Angeles sewer system, and from this system into the Pacific Ocean. Through 1970, an estimated 1,900 tons of DDT entered the ocean in this manner. Once this source was discovered, the discharge was terminated. Since 1970 the populations of brown pelicans have slowly recovered (Anderson and Gress 1983), and are once again common along the Southern California coast.

A legacy of this contamination episode remained into the 1980s in the form of an abnormal sex ratio among western gulls. Population declines of this species were associated with a high ratio of females to males at nesting colonies (Hunt and Hunt 1977). Experimental studies on the effect of DDT on development of gull eggs showed that male embryos were becoming "feminized" to a degree in their anatomical development (Fry and Toone 1981). Feminized males were believed to be abnormal in behavior, failing to establish territories and engage in normal courtship, thus creating a shortage of males in colonies and leading to the abnormal female pairings.

Even in the 1990s, DDT effects persist in southern California waters. Municipal waste water discharge continues to carry a small amount of DDT into the ocean, a quantity estimated at about 40 tons over the period from 1971 to 1983. About 36 percent of the white croaker, a popular sport fish, in San Pedro Bay, near Long Beach, possess tissue levels of DDT greater than 4 ppm, at which level they are apparently unable to spawn (Hose et al. 1989). Chlorinated hydrocarbon pesticides thus may still be contributing to the reduced populations of several sport fish along the southern California coast.

In 1990 the California State Department of Health recommended closure of the white croaker fishery. The National Oceanic and Atmospheric Administration also filed suit against the Montrose Chemical Corporation and several other industries for damages to fisheries by pollution of the coastal waters by DDT and other materials.

Another episode of industrial pollution occurred at Wheeler National Wildlife Refuge on the Tennessee River near Huntsville, Alabama (O'Shea et al. 1980). In this case, chemical wastes from a DDT plant at the Redstone Arsenal were discharged into the refuge through 1970. This discharge, believed to involve about 4,000 tons of DDT, has led to the contamination of wintering waterfowl as well as resident birds and mammals, and may be responsible for the decline or disappearance of breeding populations of bald eagles, double-crested cormorants, and several herons at the refuge. Sampling of sediments and aquatic invertebrates downstream from the discharge point in 1983–1984 revealed concentrations of DDT ranging from 12 to 2,730 ppm in sediments, and from 24 to 125 ppm in detritus-feeding invertebrates (Webber et al. 1989), so that this pollution episode will probably continue to exert effects for many decades.

Renewed Use of Chlorinated Hydrocarbons

The resurgence of malaria throughout the tropics has led to the revival of heavy DDT use against *Anopheles* mosquitos in several countries, notably India and Brazil. The World Health Organization considers DDT to be the most effective pesticide for mosquito control in homes and other buildings. In 1990, Brazil embarked on a five-year plan of spraying homes for mosquito control in Amazonia, a program estimated to require 3,000 tons of DDT.

OTHER INSECTICIDES AND WILDLIFE

Carbamates and **organophosphates** are two other chemical groups of pesticides that contain alternatives to many of the chlorinated hydrocarbons. Most of these compounds have very high toxicity, but their environmental life is short, and they do not show biological concentration or transport to distant locations. Most of their side effects are thus direct (Smith 1987), although some sublethal effects on behavior and reproduction have been noted (Grue et al. 1983).

Carbofuran, one of the carbamates, has caused heavy mortality of waterfowl feeding in treated alfalfa fields (Smith 1987). Carbofuran is often applied into the soil with seed corn at planting to control corn rootworms, leading to mortality of songbirds feeding on earthworms and insects that have become contaminated (Balcomb et al. 1984). Carbofuran may be especially toxic to burrowing owls, either through direct contact or through ingestion of insects killed by spraying (Fox et al. 1989). Carbofuran and a number of other pesticides are also commonly applied as granules, which birds sometimes ingest directly as grit (Mineau 1988). These problems have led to a recommendation by the United States Environmental Protection Agency to cancel the use of carbofuran (Thomas 1989).

Organophosphate pesticides, especially parathion, methyl parathion, and diazinon, have also occasionally caused mortality of herbivorous birds such as geese or sage grouse feeding in recently treated areas, or of other birds feeding on insects killed by application of these materials (Smith 1987, Blus et al. 1989). The use of diazinon to control insects in golf courses and sod farms was banned in 1990 by the United States Environmental Protection Agency because of the frequent mortality of Canada geese and other birds. Applications of other organophosphates for control of forest insect outbreaks have likewise been accompanied by mortality of various birds, especially insectivorous songbirds.

Recent efforts to control spruce budworm outbreaks in Canada have utilized an organophosphate pesticide, fenitrothion, rather than DDT. Application of this pesticide at even the lowest operational doses has caused extensive mortality of pollinator insects and occasional mortality of adult songbirds (Ernst et al. 1989). Light applications can also lead to more than 65 percent decline in the aquatic insect biomass in streams of sprayed areas. This change in insect food translates into about a 13 to 15 percent reduction in growth of salmon in affected streams (Muirhead-Thompson 1987).

Organophosphates are also being used heavily in the long-term effort by the World Health Organization to eradicate onchocerciasis, or river blindness, in West Africa. The vector of the disease agent is a blackfly, *Simulium damnosum*,

FIGURE 13.7

■ Phenoxy herbicides are routinely used to reduce brush growth in regenerating clearcuts in commercial forests of the Pacific Northwest.

with an aquatic larva. The main control effort involves aerial applications of pesticide to river surfaces. Begun in 1974, the plan is to continue an intensive spray program for 20 years, or long enough for most of the infected human population to die. Up to 18,000 km of river are sprayed weekly during the rainy season. The overall side effects of this program are not known, but some long-term changes in benthic invertebrate communities have been documented (Muirhead-Thompson 1987).

MAMMALIAN PESTICIDES

Compound 1080 (sodium monofluoroacetate) has been used in poisoned baits for control of rodents, coyotes, and a few other animals. This extremely toxic material may cause both direct poisoning of non-target species and secondary poisoning of animals that scavenge carcasses of animals killed by direct poisoning. This material is now permitted to be used only in collars placed on animals such as sheep, the intent being selective poisoning of predators that attack these individuals. Other vertebrate poisons, such as strychnine and sodium cyanide (used in the M-44 device, chapter 12), are restricted to relatively limited uses.

HERBICIDES

Herbicides, like insecticides, belong to several different chemical groups. Most herbicides have a short environmental life. Although not without toxicity to non-target organisms,

few herbicides have caused major direct or indirect side effects. Some indirect effects have been noted on survival of young game birds, apparently because of the reduction of food arthropods that depended on weeds in cropland (Rands 1985).

One group of herbicides, the **phenoxy herbicides,** is similar in structure to plant growth hormones. Certain of these chemicals, particularly silvex and 2,4,5-T, contain trace quantities of dioxin, an intermediate-life chemical that may be concentrated along food chains. The so-called "Agent Orange" used for tropical forest defoliation during the Vietnam war consisted of dioxin-containing phenoxy herbicides. Dioxin is both cancer-inducing and a cause of birth defects in laboratory animals, so that herbicides containing dioxin are presumably dangerous to all vertebrates.

For herbicides, a major conservation issue is their widespread use in vegetation management, such as to rid rangeland of invasive shrubs and trees or reduce shrubs that may compete with regrowing conifers on forest clearcuts (Green 1982). In the Pacific Northwest, phenoxy herbicides are used routinely to inhibit early successional shrubs such as snowbrush, *Ceanothus velutinus,* and red alder, *Alnus rubra,* which are considered to suppress the growth of seedlings and saplings of Douglas fir and other commercial conifers (Fig. 13.7). Both of these shrubs, however, have nitrogen-fixing root nodules, and some studies suggest that they are important in rebuilding the nitrogen pool of the forest, much of which is lost when an area is clear-cut and burned to eliminate dead slash material. From the ecosystem perspective, herbicide use may thus be reducing fertility over the long run.

Furthermore, successional shrublands are important habitat for mule deer and other species of animals, as well as many successional plants. From the perspective of maintaining biotic diversity, the use of herbicides is tending to create monocultures of a few tree species of commercial value.

REGULATION OF PESTICIDE USE

In the United States, use of pesticides is regulated under the **Federal Insecticide, Fungicide, and Rodenticide Act (FIFRA)** of 1947, which requires registration of pesticide chemicals, and the **Federal Environmental Pesticide Act** of 1972, which established federal authority over intrastate use of pesticides. Registration of pesticides, as well as cancellation of those found to be harmful, is carried out by the Environmental Protection Agency, which periodically published a list of pesticides that have been suspended, restricted in use, or banned (Office of Pesticides and Toxic Substances 1985).

At the international level, however, regulation of pesticide use is still very lax (Prabhu 1988). Most pesticides are produced in the United States, western Europe, and Japan—countries which, in general, have agencies that regulate pesticide use. Some 30 companies in the United States, for example, produce 45 percent of the pesticides used throughout the world. Much of this production is exported, with the exports including chemicals that are banned or severely restricted for use in the United States. Most developing countries, however, do not have agencies capable of regulating pesticide use in any detail. Several international organizations, such as the Organization for Economic Cooperation and Development (OECD), the World Health Organization (WHO), and several branches of the United Nations are attempting to develop international systems of trade in pesticides and regulation of their use. Nevertheless, the fact is that many pesticides, such as several of the chlorinated hydrocarbons that have adverse environmental impacts, are still in widespread use.

The use of pesticides in many developing countries, unfortunately, has often been subsidized heavily by international assistance agencies such as the World Bank and the United States Agency for International Development (Repetto 1985). Development loans or agricultural credit programs supported by these organizations have tended to emphasize technological approaches to agricultural production, such as chemical pest control. In addition, much of the pesticides used have been imported from industrial countries, often the countries supporting the loan programs. The result is subsidization of pesticide use to the extent of up to 80 to 90 percent of the cost of the pesticides involved.

ALTERNATIVES TO PESTICIDES

The strategy of exclusive reliance on chemical pesticides for control of agricultural pests and disease vectors has proven unsatisfactory (Cox and Atkins 1979). In large measure, this results from two facts. First, pesticides disrupt naturally ex-

FIGURE 13.8

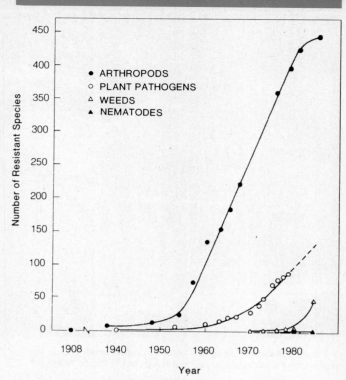

■ Genetic resistance of arthropods and weeds to pesticides has shown exponential increase since the start of widespread use of insecticides and herbicides (Modified from National Research Council 1986).

isting biological control relationships, promoting **rapid resurgence,** or quick recovery of the target pest itself. Damage to natural biological control species often permits **secondary outbreaks** of other species that were formerly not pests. Second, frequent pesticide applications strongly select for mechanisms of **pesticide resistance,** the ability to avoid, tolerate, or inactivate the pesticide chemical. More than 447 insects and mites have evolved resistance to insecticides of almost all major chemical types (Fig. 13.8), and more than 41 species of weeds are now known to have to evolved resistance to herbicides (LeBaron and Gressel 1982, National Research Council 1986). Resistance is also shown by vertebrate pests such as rats, as well as by many fungal pathogens and nematodes that attack crop plants.

The heavy use of pesticides in mechanized agriculture has largely resulted from the fact that they are a highly marketable pest control form that promises a reduction in labor costs for pest control. The fact that pesticide use often increases the pest problem itself is also a significant contributor to increased pesticide use. The economic weakness of excessive reliance on chemical control is evident to most farm producers, and has stimulated the development of new systems of **integrated pest management** that combine a variety of approaches—cultural, biological, and chemical (Cox and Atkins 1979). At the same time, much has been learned about ways to make pesticidal chemicals freer of side effects by increasing their specificity, reducing their persistence, and lessening their mobility. Further research is needed both into the design of pest management systems that are effective, and into ways of making such an approach attractive to private enterprise.

KEY CONSERVATION STRATEGIES RELATING TO PESTICIDES

1. Increased use of integrated pest management systems that incorporate selective, non-persistent pesticides.

2. Comprehensive monitoring of pesticides and their effects in ecosystems, with particular attention to animals at high trophic levels.

LITERATURE CITED

Ames, P. L. 1966. DDT residues in the eggs of the osprey in the north-eastern United States and their relation to nesting success. *J. Appl. Ecol.* **3**(Supplement):87–97.

Anderson, D. W. and F. Gress. 1983. Status of a northern population of California brown pelicans. *Condor* **85**:79–88.

Balcomb, R., C. A. Bowen, D. Wright, and M. Law. 1984. Effects on wildlife of at-planting corn applications of granular carbofuran. *J. Wildl. Manag.* **48**:1353–1359.

Bednarz, J. C., D. Klem, Jr., L. J. Goodrich, and S. E. Senner. 1990. Migration counts of raptors at Hawk Mountain, Pennsylvania, as indicators of population trends, 1934–1986. *Auk* **107**:96–109.

Blus, L. J. 1982. Further interpretation of the relation of organochlorine residues in brown pelican eggs to reproductive success. *Env. Pollution (Series A):* **28**:15–33.

Blus, L. J., C. S. Staley, C. J. Henny, G. W. Pendleton, T. H. Craig, E. H. Craig, and D. K. Halford. 1989. Effects of organophosphorus insecticides on sage grouse in southeastern Idaho. *J. Wildl. Manag.* **53**:1139–1146.

Carson, R. 1962. *Silent spring.* Houghton Mifflin, Boston.

Chapdelaine, G., P. Laporte, and D. N. Nettleship. 1987. Population, productivity and DDT contamination trends of Northern Gannets (*Sula bassanus*) at Bonaventure Island, Quebec, 1967–1984. *Can. J. Zool.* **65**:2922–2926.

Cox, G. W. and M. D. Atkins. 1979. *Agricultural ecology.* W. H. Freeman, San Francisco.

Dunlap, T. R. 1981. *DDT: Scientists, citizens, and public policy.* Princeton Univ. Press, Princeton, NJ.

Ernst, W. R., P. A. Pierce, and T. L. Pollock. 1989. *Environmental effects of fenitrothion use in forestry.* Environment Canada, Dartmouth, Nova Scotia.

Findholt, S. L. 1984. Organochlorine residues, eggshell thickness, and reproductive success of snowy egrets nesting in Idaho. *Condor* **86**:163–169.

Fox, G. A., P. Mineau, B. Collins, and P. C. James. 1989. *The impact of the insecticide carbofuran (furadan 480F) on the burrowing owl in Canada.* Technical Report Series No. 72, Canadian Wildlife Service, Ottawa, Canada.

Fry, D. M. and C. K. Toone. 1981. DDT-induced feminization of gull embryos. *Science* **213**:922–924.

Green, K. 1982. *Forests, herbicides and people.* Council on Economic Priorities, New York.

Grier, J. W. 1982. Ban of DDT and subsequent recovery of reproduction in Bald Eagles. *Science* **218**:1232–1235.

Grue, C. E., W. J. Fleming, D. G. Busby, and E. F. Hill. 1983. Assessing hazards of organophosphate pesticides to wildlife. *Trans. N.A. Wildl. Nat. Res. Conf.* **48**:200–220.

Henny, C. J., and G. B. Herron. 1989. DDE, selenium, mercury, and white-faced ibis reproduction at Carson Lake, Nevada. *J. Wildl. Manag.* **53**:1032–1045.

Hose, J. E., J. N. Cross, S. G. Smith, and D. Diehl. 1989. Reproductive impairment in a fish inhabiting a contaminated coastal environment off southern California. *Env. Pollution* **57**:139–148.

Hunt, E. G. and A. I. Bischoff. 1960. Inimical effects on wildlife of periodic DDD applications to Clear Lake. *Calif. Fish and Game* **46**:91–106.

Hunt, G. L., Jr., and M. W. Hunt. 1977. Female-female pairing in western Gulls (*Larus occidentalis*) in Southern California. *Science.* **196**:1466–1467.

LeBaron, H. M. and J. Gressel (Eds.). 1982. *Herbicide resistance in plants.* John Wiley and Sons, New York.

Lincer, J. L. 1975. DDE-induced eggshell-thinning in the American Kestrel: A comparison of the field situation and laboratory results. *J. Appl. Ecol.* **12**:781–793.

Lincer, J. L., D. Zalkind, L. H. Brown, and J. Hopcraft. 1981. Organochlorine residues in Kenya's Rift Valley lakes. *J. Appl. Ecol.* **18**:157–171.

MacGregor, J. S. 1974. Changes in the amount and proportions of DDT and its metabolites, DDE and DDD, in the marine environment off southern California, 1949–1972. *Fishery Bull.* **72**:275–293.

Matsumura, F. 1985. *Toxicology of insecticides.* 2nd ed. Plenum Press, New York.

McEwen, F. L. and G. R. Stephenson. 1979. *The use and significance of pesticides in the environment.* John Wiley and Sons, New York.

Mineau, P. 1988. Avian mortality in agro-ecosystems. 1. The case against granular insecticides in Canada. Pp. 3–12 in *Environmental effects of pesticides.* BCPC Monogr. No. 40.

Muirhead-Thompson, R. C. 1987. *Pesticide impact on stream fauna with special reference to macroinvertebrates.* Cambridge Univ. Press, Cambridge, England.

National Research Council. 1986. *Pesticide resistance: Strategies and tactics for management.* National Academy Press, Washington, DC.

Office of Pesticides and Toxic Substances. 1985. *Suspended, canceled, and restricted pesticides.* U.S. Environmental Protection Agency, Washington, D.C.

O'Shea, T. J., W. J. Fleming, and E. Cromartie. 1980. DDT contamination at Wheeler National Wildlife Refuge. *Science* **209**:509–510.

Prabhu, M. A. 1988. International pesticide regulatory programs. *Environment* **30**(9):43–45.

Rands, M. R. W. 1985. Pesticide use on cereals and the survival of gray partridge chicks. *J. Appl. Ecol.* **22**:49–54.

Ratcliffe, D. A. 1967. Decrease in egg shell weight in certain birds of prey. *Nature* **215**:208–210.

Repetto, R. 1985. *Paying the price: Pesticide subsidies in developing countries.* World Resources Institute Research Report #2.

Rudd, R. L. 1964. *Pesticides and the living landscape.* Univ. of Wisconsin Press, Madison.

Smith, G. J. 1987. Pesticide use and toxicology in relation to wildlife: Organophosphorus and carbamate compounds. *U.S. Dept. Int., Fish Wildl. Ser.,* Res. Publ. 170.

Spitzer, P. R. and A. F. Poole. 1980. Coastal ospreys between New York City and Boston: A decade of reproductive recovery, 1969–1979. *Am. Birds* **34:**234–241.

Tannock, J., W. W. Howells, and R. J. Phelps. 1983. Chlorinated hydrocarbon pesticide residues in eggs of some birds in Zimbabwe. *Env. Pollut., Ser. B.* **5:**147–155.

Thomas, L. 1989. Pesticides and endangered species: New approaches to evaluation impacts. *End. Sp. Tech. Bull.* **14(1–2):**1, 7.

Turnbull, R. E., F. A. Johnson, M. A. Hanandez, W. B. Wheeler, and J. P. Toth. 1989. Pesticide residues in fulvous whistling-ducks from south Florida. *J. Wildl. Manag.* **53:**1052–1057.

Webber, E. C., D. R. Bayne, and W. C. Seesock. 1989. DDT contamination of benthic macroinvertebrates and sediments from tributaries of Wheeler Reservoir, Alabama. *Arch. Environ. Contam. Toxicol.* **18:**728–733.

Woodwell, G. M., C. F. Wurster, Jr., and P. A. Isaacson. 1967. DDT residues in an East Coast estuary: A case of biological concentration of a persistent pesticide. *Science* **156:**821–823.

Wurster, C. F. 1969. Chlorinated hydrocarbon insecticides and the world ecosystem. *Biol. Cons.* **1:**123–129.

Aquatic Ecosystems

Aquatic ecosystems make up more than 70 percent of the biosphere. The biotic resources of these ecosystems have long been attractive to humans, and many have been exploited heavily since ancient times. Aquatic ecosystems have been the source of food, water, energy, and even minerals. They have served as routes of exploration, travel, and commerce. They have provided recreation. Not least of all, they have served as sewers to dilute and flush away the wastes of human activity. Not surprisingly, almost all major human settlements are situated on the banks or shores of streams, lakes, or oceans. Aquatic ecosystems thus have received all degrees of impact by humans. Some are so changed that we can only guess at their pristine state. Others are only now beginning to feel the impact of human activity.

CHAPTER 14

Lakes, Ponds, and Marshes

Lake, pond, and marsh ecosystems are as diverse as the full range of terrestrial biomes. They range from large, deep, and geologically ancient lakes, such as Baikal in the Soviet Union, to shallow seasonal marshes and tiny temporary pools. Their waters likewise range from fresh and nutrient deficient, as in Lake Tahoe, to fresh but nutrient rich, as for Lake Erie, or hypersaline, as for the Great Salt Lake.

Few types of ecosystems have been abused as extensively as lakes. The impacts of humans have been direct and indirect, intentional and inadvertent, biotic and abiotic. That these systems retain as much wildlife as they do is testimony to considerable resilience. In the developed nations, diversion of fresh water inflow to gain water for agricultural or urban use, and the filling or drainage of ponds and marshes to create land for urban development or farming, are destroying many of these ecosystems. Throughout the world, the watersheds of lakes, ponds, and marshes have been modified by forest clearing, agricultural activity, and urban growth. The biotic resources of these systems have been exploited carelessly, and their basins have served as convenient disposal sites for agricultural, urban, and industrial wastes. Intentionally or by accident, exotic plants and animals have been introduced. Often, these impacts have distorted basic patterns of nutrient cycling, usually by flooding the ecosystem with nutrients from sewage discharges, fertilizer runoff from farmland, or accelerated erosion of the watershed.

In this chapter we shall first examine these impacts in detail, giving special attention to enrichment of the ecosystem by increased nutrient inflow. We shall defer fuller discussion of some impacts, such as those of acid deposition and chemical pollution, to later chapters. Next, we shall consider some of the special lake, pond, and marsh ecosystems of major conservation importance. Finally, we shall consider national and international efforts to conserve wetlands ecosystems.

DIVERSION AND DRAINAGE

Diversion of water for agricultural and urban use has had a major impact on lakes in semiarid regions such as western North America and central Asia. Many such lakes are saline and fish-free, but possess distinctive invertebrate faunas that are heavily exploited by water birds. In North America, Mono Lake, Pyramid Lake, and other lakes in the Great Basin have lost much of their water inflow by diversion. In the Soviet Union, the Aral Sea has shrunk in surface area by 40 percent since 1960, due to the diversion of water for irrigation.

Mono Lake, in eastern California, has been the focus of great controversy in North America (Mono Basin Ecosystem Study Committee 1987). This lake, with a solute content nearly three times that of the ocean, is dominated by an aquatic biota of algae, brine shrimp, and the larvae of brine flies. The abundance of the brine shrimp and brine fly larvae make the lake an important stopover area for certain migratory birds (Jehl 1988). Up to 900,000 eared grebes spend three to eight months at the lake in late summer and autumn to molt and store fat for southward migration (Fig. 14.1). About 100,000 Wilson's phalaropes spend three to five weeks at the lake for the same purpose (Fig. 14.2), and about 54,000 red-necked phalaropes use the lake as a brief migratory stopover area. In addition, 61,000 or so California gulls nest on islands in the lake, which are free of terrestrial predators such as coyotes.

By the mid-1930s, Los Angeles had acquired water rights for much of the watershed of Mono Lake, and in 1941 began diverting water into the Owens Valley Aqueduct leading to Los Angeles. Until 1970, about 56,000 acre-feet of water were diverted annually. In 1970, with growth in Los Angeles water demand, the aqueduct tunnels were enlarged and diversions increased to 101,000 acre-feet annually. Increased diversion caused a drop of over 12 m in lake level, a 50 percent decrease in surface area, and more than a doubling of salinity. Dropping water level created land connections to several islands on which California gulls were nesting, making them accessible to coyotes, so that the gulls were forced to shift their nesting to other islands. Studies revealed that if this diversion were continued, the lake level would fall, and the water become saline enough to cause a serious decline in the lake invertebrates and to eliminate most nesting islands for California Gulls.

In 1988, the United States Forest Service, which manages Mono Lake and its watershed, evaluated existing studies of the Mono Lake Ecosystem and concluded that a reduction of 50 to 75 percent in diversion was needed. With this diversion, the lake is expected to stabilize at a surface elevation about 2.2 m above the 1988 level, and sufficiently high to maintain the isolation of all the major islands. Whether this plan of reduction in water diversion will be implemented remains to be seen.

Ponds and marshes have also suffered destruction. In the upper Great Plains of the United States and Canada, drainage of prairie ponds and marshes, collectively termed **potholes,**

FIGURE 14.1

■ About 900,000 eared grebes use Mono Lake as a summer and autumn staging area prior to southward migration to wintering areas at the Salton Sea in California and in the Gulf of California.

has been especially widespread (Weller 1981). Ten million or more potholes originally occupied the undulating topography left by retreat of Pleistocene glaciers (Fig. 14.3). Perhaps 50 to 60 percent of these have been drained and converted to farmland (Mitsch and Gosselink 1986). Iowa has lost about 99 percent of its original wetlands and North Dakota about 60 percent, for example (Tiner 1984). Loss of prairie potholes has been coupled with major wetland losses in other parts of the continent. Coastal Louisiana, with about 40 percent of the coastal wetlands of the coterminous 48 states, is losing wetlands at a rate of 100 to 150 square kilometers annually (Baldwin et al. 1990). Land subsidence following petroleum removal, construction of canals, and other types of development are apparently contributing to these losses. In the coterminous states, only about 40 million hectares of coastal and inland wetlands now remain, roughly 46 percent of those originally present (Tiner 1984).

Loss of wetlands, particularly in the prairie pothole region, has no doubt caused a major reduction in waterfowl populations from their original levels in North America. Disappearance of the whooping crane from most of its former breeding range (See chapter 25) was partly the result of early habitat destruction. Since the 1920s, however, waterfowl numbers have fluctuated widely, largely due to wet and dry

FIGURE 14.2

■ Wilson's Phalaropes spend several weeks at Mono Lake, where they deposit fat that serves as fuel for migration south to wintering areas on salt lakes of the South American altiplano.

FIGURE 14.3

■ Prairie potholes, largely the product of continental glaciation, are the breeding sites for many species of North American waterfowl and shorebirds.

cycles that affect breeding populations in the prairie pothole region. Severe population declines were recorded in the 1930s, early 1960s, and late 1980s. The dependence of waterfowl numbers on favorable habitat conditions now means that any further losses will certainly translate into permanent reduction of populations.

In California, **vernal pools**—temporary ponds formed in winter and spring—are a distinctive and endangered habitat (Zedler 1987). More than 100 species of plants are typical of this environment (Holland and Jain 1977), and many of these are endemics restricted to local vernal pool areas. A number of rare or endemic temporary pond invertebrates also occur in these pools. Vernal pools were originally widespread in the Central Valley and in coastal areas from Santa Barbara south. Agricultural and urban developments have eliminated about 90 percent of the original vernal pool habitat, and several species of plants are now endangered.

WATERSHED MODIFICATION

Even where the water inflow to lakes is not directly impaired, changes in watershed vegetation and land use affect lakes. Clearing or thinning of watershed vegetation typically leads to increased runoff and erosion, and thus to greater transport of dissolved materials and silt into lakes. The seasonal pattern of flow is changed, as well, and wet season flows increasing while dry season flows decreasing. Where a forest cover is removed, temperatures of streams are also altered, typically becoming warmer in summer due to increased exposure of the stream surface to sunlight. These changes modify the biota of watershed streams in a complex fashion, which in turn influences the biota of the lakes into which they flow.

OVEREXPLOITATION AND INTRODUCTION OF EXOTICS

Overexploitation by commercial fisheries in large lakes has been nearly universal, as illustrated by salmonid fisheries in many lakes of northern Eurasia and North America. The original fisheries of most such lakes were exploited heavily before modern techniques of assessing fish stocks were available. Large catches during the early phase of exploitation nearly always led to harvests, and even to scientific estimates of sustainable harvest, that exceeded the long-term potential of the ecosystem (Regier and Loftus 1972).

The decline in yield that tends to follow overexploitation favors other biotic changes, such as the establishment of exotic species. Some exotics arrive by natural dispersal or are introduced accidentally. Species purported to be better game or commercial fish, however, are often introduced deliberately. Few lakes have escaped such meddling. Introduced species may completely restructure the lake ecosystem, often to the detriment of valuable, endemic species.

The introduction of the peacock bass, *Cichla ocelaris,* to Gatun Lake in Panama illustrates the extent of ecosystem restructuring that can occur. A large, predatory fish native to the Amazon River, the peacock bass was introduced as a sport fish to ponds in the upper drainage of the Chagres River in 1967. Floods carried it into Gatun Lake, where it gradually spread, eliminating or reducing the populations of native fish (Zaret and Paine 1973). Decline of the smaller fish has led to the disappearance of several fish-eating water birds. Loss of these fish also permitted increase in zooplankton and insect larvae, including those of *Anopheles* mosquitos, which are carriers of malaria.

In similar fashion, introduction of the large opossum shrimp, *Mysis relicta,* to lakes in the western United States has sometimes decimated smaller zooplankton populations. In turn, this has led to decline of fish the opossum shrimp were supposed to increase, and sometimes even to decline of fish-eating birds such as the bald eagle (Spencer et al. 1991). Introductions of fish to previously fish-free lakes have also disrupted these ecosystems, often reducing or eliminating invertebrates that are intolerant of fish predation.

EUTROPHICATION

Eutrophication, the increase in fertility and productivity of an ecosystem due to an increased rate of nutrient input, is perhaps the most pervasive impact of human activities on lake ecosystems. Eutrophication may be natural, due to the slow increase in fertility of a lake watershed through geological time. Almost all examples of recent eutrophication, however, are due to human activity. Sewage discharges, organic waste discharges by industries, runoff of fertilizers from agricultural lands, and increased leaching and erosion of watersheds disturbed by farming and construction all increase nutrient inflows to lakes. For most fresh waters, phosphorus is the limiting nutrient for primary production, and, excess inputs of phosphorus the primary culprit in eutrophication due to human activity (Schindler 1974).

To understand the impacts of eutrophication, we must first recognize that deep lakes in the temperate zone become thermally stratified in summer. Solar heating creates a surface zone of warm, low-density water—the **epilimnion**—overlying the cold, high-density bottom water—the **hypolimnion** (Fig. 14.4). These layers are separated by the **thermocline,** a relatively thin zone where temperature and density change rapidly with depth. Summer stratification prevents exchanges between the epilimnion and hypolimnion. Wind-induced currents circulate only within the epilimnion, and oxygen is not carried downward, nor are nutrients released by decomposition carried upward. In cooler periods of autumn and spring, this stratification breaks down, and currents circulate through the entire lake profile.

Lakes vary greatly in their fertility and productivity. A lake with low concentrations of nutrients and a low rate of nutrient input is termed an **oligotrophic** lake (Henderson-Sellers and Markland 1987). Oligotrophic lakes have a low abundance of phytoplankton in the epilimnion, and thus tend to be clear—the abundance of phytoplankton largely determines the clarity of lakes free of massive sediment pollution

FIGURE 14.4

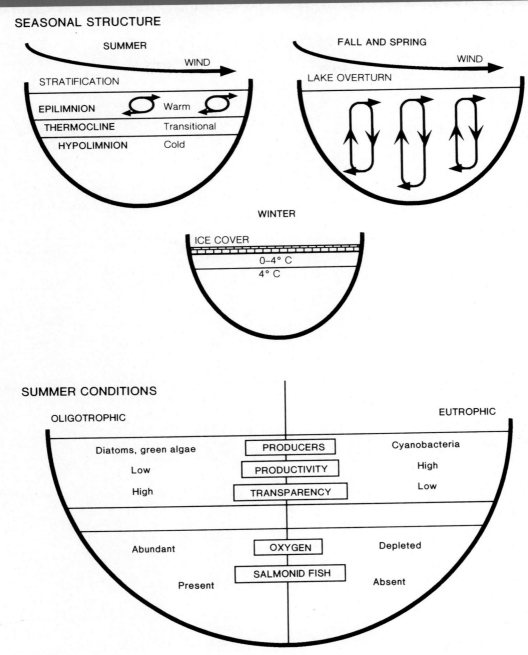

SEASONAL STRUCTURE

SUMMER — WIND

STRATIFICATION

EPILIMNION	Warm
THERMOCLINE	Transitional
HYPOLIMNION	Cold

FALL AND SPRING — WIND

LAKE OVERTURN

WINTER

ICE COVER
0–4° C
4° C

SUMMER CONDITIONS

OLIGOTROPHIC — EUTROPHIC

Diatoms, green algae	PRODUCERS	Cyanobacteria
Low	PRODUCTIVITY	High
High	TRANSPARENCY	Low
Abundant	OXYGEN	Depleted
Present	SALMONID FISH	Absent

■ The vertical structure and seasonal cycle of oligotrophic and eutrophic lakes in the Temperate Zone.

(Fig. 14.4). The low abundance of producers—mainly diatoms and green algae—is coupled with a low primary productivity, and this, in turn, means that the amount of dead organic matter ultimately sinking into the hypolimnion is small. Thus, in summer, bacterial decomposition of organic matter is rarely great enough to deplete the oxygen supply of the hypolimnion. Oligotrophic lakes are thus favorable to salmonid fish such as lake trout, whitefish, and other deep-water fish that demand cold, high-oxygen waters, and, of course, to invertebrates such as mayfly larvae and freshwater shrimp, on which these fish feed. In general, the dominant fish of oligotrophic lakes are high-quality game fish and table fish, so that even though their productivity is not great, their dominance is highly desirable.

The North American Great Lakes were originally all oligotrophic lakes dominated by faunas of salmonid fish such as lake trout, the Atlantic salmon (Lake Ontario), and various whitefish such as the ciscos and the lake herring (Regier 1979). Lakes Ontario, Michigan, Huron, and Superior possessed 10 to 13 species of salmonids, and Lake Erie, the shallowest of the lakes, 4 species. All of the lakes have been profoundly changed by the impacts of human activity.

Overfishing contributed to the decline of several whitefish, as well as other prize species such as the lake sturgeon, in the 19th and early 20th centuries. In 1879, for example, 1.78 million kg of sturgeon were harvested in Lake Michigan. In only 20 years, this harvest had declined to less than 0.05 million kg.

The watersheds of the lakes were modified extensively by clearing of forests, farming, and urban growth. Deforestation raised water temperatures of the streams flowing into the lakes, making them less favorable for spawning by native sturgeon and salmonids, and more favorable for spawning of introduced species, particularly the sea lamprey. Several attempts to reestablish Atlantic salmon in Lake Ontario have failed because streams flowing into the lake are too warm and shallow to provide the well oxygenated gravel bottoms required for spawning (Moss 1988). Drainage of the Great Black Swamp at the western end of Lake Erie led to massive siltation in the western end of the lake by the Maumee River, contributing to major ecological change (Egerton 1987).

Overfishing and watershed changes, combined with other factors, permitted the invasion of the lakes by several exotic species. The construction of the Erie Canal linking the Hudson River and Lake Ontario allowed the sea lamprey to invade Lake Ontario in the 1860s. Once in the lake, the lamprey found suitably warm spawning streams in its watershed. Later, with expansion and reduction in number of locks in the Welland Canal that allowed shipping to pass around Niagara Falls, the lamprey reached the upper lakes, establishing populations in these lakes between 1921 and 1946. The sea lamprey is a predator on other fish, feeding on them by attaching to the body surface with a sucker-like mouth, rasping through the body wall with horny teeth, and draining the body fluids of the prey (Fig. 14.5). Im-

FIGURE 14.5

■ The sea lamprey, introduced to the Great Lakes, caused the decline of many species of native salmonid fish. These lake trout were severely wounded by lamprey attacks.

pacts of the lamprey on the native salmonids were severe in Lakes Ontario, Michigan, Huron, and Superior. The lamprey first attacked the largest deep water salmonids, such as the lake trout. When their populations declined, it shifted to the next largest species. The lake trout and Atlantic salmon were driven to extinction in Lake Ontario before 1900, due to the combined effects of overfishing, watershed change, and sea lamprey predation. In the upper lakes, sea lamprey predation caused catastrophic declines in salmonid populations in the 1930s and 1940s, driving several species to extinction.

FIGURE 14.6

■ Electrical barriers are used to prevent adult lampreys from entering tributary streams of the Great Lakes to spawn.

The faunal vacancies created by decimation of the salmonids permitted the establishment of the alewife and rainbow smelt. The alewife followed the lamprey into Lake Ontario in the 1870s, and into the upper lakes between 1931 and 1953, filling the plankton-feeding niche of the lake herring, one of the lake whitefish. The smelt first entered Lake Michigan by escaping from a reservoir on a tributary stream, into which it had been stocked. Smelt spread throughout the lake system between 1923 and 1935.

In Lake Erie, the most severe impact has been eutrophication. Because it is the shallowest of the lakes, eutrophication has caused the greatest changes in conditions of the hypolimnion (Regier and Hartman 1973). Oxygen concentrations below 3 ppm occur throughout more than 70 percent of the lake during summer stratifica-tion. Eutrophic conditions have led to changes in the communities of phyto-plankton, zooplankton, and invertebrates. Mayflies have disappeared from large areas of the lake bottom, being replaced by midge larvae and tubificid worms tolerant of low oxygen. These invertebrates, how-ever, are poor food for salmonid fish, which have also been excluded from deep water areas by lack of oxygen. Two salmonids, the lake trout and the longjaw cisco, have become extinct. The original salmonid fishery, which yielded about 22.7 million kg of high quality table fish annually, has now been replaced by a fishery of roughly equal volume, but one that yields pri-marily trash species that are used for pro-duction of fish meal.

Recently, efforts have been made to counteract these impacts. Control of sea lampreys has been attempted in several ways. Dams and electrical barriers have been installed at the mouths of tributary streams used for spawning by lampreys (Fig. 14.6). A specific **lampricide**, TFM (3-trifluoromethyl-4-nitrophenol) has been used to kill lamprey larvae during their early life in streams. Substantial control has been achieved in the watersheds of Lakes Michigan, Huron, and Superior, leading to increases in the harvest of some of the lake whitefishes (Dahl and McDonald 1980, Smith and Tibbles 1980).

A somewhat different response—some might say a form of ecological "tin-kering"—has been the introduction of exotic salmonids to the Great Lakes system. Coho, chinook, pink, and kokanee salmon from the Pacific Coast have been intro-duced to the upper lakes in an effort to re-establish a salmonid sport fishery. Initially, these populations depended entirely on in-troduced fry. Chinook and pink salmon, however, now spawn in the upper lakes, and appear to have established self-sustaining populations.

With an increase in fertility and productivity, a lake becomes **eutrophic.** As populations of phytoplankton become more dense, the clarity of the water declines. The composition of the producer community shifts markedly, often to the dominance of blue-green algae, many of which are incompletely harvested by herbivores. Often, scums of these algae accumulate on the surface, blow to shore, and pile up in windrows, where they rot and create a vile stench. Rotting blue-green algae often taint the water with foul-tasting chemicals that are difficult to remove by municipal water treatment. With high productivity, large quantities of dead organic matter also sink to the bottom waters of the lake. There, bacterial decomposition may exhaust the supply of oxygen during summer stratification, leading to mortality of invertebrates and fish. Ultimately, these physical and chemical changes cause the replacement of animals typical of oligotrophic lakes with invertebrates and fish tolerant of low oxygen. The fish fauna of eutrophic lakes in the temperate zone is dominated by sunfish, bass, yellow perch, carp, catfish, and other forms, some of which are neither good sport fish nor good table fish. Although the fish harvest from such lakes may be much higher than that of the pre-existing oligotrophic lake, its value may be primarily for production of fish meal and fertilizer.

RECOVERY OF LAKES FROM EUTROPHICATION

Eutrophication has often been likened to the "aging" of lakes, based on the idea that it is a natural trend that occurs over the geological lifetime of a lake basin. By this analogy, recovery from eutrophication—"de-aging"—is not possible. Fortunately, cultural eutrophication does not constitute an irreversible change in lake ecology, and there is now abundant evidence that lakes that have become eutrophic from human activities can be restored to an oligotrophic condition.

Lake Washington, in the Seattle, Washington, metropolitan area, provides a good example of such recovery (Edmondson and Lehman 1981). During and after World War II, Seattle and its inland suburbs grew rapidly. The sewage systems constructed to serve these communities used the most expeditious approach to disposal of secondary effluent—discharge into Lake Washington. By 1955, the lake was receiving 24,200 m³ of secondary effluent from 10 systems. At this point, sewage contributed 56 percent of total phosphorus input to the lake ecosystem. The result was severe eutrophication of a large lake lying in the center of a major metropolitan area. Water clarity, which in the mid-1930s enabled a black-and-white Secchi disk to be seen at a depth of over 6 m, declined to about 1 m. Massive blooms of blue-green algae made the lake a nuisance to nearby residents.

Fortunately, studies by limnologists at the University of Washington had not only documented the deterioration of the lake, but established the cause—phosphorus input in

FIGURE 14.7

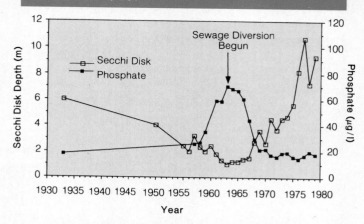

■ Changes in phosphate (annual mean concentration) and maximum Secchi disk depth in July and August in Lake Washington from 1933 to 1979. Source: Data from W. T. Edmondson and J.T. Lehman, "The Effect of Changes in the Nutrient Income on the Condition of Lake Washington" in Limnology and Oceanography, 26:1–29, 1981.

secondary sewage effluent. Based on this knowledge, Seattle undertook the construction of a comprehensive sewage disposal system that now discharges effluent into Puget Sound. Diversion of effluent discharges into Lake Washington was accomplished in stages, with almost complete diversion finally occurring in 1968. Recovery of the lake was rapid, and by 1978 algal blooms had ceased to occur, water clarity was again at mid-1930s levels, and the phosphorus concentration in the lake water was essentially equal to that measured in 1933 (Fig. 14.7).

ULTRAOLIGOTROPHIC LAKES: A SPECIAL CASE

Several highly oligotrophic lakes have exceptional aesthetic or biotic values. Among these are Lake Tahoe in the western United States and Lake Baikal in the Soviet Union.

Because of its setting and water clarity, Lake Tahoe is one of the most beautiful lakes in the world. The Secchi disk depth of Lake Tahoe, once about 40 m, is now about 25 m, and is declining about 0.4 m per year (Goldman 1989). Growth of attached green algae in shallow water is conspicuous, and the primary productivity level of the lake has nearly trebled since 1958. Extensive development of resort communities around the lake is the main cause of this eutrophication. Even though sewage effluent from communities in the Tahoe Basin is exported, disturbance of the land surface by construction activity has increased nutrient inflow, particularly that of phosphorus, by erosion and leaching. Runoff from paved surfaces and rooftops also carries nutrients deposited by atmospheric fallout into the lake.

Introductions of kokanee salmon, *Onchorhynchus nerka,* and opossum shrimp to improve sport fishing have led to the decline of the native zooplankton of Lake Tahoe. The loss of these zooplankters, and their grazing on open-water phytoplankton, may also be contributing to increase in algal populations in the lake waters.

Lake Baikal (Fig. 14.8) is the most distinctive lake ecosystem on earth (Nijhoff 1979). The lake, which has existed for about 25 million years, occupies a basin formed by a rift in the Asian continental plate. Its depth of 1,620 m, combined with a surface area of 31,500 square kilometers, gives it the greatest volume of any lake on earth. Its volume equals, in fact, one-fifth of the earth's fresh water. The lake is highly oligotrophic, with a Secchi depth of 36 m.

Even more unusual is the biota of the lake. About 2,700 species of aquatic plants and animals are known from the lake, 84 percent of these being endemic. The lake contains hundreds of species of planktonic diatoms, and over 100 species of ostracods, 250 species of amphipods, and 70 species of molluscs. An endemic family of sponges, a group of organisms rare in fresh water, occurs in the lake. Among vertebrates, the lake possesses an endemic seal, the nerpa, and an important whitefish, the Baikal omul. Both the seal and the omul have been exploited heavily.

The greatest threat to Lake Baikal's health is pollution from industries and deforestation. Several wood processing plants are located at the south end of the lake, and rafting of logs to these plants has polluted the lake with sunken timber. The plants, which produce a wide range of lumber, cellulose, and paper products, discharge treated wastes into the lake, although alternative means of disposal exist. On the east side of the lake, a large industrial complex on the Selenga River produces petroleum, sulfates, chlorides, and thermal effluent. Wastes from these plants are discharged into the river and carried into Lake Baikal. Completion of the Baikal-Amur railroad to the north end of the lake has led to tanker transport of petroleum and other chemicals from production areas to railway loading facilities. Due to these forms of pollution, the Baikal omul reportedly has ceased to spawn in nature (Stewart 1990). Recently, however, a governmental commission has developed plans to eliminate industrial pollution, log rafting, and watershed deforestation during the 1991–1995 period, and an International Centre for the Ecological Study of Lake Baikal is being established (Galazii 1991).

AFRICAN RIFT VALLEY LAKES

Several of the Rift Valley lakes of East Africa possess rich, highly endemic faunas of great significance. Of particular importance are Lakes Victoria, Tanganyika, and Malawi. These lakes are ancient and the last two very deep. Through climatic cycles that have isolated and reconnected portions of their basins, speciation and adaptive radiation have produced rich endemic biotas. These species now face the varied challenges of growing human exploitation and impact.

FIGURE 14.8

■ Soviet and American scientists are cooperating in studies of Lake Baikal. Here, Soviet seal biologist Eugene Petrov is measuring a nerka, or freshwater seal, before attaching radio-telemetry instruments.

Lake Victoria, bordering Kenya, Tanzania, and Uganda, has a rich, highly endemic fish fauna. More than 200 species of small cichlids, known as haplochromines, occur in the lake. Many haplochromines are brightly colored fish prized in the aquarium trade. Haplochromines, native and introduced tilapias, and native catfish supported local fisheries that used traditional techniques and preserved fish by sun-drying.

In the late 1950s, the Nile perch, *Lates niloticus,* a large predatory fish (Fig. 14.9), was introduced to Lake Victoria. The Nile perch, which reaches a weight of 250 kg, has almost completely restructured the fish fauna of the lake, and severely affected the traditional fisheries (Payne 1987). About 80 percent of the fish biomass of the lake is now made up of Nile perch. The perch cannot be sun-dried, and must either be dried over fires, using hard-to-get firewood, or frozen. The

FIGURE 14.9

■ The Nile perch, introduced to Lake Victoria in 1962, has largely restructured the largely endemic fauna of native fish and reduced the harvest of the more preferred native species by African fishermen.

Nile perch has become a major commercial fish, with the Kenyan harvest alone being about 100,000 tons annually, but many of the haplochromines have become extinct.

Lake Tanganyika, 1470 m deep, has existed for between 1.6 and 6.0 million years, and completely lacks oxygen in its deepest layers (Burgis and Morris 1987). The lake has a highly endemic fauna, including 60 species of aquatic snails, 30 of them endemic. The fish fauna is dominated by a group of endemic herrings—a typically marine group of fish—commonly known as "Tanganyika sardines." Lake Malawi, 704 m deep and lacking in oxygen in its bottom waters, is also several million years old (Burgis and Morris 1987). The fish fauna—almost completely distinct from that of Lake Tanganyika—is richer than that of any other lake in the world. Some 245 species are known from Lake Malawi, 219 of them endemic. Of the endemics, 190 are cichlids, which have undergone an astounding evolutionary radiation. The cich-

lids have diversified greatly in feeding ecology, to the point that up to 30 species sometimes occur together. The biotas of Lakes Tanganyika and Malawi have not been damaged by introduced species such as the Nile perch, and must be rigorously protected against such an action.

THE EVERGLADES

Southern Florida originally possessed a complex of over 23,000 square kilometers of wetlands extending south from Lake Kissimmee, near Orlando, to Florida Bay (Tiner 1984). These wetlands include Lake Okeechobee, Big Cypress National Preserve, and Everglades National Park (Fig. 14.10). They are home to the American alligator, Florida panther, Florida manatee and many breeding and wintering water birds. The fresh water flow from this wetland also feeds an extensive estuary environment along the southwest Florida coast.

FIGURE 14.10

■ The Florida Everglades are a vast wetland consisting of open marsh, wooded hammocks, cypress swamps, and coastal mangroves.

Beginning in the 1920s, drainage and channelization projects, many of them federally supported, have destroyed much of the original northern and eastern parts of the Everglades, converting them to farmland and urban land. Between present agricultural areas and Everglades National Park lie some 3,400 square kilometers of Water Conservation Areas, in which water is stored and regulated for discharge to agricultural, municipal, and park areas. The canals and dikes of these projects, however, have reduced the volume of fresh water flowing southward into the everglades proper. They have also created an unnatural stabilization of the remaining flows, largely eliminating the flood flows that once occurred after large summer storms.

These developments have had major impacts on both breeding and non-breeding water birds (Bancroft 1989). Large shifts of breeding birds from the park into the Water Conservation Areas occurred when the latter were created in the 1960s. Although breeding populations have fluctuated greatly due to water conditions, the overall population trend has been downward. In 1975–1976, for example, 23,500 to 26,000 pairs of wading birds nested in Everglades National Park and the Water Conservation Areas; in the late 1980s the highest numbers nesting were 16,500 pairs (Bancroft 1989).

Pollution from agricultural and urban sources is also a problem. Mercury, of uncertain origin, is a major contaminant of Everglades food chains. Phosphorus from agricultural fertilizers has also caused eutrophication of large areas of marsh, causing replacement of sawgrass by cattails, which are poorly utilized by water birds. Clearly, much remains to be done to devise a plan that protects the quantitative and qualitative character of the water supply to this unique ecosystem.

LAKE ECOSYSTEMS AND GLOBAL WARMING

Global warming will likely have great impacts on lake, pond, and marsh ecosystems, although the nature of these impacts is still difficult to predict. Based on long-term studies of lakes in the boreal forest region of Ontario, Canada, Schindler et al. (1990) suspect that northern lakes will become ice-free for a greater portion of the year, increase in solute content, and develop a deeper layer of warm epilimnetic water in summer. Thus, the suitability of these lakes for species typical of cold, oligotrophic waters may be reduced considerably.

PROTECTION OF WETLANDS

In the United States, under the **Clean Water Act** (Federal Water Pollution Control Act) as reauthorized in 1987, lakes, ponds, marshes, and other wetlands receive special protection. This act requires that filling or dredging actions minimize detrimental impacts, or, if such impacts are unavoidable, mitigate them by restoration of degraded wetlands or creation of artificial wetlands (Barton 1987). This act has been subject to somewhat different interpretation by the administrative agencies and by conservation groups, however.

The United States and Canada cooperate in protection of the Great Lakes through activities of the **International Joint Commission,** established by the **Boundary Waters Treaty** of 1909 to regulate matters such as lake levels and water quality, and the **Great Lakes Fishery Commission,** established in 1955 by the **Convention on Great Lakes Fisheries** to control the sea lamprey and coordinate other aspects of fisheries research and management. Only since about 1976 have these complementary commissions been actively coordinating efforts, and an integrated approach to regulation of the Great Lakes Ecosystem is far from being implemented.

The **Convention on Wetlands of International Importance,** developed at Ramsar, Iran in 1971, provides a framework for the preservation of wetlands, especially those of importance to waterfowl. More commonly termed the **Ramsar Convention,** this agreement provides for member nations to designate specific areas for inclusion on a *List of Wetlands of International Importance.* International cooperation in research and protection of listed wetlands is provided for under the convention. By 1989, some 54 nations, including the United States, had become parties to the Ramsar Convention, and more than 400 wetland sites had been placed on the list, including Everglades National Park, Florida. A priority for the Ramsar organization is the addition of additional southern hemisphere members such as Botswana, Argentina, and Brazil. The wetlands of the Okavango Delta of Botswana constitute one of the most important wildlife areas of southern Africa (See chapter 15). Those of Argentina and Brazil are important both as breeding areas for resident wildlife and as wintering areas for North American shorebirds.

KEY CONSERVATION STRATEGIES FOR LAKE, POND, AND MARSH ECOSYSTEMS

1. Maintenance of normal quantity and pristine quality of water inflow from the watershed and from urban-industrial sources.

2. Prevention of exotic introductions or invasions and the control of problem exotics.

LITERATURE CITED

Baldwin, M. F., M. Leslie, and E. H. Clark II. 1990. Government programs inducing wetlands alterations. Pp. 191–210 *in* G. Bingham, E. H. Clark II, L. V. Haygood, and M. Leslie (Eds.), *Issues in wetland protection: Background papers prepared for the National Wetlands Policy Forum.* The Conservation Foundation, Washington, D.C.

Bancroft, G. T. 1989. Status and conservation of wading birds in the Everglades. *Amer. Birds* **43:**1258–1265.

Barton, K. 1987. Federal wetlands protection programs. Pp. 179–198 *in* R. L. Di Silvestro (Ed.), *Audubon wildlife report 1987.* Academic Press, Orlando, FL.

Burgis, M. J. and P. Morris. 1987. *The natural history of lakes.* Cambridge Univ. Press, Cambridge, England.

Dahl, F. H. and R. B. McDonald. 1980. Effects of control of the sea lamprey (*Petromyzon marinus*) on migratory and resident fish populations. *Can J. Fish. Aq. Sci.* **37:**1886–1894.

Edmondson, W. T. and J. T. Lehman. 1981. The effect of changes in the nutrient income on the condition of Lake Washington. *Limnol. Oceanogr.* **26:**1–29.

Egerton, F. N. 1987. Pollution and aquatic life in Lake Erie: Early scientific studies. *Env. Review* **11:**189–205.

Galazii, G. 1991. Lake Baikal reprieved. *Endeavour* **15:**13–17.

Goldman, C. R. 1989. Lake Tahoe: Preserving a fragile ecosystem. *Environment* **31:**6–11, 27–31.

Henderson-Sellers, B. and H. R. Markland. 1987. *Decaying lakes.* John Wiley and Sons, New York.

Holland, R. and S. Jain. 1984. Vernal Pools. Pp. 515–533 *in* M. G. Barbour and J. Major (Eds.), *Terrestrial vegetation of California.* John Wiley and Sons, New York.

Jehl, J. R., Jr. 1988. Biology of the eared grebe and Wilson's phalarope in the nonbreeding season: A study of adaptations to saline lakes. *Studies in Avian Biology* No. 12.

Mitsch, W. J. and J. G. Gosselink. 1986. *Wetlands.* Van Nostrand Reinhold, New York.

Mono Basin Ecosystem Study Committee. 1987. *The Mono Basin ecosystem.* National Academy Press, Washington, D.C.

Moss, B. 1988. *Ecology of fresh waters: Man and medium.* 2nd ed. Blackwell Scientific Publications, Oxford, England.

Nijhoff, P. 1979. Lake Baikal endangered by pollution. *Env. Conserv.* **6:**111–115.

Payne, I. 1987. A lake perched on piscine peril. *New Scientist* **115(1575):**50–54.

Regier, H. A. 1979. Changes in species composition of Great Lakes fish communities caused by man. *Trans. N.A. Wildl. Nat. Res. Conf.* **44:**558–566.

Regier, H. A. and W. L. Hartman. 1973. Lake Erie's fish community: 150 years of cultural stress. *Science* **180:**1248–1255.

Regier, H. A. and K. H. Loftus. 1972. Effects of fisheries exploitation on salmonid communities of oligotrophic lakes. *J. Fish. Res. Bd. Canada* **29:**959–968.

Schindler, D. W., K. G. Beaty, E. J. Fee, D. R. Cruikshank, E. R. DeBruyn, D. L. Findley, G. A. Linsey, J. A. Shearer, M. P. Stainton, and M. A. Turner. 1990. Effects of climatic warming on lakes of the central boreal forest. *Science* **250:**967–970.

Schindler, D. W., K. G. Beaty, E. J. Fee, D. R. Cruikshank, E. R. DeBruyn, D. L. Findley, G. A. Linsey, J. A. Shearer, M. P. Stainton, and M. A. Turner. 1990. Effects of climatic warming on lakes of the central boreal forest. *Science* **250:**967–970.

Smith, B. R. and J. J. Tibbles. 1980. Sea lamprey (*Petromyzon marinus*) in Lakes Huron, Michigan, and Superior: History of invasion and control, 1936–1978. *Can. J. Fish. Aq. Sci.* **37:**1780–1801.

Spencer, C. N., B. R. McClelland, and J. A. Stanford. 1991. Shrimp stocking, salmon collapse, and eagle displacement. *BioScience* **41:**14–21.

Stewart, J. M. 1990. The great lake is in great peril. *New Scientist* **126(1723):**58–62.

Tiner, R. W., Jr. 1984. *Wetlands of the United States: Current status and recent trends.* U.S. Dept Interior, Fish and Wildl. Serv., Washington, D.C.

Weller, M. W. 1987. *Freshwater marshes.* 2nd ed. Univ. of Minn. Press, Minneapolis.

Zaret, T. M. and R. T. Paine. 1973. Species introduction in a tropical lake. *Science* **182:**449–455.

Zedler, P. H. 1987. The ecology of southern California vernal pools: A community profile. *U.S. Fish Wildl. Serv. Biol. Rep.* 85(7.11).

CHAPTER 15

Rivers and Streams

Rivers and streams hold only about 1.3 percent of the water in all freshwater ecosystems combined (Berner and Berner 1987). Nevertheless, they play an important role in the flux of water through the biosphere, and are an important habitat for many types of organisms. The total annual flow of rivers to the ocean is estimated to be 39,000 km³, of which 12,000 km³ represent the stable base flow (reaching river channels through the water table) and 27,000 km³ flood runoff (entering channels as surface runoff).

Rivers also transport large quantities of dissolved and suspended organic and inorganic materials to the oceans. The annual transport of inorganic sediment is estimated to be 13.5 billion tons, or about 152 tons for each square kilometer of watershed. Human-induced changes in watersheds, primarily deforestation, livestock grazing, and crop cultivation, have about doubled the natural sediment discharge of rivers to the ocean.

The major rivers of the world are concentrated in two latitudinal zones: the moist temperate to subarctic belt and the humid tropics. Most rivers at high latitudes in the Northern Hemisphere were heavily disturbed in recent geological time by Pleistocene glaciation. The fish faunas of these river systems are not very rich, but they show distinctive patterns of adaptation, foremost among which is migration between fresh and salt water. Many species are **anadromous,** entering these rivers to spawn. Anadromous fish include salmon, shad, striped bass, sea lampreys, and others. Other species, such as the American eel, are **catadromous,** living in fresh water but returning to the sea to spawn.

The largest rivers of the world are in the tropics, and like many other tropical ecosystems, they are biotically rich (Davies and Walker 1986). The Amazon River,

for example, possesses more than 1,300 species of fish, together with endemic and highly distinctive species of crocodilians, turtles, and a freshwater dolphin. The Mekong River of southeast Asia probably has more than 850 fish species and the Zaire River of Africa about 700 species. The faunas of all three rivers are still not fully described. Most of these species are also endemic, so that preserving biodiversity in these rivers is a major conservation objective.

Rivers are integrally tied to the riparian ecosystems of their banks and floodplains (Hunt 1988). Riparian ecosystems are among the most threatened environments on earth, and riparian forms are numerous on lists of endangered species. In the Sacramento Valley of California, for example, only 4,860 ha of riparian woodland remain of the more than 313,000 ha originally present (Hunt 1988). Several river systems are linked with major wetland areas of enormous conservation significance. In southern Africa, the Okavango River, arising in Angola, flows south into Botswana, forming an immense interior delta in the Kalahari Desert. The Okavango Delta, consisting of papyrus bordered channels and vast areas of seasonal marshland (Fig. 15.1), contains a distinctive fish fauna, and a rich mammal and bird fauna. In Brazil, the Paraguay River floods seasonally to create what is perhaps the world's largest wetland complex, the Pantanal (Alho et al. 1988). This region is now beginning to experience major impacts of exploitation and development (Mittermeier et al. 1990). The deltas of major rivers such as the Mississippi in the United States, and the Rhone and Danube in Europe are also wetlands rich in wildlife.

TABLE 15.1

Changes in Stream pH and Biota Due to the Influence of Acid Mine Drainage in the Beaver Creek Watershed, Eastern Kentucky.

	CANE BRANCH (ACID IMPACTED)	HELTON BRANCH (UNIMPACTED)
pH[1]	3.5	6.9
Sulfate Concentration[1] (ppm)	242	0.4
Benthic Insects[2] (No./m²)		
Ephemeroptera	0	365.5
Plecoptera	0	367.6
Trichoptera	0	185.0
Megaloptera	19.3	8.6
Coleoptera	2.1	107.5
Diptera	301.0	814.8
Odonata	tr	37.6
TOTAL	322.4	1886.6
Fish Biomass[1] (kg/ha)	0	18.4

[1]Mean for 1956–1966. [2]Mean for 1959–1965.
Source: Data from C. R. Collier, et.al., "Influences of Strip Mining on the Hydrologic Environment of Parts of Beaver Creek Basin, Kentucky, 1955–66" in *Geological Survey Professional Paper 427–C*, viii + 80 pp., 1970.

FIGURE 15.5

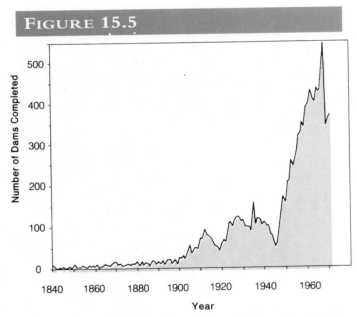

Rate of inauguration of new dams on the world's rivers since 1840 (Modified from Berner and Berner 1987).

Amazon and Parana Rivers of South America and the Mekong River of southeastern Asia (Davies and Walker 1986, Barthem et al. 1991), and damming thus constrains their movements.

Dams, as they are designed to do, also modify the seasonal pattern of flow volume, in particular, reducing flood flows downstream. This alone changes the character of the river system. On the Platte River in Nebraska, for example, the floods that formerly scoured the channel, clearing it of woody vegetation and maintaining a broad complex of open sandbars and channels, no longer occur. The river channel has become more confined, and much of the original channel has grown up in riparian cottonwood woodland. Likewise, below Glen Canyon Dam on the Colorado River, woody vegetation has invaded areas formerly scoured by flood flows.

Stream temperatures, sediment loads, flow volume, and salinity are all changed dramatically by damming, as well. The general result of damming is to increase evaporation from reservoir and irrigated land surfaces, increase the concentration of dissolved materials, trap sediments in reservoir basins, and decrease terminal water and sediment flows into the ocean.

Channelization

In some situations, such as highly urbanized or intensively farmed areas, humans sometimes view rivers primarily as channels to carry flood runoff. In such places, the idea of straightening a meandering stream channel to speed flood water outflow, or **channelization,** often arises. From the standpoint of stream and riparian wildlife, and often from a broad aesthetic viewpoint, channelization can be disastrous (Simpson et al. 1982).

entire stream and riparian zone is destroyed by inundation. At the extreme, the river is converted into a set of warm, quiet pools between which upstream movement of organisms is impossible. Damming thus interferes with fish which migrate between the ocean and fresh water, such as species of salmon, which spawn in fresh water, and certain eels that spawn in the ocean. Many other strictly freshwater fish also migrate extensively within major river systems, such as the

The Columbia River is the longest North American river flowing into the Pacific Ocean, and the second largest river, in volume of flow, in the United States. Development in the Pacific Northwest has been recent, and impacts on the river have occurred mostly since World War II.

The first dam on the main channel of the Columbia, at Rock Island, was constructed in 1933. Today 86 dams exist on the Columbia and its major tributaries (Fig. 15.6), and much of the river is simply a chain of reservoirs. These dams have had a profound effect on the river ecosystems.

Several species of anadromous fish migrate into the Columbia River to spawn. Originally, the most important of these were salmonids: the steelhead, chinook, coho, sockeye, and chum salmon. The shad, an anadromous fish from the Atlantic coast of North America, was also introduced to the Columbia. In the early 1900s, river fisheries for the salmonids yielded over 22,000 tons annually. Today, due to the changes in the river environment and the exploitation of salmon both in the river and at sea, the annual harvest is only about 6,800 tons (Trefethen 1972). Commercial exploitation is now restricted to the lower 225 km of the Columbia channel, with exploitation of the fisheries in the next 210 km, between Bonneville and McNary dams, being reserved for American Indians.

Several changes related to damming have adversely affected the migratory fish of the Columbia system. Although all of the major dams on the lower river have fish ladders that enable adults to pass the dams on their upstream migration (Fig. 15.7), not all of the upper dams allow such passage. The Columbia above the Grand Coulee Dam and the Snake River above Brownlee Dam, together with smaller areas of watershed elsewhere in the system, are now unavailable to migratory fish (Fig. 15.8).

Several formidable challenges are faced by the young salmon on their downstream migrations. Conversion of the river from a swiftly flowing system to a series of reservoirs has increased the downstream migration time between the mouth of the Salmon River, in eastern Washington, to the John Day Dam, in central Washington, by about one month. In passing the various dams, young salmon risk being trapped in the flows diverted into hydroelectric turbines, where they are killed by

FIGURE 15.6

■ Bonneville Dam on the lower Columbia River, completed in 1938, was originally designed without provision for passage of anadromous fish.

FIGURE 15.7

■ Fish ladders were added to the design of Bonneville Dam to allow anadromous salmon and other fish to migrate upstream to spawn in the upper Columbia River system.

the intense turbulence. Ponding of the river has also raised its summer temperatures, which in August now rise above 20 degrees C, well above the range of 5.8 to 12.8 degrees C that is optimum for salmonid fish, and just below the lethal limit of 22.5 degrees C for juvenile salmon. Where thermal plumes from power plants occur,

these limits of tolerance are exceeded. Entrainment in the cooling systems of power plants is still another risk.

FIGURE 15.8

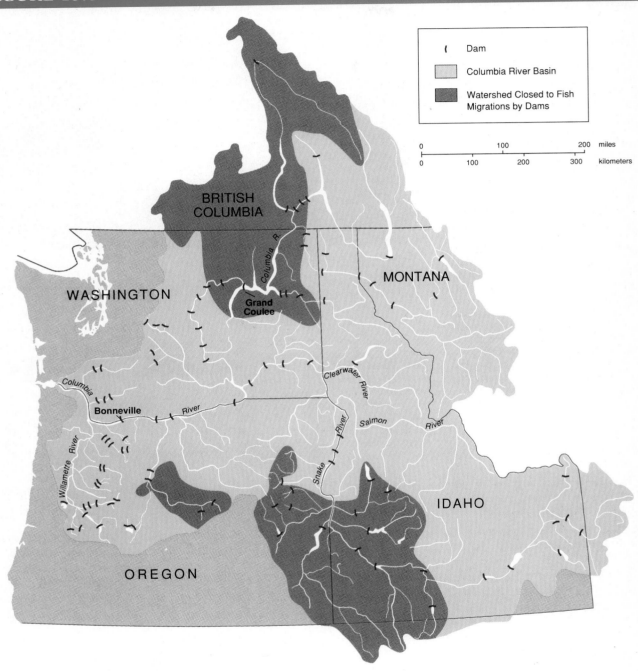

Dam

Columbia River Basin

Watershed Closed to Fish Migrations by Dams

BRITISH COLUMBIA

Columbia R.

Grand Coulee

WASHINGTON

MONTANA

Columbia

Bonneville

River

Clearwater River

Willamette River

Snake River

Salmon

River

IDAHO

OREGON

■ Portions of the upper Columbia River system are no longer accessible to migratory fish because of the lack of fish ladders on dams.

For many years, nitrogen supersaturation in the plunge pools below several dams was a major problem. Water, mixing with air as it shot down the spillway, plunged to depths at which pressures forced large amounts of nitrogen into solution. The blood of fish resting in these pools also became supersaturated. When they moved into shallower or warmer water, the excess gas came out of solution as bubbles, the equivalent of "bends" in human divers who return to the surface too quickly (Weitkamp and Katz 1980). This problem has been minimized in recent years by the construction of spillway deflectors that direct water from the spillway outward onto the surface of the pool below the dam.

The loss of access to spawning areas and slowed return of young salmon to the ocean have been attacked in several ways. Eggs and sperm are collected from adults returning to spawn and are used to produce fertile eggs in hatcheries. The fry are then reintroduced to the river below Bonneville Dam, the lowest dam on the river. Young steelhead and salmon are also collected at John Day Dam and transported by barge to Bonneville Dam, enabling them to pass this stretch of the river without high mortality. Although these programs may be beneficial to steelhead fry, they do not seem to have benefitted other salmon species.

FIGURE 15.9

■ Several stocks of chinook salmon in the Snake River portion of the Columbia River system have recently declined to the brink of extinction.

Overall, these efforts have not restored the abundant fisheries of the early 1900s. In fact, the release of fry reared in hatcheries may have further damaged some wild populations. In particular, Chinook salmon stocks in the Snake River have continued to decline toward the brink of extinction (Fig. 15.9). Future developments on the river, and the possibility of large diversions of water for irrigation and interbasin transfer, are likely to intensify the challenges facing migratory fish.

When a stream channel is straightened, it is shortened and its downhill gradient is steepened. The result is swift outflow of water, which affects not only the channelized section, but areas both upstream and downstream. Within the channelized section erosion is usually accelerated, widening and deepening the channel. Channelization of the Blackwater River in Missouri (Emerson 1971), for example, led to a deepening of the channel by about 3.5 m. Widening of the channel has necessitated replacement of bridges 15 to 30 m long by structures 60 to 124 m in length. Often, channel widening increases the total area occupied by the channel, in spite of the shortening that occurred. From the standpoint of aquatic wildlife, the microhabitats important to many organisms are eliminated by channelization, so that the biomass of life per unit area of stream declines. In the Blackwater River, fish biomass declined from 632 to 126 kg/ha of stream area as a result of channelization (Emerson 1971).

Channel stabilization on the Missouri River by damming, dredging, and various forms of bank protection illustrate the extent of impacts (Hunt 1988). The Missouri River Bank Stabilization and Navigation Project has resulted in large losses of island and sandbar habitat, riparian woodland and wetland, and total water surface area of the river itself. These changes have led to the decline of riparian wildlife such as river otter, mink, and many other species.

Combined Impacts

All of these changes affect the wildlife of streams, especially migratory and wide-ranging species. In the Pacific Northwest, some 90 out of 195 spawning populations of native anadromous salmonid fish were recently judged to be at a high risk of extinction, with 52 other populations having a moderate extinction risk. Some of the world's most endangered

FIGURE 15.10

■ Many fish in the Amazon River system, such as this freshwater croaker of the genus *Plagioscion,* are adapted to feed on tree fruits that they obtain in the flooded *varzea* forest zone during the high water stage of the river's annual cycle.

large animals also inhabit tropical and subtropical rivers. Dams, boat traffic, and fishing activities are implicated in the decline of the endemic river dolphins of the Ganges, Indus, and Yangtze Rivers of Asia (See chapter 18). Of these, the Chinese river dolphin, *Lipotes vexillifer,* is most endangered, having been reduced to less than 200 individuals (Peixun and Ding 1988).

THE MAJOR TROPICAL RIVERS

Occurring in the humid tropics, rivers such as the Amazon, Mekong, and Zaire show large seasonal variations in their flow. During flood seasons, large areas of their floodplains become functional units of the river ecosystem. In the Amazon Basin, seasonally flooded forests of the varzea zone bordering the river channel (See chapter 6) form a major feeding habitat

for many species of fish (Fig. 15.10), some of which are highly specialized for feeding on tree fruits (Goulding 1980, 1985). In some cases the fish are predators on the seeds of these fruits, in other cases the seeds pass through the digestive tracts of the fish, which act as their dispersal agents. Perhaps 75 percent of the fisheries productivity of the Amazon River system is believed to depend on this relationship. Because the varzea forests are accessible from the river channels and logs can easily be floated out from them during flood periods, logging of these forests has been heavy.

Many distinctive reptiles, birds, and mammals are also associated with these tropical rivers. In the Amazon Basin, several of these have been exploited heavily for food, skins, and other purposes. The giant otter, two species of giant river turtles, and two species of caimans have been reduced to endangered status (Johns 1988).

The Colorado River has experienced even more severe impacts than the Columbia River from water diversion and damming (Carothers and Johnson 1983, Hickman 1983). Diversion of water for agricultural and urban use consumes the entire flow of the river. The legal allocation of Colorado River water to various states and Mexico equals 14.84 million acre-feet (Miller et al. 1986). The Colorado system, which now loses about 1.89 million acre-feet annually by evaporation, has an annual flow of about 11.0 to 13.8 million acre-feet (Hunt 1988). The only water regularly entering the Gulf of California is now waste irrigation water from southern Arizona and the Mexicali Valley of Mexico. In Colorado, water is diverted to the east side of the continental divide, as well as to irrigated land in the western part of the state. The Central Arizona Project now carries water to Phoenix and Tucson for urban and agricultural use. In California, water is diverted to Los Angeles and San Diego, where it is ultimately

discharged into the Pacific Ocean, and to the Imperial Valley, where waste irrigation water flows into the Salton Sea. Mexico takes the remaining water, which is used for irrigation in the Mexicali Valley and to supply the cities of Mexicali and Tijuana.

Damming of the Colorado River has also changed physical and chemical conditions of this river (Carothers and Dolan 1982). From a free-flowing river with enormous spring floods, the lower Colorado has largely become a series of lakes without seasonal change in flow. Once a stream so laden with sediment that it was named, in Spanish, the "Red River," it is now clear for some distance below Glen Canyon Dam, due to deposition of the sediment in Lake Powell. In the free-flowing section of river in the Grand Canyon, the water now remains icy cold in summer because the water in this section comes from below the thermocline of Lake Powell. The release of water through hydroelectric generators is also timed to provide power at peak demand times, and this caused the river below the dam to fluctuate as much as 4 m in depth during the

day (Elfring 1990). This enormous variation in flow can alternately flood and expose the spawning areas of fish. In the hot, arid climate of the Southwest, evaporation form the various reservoirs and the inflow of salts in waste irrigation water have doubled the salinity of the water.

Biotically, the Colorado River has a native fish fauna with a greater degree of endemicity than that of any other North American river (Minckley and Deacon 1968). Loss of the river habitats to which these species were adapted along with competition from species introduced into the reservoirs have endangered this fauna almost in its entirety. Nine species of larger native fish, including eight species of river chubs and the squawfish, are now designated as threatened or endangered (Hickman 1983). In the colder portions of the river, these natives have largely been replaced by introduced trout, and in warmer regions by channel catfish.

CONSERVATION MANAGEMENT OF RIVER ECOSYSTEMS

Central to management of flowing water ecosystems is preserving a flow adequate to maintain the basic features of the aquatic ecosystem. The rights to water in rivers and streams, however, are governed by complex laws that vary from country to country and even state to state (Cooper 1990). Versions of the **riparian doctrine** derived from English common law govern water rights in most of Europe and eastern North America. This doctrine guarantees owners of land along streams the right to an undiminished and undefiled flow, a guarantee that is compatible with preservation of stream wildlife. Under the riparian doctrine, the rights to the use of water cannot be bought or sold separately from the stream-side land, nor can they be lost because water is not used in some active fashion.

In the arid and semiarid western United States, however, a very different type of water law, the **prior appropriation doctrine,** evolved. This doctrine permits diversion of water for beneficial uses, with the rights to the water being governed by the dictum "first in time, first in right." Under this doctrine water rights can be sold, and, if they are not used for beneficial purposes *involving diversion,* they can be

lost (Meyer 1989, Wiley 1990). The appropriation doctrine does not recognize an intrinsic beneficial use of water *in the stream itself.*

Several strategies are now being used to assure the maintenance of basic flows in important wildlife streams in the western United States. Acquisition of water rights by conservation organizations, coupled with legal recognition of instream flow as constituting a beneficial use, is one approach (Wiley 1990). The concept of **reserved water rights,** or implicit rights to the maintenance of basic flows that existed in streams within national forests, wildlife refuges, and recreation areas at the time of their establishment has been supported by a number of court decisions (Hunt 1988). Application of the **public trust doctrine** is another. This doctrine states that the public's right to natural features such as streams and lakes overrides excessive allocations that could destroy these features (Cooper 1990). These emerging doctrines are aided by new approaches to the economic valuation of ecological resources such as free-flowing streams and their wildlife (See chapter 22).

In addition to maintaining adequate flows, the quality of the water in rivers and streams must be preserved. The United States Geological Survey monitors water quality at a series of 388 stream sampling stations (Smith et al. 1987). The

Clean Water Act of 1972, together with several subsequent acts, requires the United States Environmental Protection Agency to establish **national effluent standards** that limit the quantities of polluting substances that are discharged into surface waters. In addition, the Environmental Protection Agency is responsible for obtaining compliance with these standards by sewage treatment plants, industries, and other sources of pollutants.

KEY MANAGEMENT STRATEGIES FOR STREAM ECOSYSTEMS

1. Preservation of free-flowing conditions or mitigation of the detrimental impacts of damming.

2. Maintenance of adequate water flow volume and original quality, including sediment load, salinity, acidity, temperature, and chemistry.

LITERATURE CITED

Alho, C. J. R., T. E. Lacher, Jr., and H. C. Goncalves. 1988. Environmental degradation in the Pantanal ecosystem. *BioScience* **38**:164–171.

Barthem, R. B., M. C. L. de Brito Ribeiro, and M. Petrere, Jr. 1991. Life strategies of some long-distance migratory catfish in relation to hydroelectric dams in the Amazon Basin. *Biol. Cons.* **55**:339–345.

Berner, E. K. and R. A. Berner. 1987. *The global water cycle.* Prentice-Hall, Englewood Cliffs, NJ.

Biggins, D. 1989. Coal mining and the decline of freshwater mussels. *End. Sp. Tech. Bull.* **14(5)**:5.

Carothers, S. W. and R. Dolan. 1982. Dam changes on the Colorado River. *Nat. Hist.* **91(1)**:75–83.

Carothers, S. W. and R. R. Johnson. 1983. Status of the Colorado River Ecosystem in Grand Canyon National Park and Glen Canyon National Recreation Area. Pp. 139–160 *in* V. D. Adams and V. A. Lamarra (Eds.), *Aquatic resources management of the Colorado River Ecosystem.* Ann Arbor Science Publishers, Ann Arbor, MI.

Collier, C. R., R. J. Pickering, and J. J. Musser (Eds.) 1970. Influences of strip mining on the hydrologic environment of parts of Beaver Creek Basin, Kentucky, 1955–66. *Geol. Surv. Prof. Paper 427-C,* viii + 80 pp.

Cooper, C. F. 1990. Recreation and wildlife. Pp. 329–339 *in* P. E. Waggoner (Ed.), *Climate change and U.S. water resources.* John Wiley and Sons, New York.

Davies, B. R. and K. F. Walker (Eds.). 1986. *The ecology of river systems.* W. Junk, Dordrecht, The Netherlands.

Elfring, C. 1990. Conflict in the Grand Canyon: How should Glen Canyon Dam be operated? *BioScience* **40**:709–711.

Emerson, J. W. 1971. Channelization: A case study. *Science* **173**:325–326.

Goulding, M. 1980. *The fishes and the forest.* Univ. of California Press, Berkeley.

Goulding, M. 1985. Forest fishes of the Amazon. Pp. 267–276 *in* G. T. Prance and T. E. Lovejoy (Eds.), *Amazonia.* Pergamon Press, Oxford, England.

Hickman, T. J. 1983. Effects of habitat alteration by energy resource developments in the upper Colorado River basin on endangered species. Pp. 537–550 *in* V. D. Adams and V. A. Lamarra (Eds.), *Aquatic resources management of the Colorado River Ecosystem.* Ann Arbor Science Publishers, Ann Arbor, MI.

Hunt, C. E. 1988. *Down by the river.* Island Press, Washington, DC.

Johns, A. D. 1988. Economic development and wildlife conservation in Brazilian Amazonia. *Ambio* **17**:302–306.

Junk, W. J. 1984. Ecology, fisheries and fish culture in Amazonia. Pp. 443–476 *in* H. Sioli (Ed.), *The Amazon: Limnology and landscape ecology of a mighty tropical river and its basin.* W. Junk, Dordrecht, The Netherlands.

Kohlhorst, D. W. 1980. Recent trends in the white sturgeon population in California's Sacramento-San Joaquin estuary. *Calif. Fish and Game* **66**:210–219.

Limburg, K. E. 1986. PCBs in the Hudson. Pp. 83–130 *in* K. E. Limburg, M. A. Moran, and W. H. McDowell (Eds.), *The Hudson River ecosystem.* Springer-Verlag, New York.

Meyer, C. H. 1989. Western water and wildlife: The new frontier. Pp. 59–91 *in* W.J. Chandler (Ed.), *Audubon wildlife report 1989/90.* Academic Press, San Diego, CA.

Miller, T. O., G. D. Weatherford, and J. E. Thorson. 1986. *The salty Colorado.* The Conservation Foundation, Washington, D.C.

Minckley, W. L. and J. E. Deacon. 1968. Southwestern fishes and the enigma of "endangered species." *Science* **159**:1424–1432.

Mittermeier, R. A., I. de G. Camara, M. T. J. Padua, and J. Blanck. 1990. Conservation in the pantanal of Brazil. *Oryx* **24**:103–112.

Peixun, C. and W. Ding. 1988. The Chinese river dolphin, *Lipotes vexillifer. Endeavour* **12**:176–178.

Petts, G. E., H. Moller, and A. L. Roux (Eds.). 1989. *Historical change of large alluvial rivers: Western Europe.* John Wiley and Sons, New York.

Simpson, P. W., J. R. Newman, M. A. Keirn, R. M. Matter, and P. A. Guthrie. 1982. *Manual of stream channelization impacts on fish and wildlife.* Biological Services Program, U.S. Fish and Wildlife Service (FWS/OBS-82/24), Washington, D.C.

Smith, R. A., R. B. Alexander, and M. G. Wolman. 1987. Water-quality trends in the nation's rivers. *Science* **235**:1607–1615.

Smith, R. L. 1973. Strip mining impacts and reclamation efforts in Appalachia. *Trans. N.A. Wildl. Nat. Res. Conf.* **38**:132–142.

Thomann, R. V. 1972. The Delaware River—a study in water quality management. Pp. 99–129 *in* R. T. Oglesby, C. A. Carlson, and J. A. McCann (Eds.), *River ecology and man.* Academic Press, New York.

Trefethen, P. 1972. Man's impact on the Columbia River. Pp. 77–98 *in* R. T. Oglesby, C. A. Carlson, and J. A. McCann (Eds.), *River ecology and man.* Academic Press, New York.

Weitkamp, D. E. and M. Katz. 1980. A review of dissolved gas supersaturation literature. *Trans. Amer. Fish. Soc.* **109**:659–702.

White, G. F. 1988. The environmental effects of the high dam at Aswan. *Environment* **30(7)**:5–11, 34–40.

Wiley, K. 1990. Untying the western water knot. *Nature Conservancy Magazine* **40(2)**:5–13.

CHAPTER 16

Oceanic Ecosystems

Oceans cover almost 71 percent of the earth's surface, and their behavior thus plays a major role in the dynamics of the biosphere. Scientists once believed that hectare for hectare, the marine environment was equal in productivity to the environment of the continents. This belief supported the view that oceans were a virtually inexhaustible resource of food for meeting the needs of the growing human population.

Oceanographers have now learned much more about the productivity of various ocean realms. The realities are that productivity varies enormously in different situations, and that the average productivity is much below that of the land. The biological resources of the oceans, far from being inexhaustible, are subject to rapid overexploitation, and must be managed with great care—a task that is proving extremely difficult in the international marine environment.

Productivity of Oceanic Regions

For purposes of analysis of basic productivity, the oceans can be divided into four zones: the open ocean, upwelling zones, continental shelf waters, and reef-estuary systems (Table 16.1). These regions differ in their extent, basic primary productivity, and food chain characteristics (Ryther 1969, Cox and Atkins 1979).

The **open ocean,** beyond the edge of the continental shelf, constitutes about 92.1 percent of the total marine environment. The net primary production of this ocean region is very low, corresponding to that of desert-edge ecosystems of the continents. The basic reason for low productivity in this region is severe nutrient limitation. Dead organisms and particulate wastes sink to the ocean depths, causing a constant loss of nutrient-containing matter from surface waters. These nutrients return only by upwelling at distant locations. The characterization of the open ocean as a "marine desert" is thus not far from the truth. In addition, food chains of this region are unusually long. They begin with microscopic phytoplankton, and involve several food chain links before species of fish that are practical to harvest are reached. These are primarily the tunas, which travel in large schools and migrate over great distances. Squid are also important fishery species in the open oceans. Some dolphins also forage in herds in the open oceans. In the eastern Pacific, these dolphins frequently travel and feed in association with schools of yellowfin tuna, an association that has led to high mortality in fishing operations (See chapter 18). Of the great whales, the sperm whale is most at home in the open oceans, where it feeds largely on squid. Among seabirds, the wide-ranging albatrosses are major foragers of the open oceans.

Upwellings occur at scattered locations near the edges of continents and around Antarctica. Along continental coastlines, two special circumstances are required for upwellings to occur: deep waters that lie close to the continental margin and prevailing winds that push surface waters away from the coast. Where such conditions are met, commonly along the west coasts of continents, cold, nutrient-rich waters rise to the surface. The slow, nutrient-rich bottom currents of the Atlantic, Pacific, and Indian Oceans converge in antarctic waters and well up at the antarctic divergence, a complex and variable zone encircling Antarctica. Upwelling areas are very localized, and amount to perhaps 0.1 percent of the area of the oceans (Fig. 16.1). Of the offshore regions of the oceans, upwelling zones are by far the most productive. Despite the coldness of the water, the abundant nutrients support a moderately high level of net primary production. In addition, the primary producers are much larger than those of the open ocean. Food chains only one or two links long lead to species such as anchovies, sardines, and invertebrates that can be harvested profitably when they are abundant. Most of the catch of sardines and anchovies is converted to oil and fish meal. **Krill** are large marine zooplankton, primarily the small shrimp, *Euphausia superba,* that abound in cold upwelling waters at high latitudes (Fig. 16.2). Krill are abundant in the antarctic upwelling zone. Marine mammals, such as baleen whales and seals are abundant in upwelling zones. Diving marine birds, such as auks, penguins, cormorants, boobies, and sea ducks are also abundant (Fig. 16.3).

Continental shelf waters, the second most extensive ocean region (Fig. 16.1), have an intermediate net primary productivity, largely supported by the inflow of nutrients from adjacent continental areas. Food chains here are also shorter,

Table 16.1

Net Primary Productivity (NPP) of Various Marine Environments, and the Potentially Sustainable Yields of Traditional Fisheries.

Characteristic	Estuary/Reef	Upwellings	Coastal Waters	Open Ocean
Area (percent of ocean area)	0.6	0.1	7.2	92.1
NPP (kcal m^{-2} yr^{-1})	9000	2500	1600	600
Mean efficiency of transfer between trophic levels (%)	10	20	15	10
Number of trophic transfers up to harvestable species	3	1.5	3	5
Production of harvestable material (kcal m^{-2} yr^{-1})	9.0	225.0	5.4	0.006
Total harvestable production (10^6 tons yr^{-1})	1.9	81.1	141.2	2.0

From G. W. Cox and M. D. Atkins, *Agricultural Ecology.* Copyright © 1979 W. H. Freeman and Company Publishers, New York, NY. Reprinted by permission of the authors. Data from Ryther 1969.

FIGURE 16.1

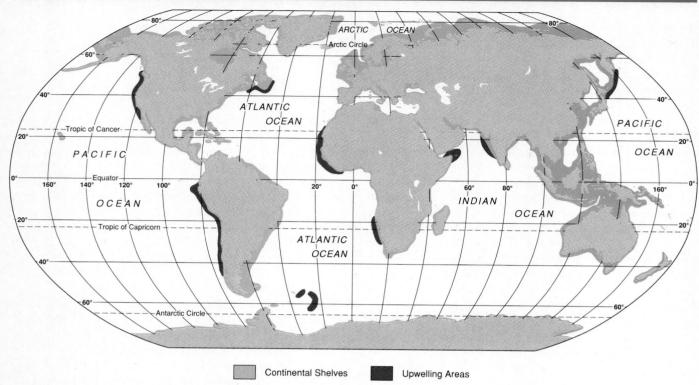

■ Continental shelf and upwelling zones of the oceans.

FIGURE 16.2

■ Euphausid shrimp, or krill, are large zooplankton that are abundant in the antarctic upwelling zone.

FIGURE 16.3

■ King penguins at this breeding colony on South Georgia are dependent on the high productivity of krill in the antarctic upwelling ecosystem.

averaging about three links from the producers to fish and shellfish of harvestable size. Many types of fish are harvested in this region of the oceans. The most important food fish are pelagic species such as mackerel, salmon, and swordfish, and various bottom-dwelling groundfish, such as cod, flounder, pollock, hake, and croaker. Herring and menhaden, small fish taken primarily for oil and fish meal production, are also harvested in this zone. Many types of shellfish, including clams, shrimp, crabs, and lobsters are also harvested in shallow continental shelf waters. Marine mammals, including seals, sea lions, walruses, porpoises, dolphins, small whales, and baleen whales, concentrate their activities in this zone. Diving marine birds are also abundant in this zone.

The net primary productivity of certain other marine environments, such as **estuaries** and **coral reefs,** is even higher than that of upwelling areas, but food chains are somewhat longer. Estuaries and reefs are major production areas for finfish and shellfish of many kinds (See chapter 17). Gulls, terns, shorebirds, and wading birds are abundant in estuaries and reefs, but few marine mammals frequent these environments.

These characteristics, coupled with efficiency of transfer of energy from one species in the food chain to the next, determine the potential productivity of the various ocean regions. Food chain efficiency tends to be highest, about 20 percent, in upwelling areas, and lowest, about 10 percent, in the open ocean. These efficiencies mean that only 20 percent and 10 percent, respectively, of the energy reaching one level in the food chain is passed on to the next level. When net primary production, food chain length, and food chain efficiency are combined, an estimate of the production of harvestable marine fish per unit ocean area is obtained (Table 16.1). This, weighted by the areas of the various regions, gives an estimate of the total production of harvestable fish in the world oceans.

The data in table 16.1 show several interesting features. First, despite their small areas, upwelling zones and continental shelf waters contribute over 98 percent of the total harvestable production. Second, the total harvestable production is only about 226 million tons, far below the billions of tons that some have proposed as the potential harvest from the oceans.

How much of this can be harvested without overexploiting the target species themselves, and disrupting food chain relationships with other members of the ecosystem? The answer to this question is not known with any real certainty. Ryther (1969), one of the most knowledgeable biological oceanographers of our time, estimated that perhaps 40 percent of this could be taken, or roughly 100 million tons, if it were harvested in a carefully controlled manner. This estimate is the same as that developed for conventional marine fisheries by the United Nations Food and Agriculture Organization (World Resources Institute 1988).

TRENDS IN MARINE FISHERY HARVESTS

At the end of World War II, world fishery harvests were just over 20 million tons annually (Fig. 16.4), and some major fisheries, such as that for Peruvian anchoveta, had not even been tapped. Improved marine technologies, however, led to rapid growth of marine fishery harvests, and total fishery yields climbed rapidly. Between 1950 and 1970 nearly a three-fold increase in total harvest was realized. These developments led to highly optimistic predictions of the potential fishery harvest from the world oceans, and to the view that such harvests could solve the protein shortages afflicting much of humanity. Predictions of ultimate annual harvests of up to one billion tons were made.

In the 1970s and 1980s, however, fishery yields began to plateau. In 1987 the United Nations Food and Agriculture Organization (UN FAO a, b) concluded that the era of large, sustained increase in marine fishery harvest was over. Only 25 of 280 fish stocks that the FAO monitored were still only slightly to moderately exploited, whereas 42 were overexploited or depleted.

Overexploitation of marine fisheries has been frequent and widespread. In the northwest Atlantic, the cod fishery crashed in the first decade of the 1900s and haddock in the 1930s (Alverson 1978). Overexploitation of Pacific halibut by United States and Canadian fleets in the early 1900s led to the crash of this fishery in 1916, forcing the creation of an international commission to oversee its harvests (Weber 1986). Other fisheries that are now badly overexploited include the Atlantic salmon, the Atlantic striped bass, New England lobster, Gulf of Mexico king mackerel, Pacific albacore, Pacific rockfish, Alaska pollock, Alaska king crab, and others (Weber 1986). Many other stocks are on the verge of overfishing, including sharks (Manire and Gruber 1990), Gulf of Mexico menhaden, Atlantic and Pacific billfishes, and Pacific yellowfin tuna (Weber 1986).

In spite of scientific fishery research programs in most major regions of the oceans, even newly exploited fisheries have proven almost impossible to manage for sustained yield. In Antarctic waters, for example, exploitation of the marbled notothenia and other species near the island of South Georgia began in 1972–1973 with a yield of about 4,000 tons. The yield of this fishery grew to 500,000 tons in 1979–1980, but subsequently crashed to 75,000 tons in 1984–1985, in spite of efforts by the International Commission for the Conser-

FIGURE 16.4

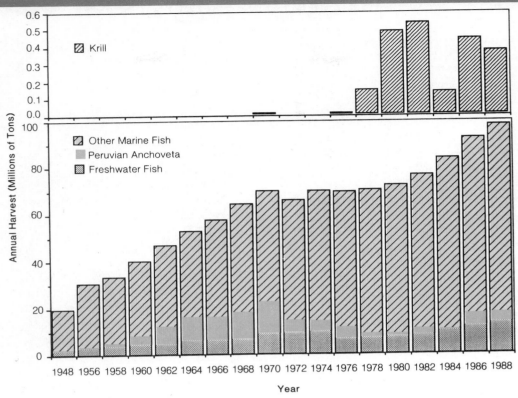

■ Global harvests of marine and freshwater fish and shellfish for
selected years since World War II (Source:
Data from annual volumes of the FAO
Yearbook of Fishery Statistics).

vation of Antarctic Marine Living Resources, whose membership includes all 12 of the nations engaged in major fishing activities there. In 1984, the commission was forced to close the fishery for the marbled notothenia because of the rapid decline in catches.

Fisheries of the upwelling zones have proven especially difficult to exploit in a sustained manner. The foremost example of this problem is the Peruvian anchoveta fishery, discussed in detail below. The Pacific sardine, once abundant from central Baja California to British Columbia, is another example (Browning 1980). Beginning in the 1920s, exploitation of the species by United States and Canadian fishermen increased until, in 1936–1937, over 800,000 tons were harvested. In the years immediately following, the fishery crashed. The factors contributing to this crash are hotly argued. Some biologists believe that overexploitation was the principal culprit, others that changes in currents and water temperature

were responsible, and still others that the abundance of the sardine was the high point of a natural, long-term cycle (Browning 1980). In over 50 years, however, the Pacific sardine fishery still has not recovered. The estimated stock of this species in 1989 was 0.5 percent of its early 1930s level.

As a result, by 1988 commercial marine fishery harvests had reached about 84.6 million tons (Fig. 16.4). Since non-commercial marine catches are about 24 million tons (World Resources Institute 1988), the total marine harvest for 1988 may have exceeded the 100 million tons estimated as being sustainable. In recent years, freshwater fish harvests, much from aquaculture, have grown at a faster rate than those of marine fish, and they now constitute about 14 percent of all world fishery harvests. Thus, it is likely that many more traditional fish stocks will begin to show signs of decline, and that more efforts will be expended on aquaculture and on harvesting non-traditional resources.

FIGURE 16.5

■ The crew of a Peruvian fishing boat prepares to bring aboard anchoveta captured in a purse seine in the coastal upwelling zone.

One of the factors contributing to the rapid growth of marine fishery harvest in the 1960s was the emergence of the anchoveta fishery in the Peruvian upwelling zone. Anchoveta are harvested by small purse seiners, boats that enclose schools of fish by a long net that extends downward from the surface, and then purse the deep edge of the net so that it closes, trapping the fish (Fig. 16.5). This fishery began in about 1956 with a harvest of 0.1 million tons and reached 13.1 million tons in 1970. Between 1964 and 1971, anchoveta harvests exceeded 9.5 million tons annually, the level estimated by fishery biologists to be a maximum sustainable harvest (Brown 1985). In 1970, anchoveta constituted almost 20 percent of the total world fishery harvest, and fish meal manufactured from the anchoveta earned $340 million for Peru, about a third of the country's foreign exchange (Fig. 16.6).

In the Peruvian upwelling zone, occasional incursions of warm tropical water suppress the upwelling phenomenon, killing and dispersing the anchoveta. This event, which is known as El Niño because it tends to occur at Christmas (El Niño refers to the Christ child), recurs at intervals of about five to ten years (Philander 1989). In most years, strong trade winds drive equatorial currents westward from the coast of South America, stimulating the upwelling of cold, nutrient-rich water that supports the high productivity of the system. At the peak of this process (now often termed "La Niña") warm surface water becomes concentrated in the western Pacific, where a very strong tropical low-pressure system simultaneously develops. Eventually, this tropical low spreads eastward, reducing the strength of the trade winds and permitting strong ocean currents to carry warm water eastward. When these currents reach South America, they seal off the upwelling, creating an El Niño. A similar but weaker system of this sort occurs in the tropical Atlantic Ocean.

Prior to the development of the anchoveta fishery, the impact of El Niños was mainly on seabirds such as brown pelicans, guanay Cormorants, and Peruvian boobies, all of which depend on the anchoveta. Original populations of these birds along the Peruvian coast, during normal years, are estimated to have been 28 million. Periodic El Niños, however, caused these populations to crash to levels as low as five million.

At the peak of the anchoveta fishery in 1972, an El Niño struck. In an effort to sustain the profitable catch, fishing effort was intensified, the result being that the decimated anchoveta stock was depleted even further. In 1973, instead of showing recovery, the catch fell to less than two million tons. The fishery showed evidence of some recovery in 1974–1976, but exploitation and subsequent El Niños have apparently caused a basic disruption of the breeding stock of the anchoveta. In the 1980s, this fishery has remained far below its yields in the early 1970s. In 1988, for example, the anchoveta harvest was 3.6 million tons. Seabird populations along the Peruvian coast are also at an all-time low.

FIGURE 16.6

■ Bagged fish meal produced from anchoveta is the major product of the fishery in the Peruvian coastal upwelling zone.

Figure 16.7

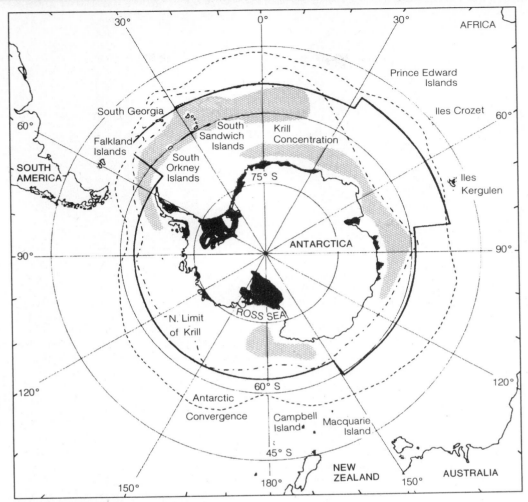

■ The antarctic marine ecosystem, showing the prevailing locations of major concentrations of krill and their northern limit of distribution in relation to the antarctic convergence (Modified from Siegfried 1985).

DROPPING DOWN THE FOOD CHAIN

The harvest of marine fisheries could obviously be increased considerably if the harvest could be taken at lower levels in the food chain. In effect, this is the strategy behind efforts to develop a fishery for krill. The effort to develop a krill fishery has been concentrated in the antarctic upwelling zone (Fig. 16.7).

Estimates of the biomass and productivity of antarctic krill are highly uncertain (Knox 1984). Euphausid krill tend to occur in dense swarms, which may extend to a depth of 100 m over many square kilometers of ocean, and in which animals may reach a density of 10 to 16 kg m^{-3}. The true extent of such swarms is difficult to measure, and estimates of standing crop biomass range between 55 million and seven billion tons (Miller and Hampton 1989), the most probable values lying in the range of several hundred million tons. Surveys in 1981, using sonar techniques, gave a krill biomass estimate of 250 to 600 million tons (El-Sayed 1988). More recently, other workers have suggested that the sonar echoes from krill swarms are much weaker than the values used in this survey, and that krill biomass might be much greater (Everson et al. 1990). Annual production is even more uncertain, due to inadequate knowledge of the life history of euphausid shrimp. If krill require only two years to reach maturity, the annual productivity would be about half the standing crop biomass, and thus possibly a few hundred million tons (Sahrhage 1989). Most fisheries biologists now believe that krill have a life-span of 6 to 10 years, so that annual production would be a much smaller fraction of standing biomass (Nicol

FIGURE 16.8

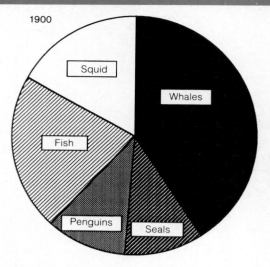
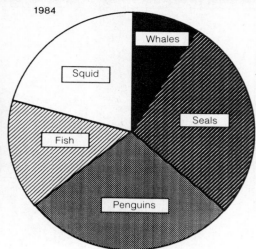

■ Major changes in the biomasses of the major krill-feeding animals have occurred in the antarctic marine ecosystem, correlated with human harvests of whales and fish Source: Data form R. M. Laws, "The Ecology of the Southern Ocean" in *American Scientist*, 73:26–40, 1985.

1990). How much a sustainable, economically sound krill fishery could add to marine fishery harvest is therefore very uncertain.

Experimental krill harvests have been carried out in antarctic waters since 1964 by several countries, notably the Soviet Union. The maximum annual harvest to date is about 530 thousand tons. Krill must be processed immediately after capture to prevent enzymatic deterioration. With limited processing, krill can be converted to protein meal for use in poultry and livestock diets. With more extensive processing, a protein additive suitable for use in human foods can be produced. Altogether, the economics of harvesting and processing krill in remote antarctic waters have not yet become positive, and krill harvests still remain an experimental enterprize.

HARVESTING INFLUENCES IN COMPLEX MARINE ECOSYSTEMS

The exploitation of marine fisheries involves the harvest of different species, often belonging to different trophic levels, that are interacting members of a single ecosystem (Beddington and May 1982). Harvesting strategies for different fishery species have usually been formulated independently, and usually from theory about the sustained yield harvest of one species (See chapter 21). Although such strategies may be suitable for species at the ends of marine food chains, for which humans are the primary predator, they are inadequate for most marine fisheries. Most harvested species are linked with competitors, predators, and prey in a complex system which can respond in unexpected ways to the exploitation of one or more members.

As a result of ecological linkages, heavy exploitation of some fisheries has caused major shifts in the composition of marine fish communities. In the Georges Bank fishery in the northwestern Atlantic, heavily fished species such as herring, mackerel, and silver hake are being replaced by sand lance, dogfish, and squid (Weber 1986). Decline of the Pacific sardine fishery off the west coast of North America was coupled with increase in the northern anchovy (MacCall 1986). In the North Sea, even though the overall fisheries harvest remained relatively constant from the mid-1960s through the late 1970s, the fraction of the harvest made up of herring and mackerel, the most desirable species, declined from about two-thirds to less than one-third (Beddington and May 1982).

The kinds of tradeoffs that involve marine fisheries harvests by man are well illustrated by the antarctic marine ecosystem. This ecosystem is one of the largest, most productive, and most sharply defined ecosystems of the world oceans. Although many uncertainties exist about rates of productivity at the level of primary producers and herbivores, it is clear that euphausid shrimp are the key consumer in the system. They constitute what some ecologists term a **foundation species,** a lower trophic level species whose abundance determines much of the upper trophic level structure of the ecosystem. They form the primary food of baleen whales, crabeater seals, penguins, and the larger fish and squid. The populations of these species are ultimately determined by krill production.

Populations of krill-eating species have changed markedly in the last century (Laws 1985), due to selective harvest of certain of these animals by humans (Fig. 16.8). For example, in the early 1900s, this antarctic region was home to perhaps 1.1 million baleen whales, with a total biomass of

FIGURE 16.9

■ Crabeater seals, the most abundant pinniped in the world, are one of the major krill predators of antarctic waters.

about 45 million tons. Whaling has reduced the biomass of whales to about nine million tons. In recent decades, heavy exploitation of several fish stocks has reduced their abundance, as well. These reductions have meant more krill for the remaining species, allowing major increases in their populations. Populations of seals and penguins are probably 2 to 3 times those of the early 1900s, and squid are also believed to have increased in abundance somewhat. The crabeater seal (Fig. 16.9), with a population of 15 to 30 million, is now the most abundant pinniped in the world (Siniff 1991). Sei and minke whales, two small species less intensively exploited than the larger whales, may also have increased in numbers.

These changes indicate that a major fishery for krill would create substantial impacts on several or all of the krill-eating species. Depletion of krill stocks would almost certainly lead to declines in populations of whales, seals, birds, and squid. Some scientists fear that reduced numbers of nesting penguins in certain antarctic colonies reflect the effect of increased krill harvests since 1986 (R. Hewitt, pers. comm.). Eventually, the krill might even be replaced as herbivores by copepods or other invertebrates that would channel their food energy along new food chains, perhaps involving various species of fish and fish predators, so that a major change in community structure might occur.

Similar tradeoffs are evident in the Peruvian upwelling ecosystem, where intense exploitation of anchoveta and occasional El Niños have depleted the anchoveta, the foundation species of this system. In the mid-1950s, before anchoveta harvests began, about 28 million seabirds—Peruvian Boobies, Guanay Cormorants, and Brown Pelicans—bred on islands along the coast of Peru. These species created the remarkable deposits of guano that were mined on these islands for fertilizer for many years. During El Niño years, shortages of anchoveta would cause their populations to crash, as in 1957, when populations dropped to about five million. Normally, however, recovery of seabirds was rapid. With the growth of the anchoveta fishery, however, bird numbers have never returned to their original levels, reaching only about 16 million in the mid 1960s. With collapse of the anchoveta stock in the 1970s, peak numbers of seabirds declined to only about

FIGURE 16.10

■ By-catch of fish, sea turtles, and other animals is heavy in shrimp trawling.

four million. In 1983, after an El Niño reduced the anchoveta harvest to its lowest level since the fishery began, less than a million seabirds survived along the Peruvian coast.

DISCARDED CATCHES OF COMMERCIAL FISHING

Another serious problem of commercial fishing is the capture, injury, and discard of non-target organisms. These include non-target species of finfish or shellfish, undersize individuals, and animals such as sea turtles, mammals, and birds. This by-catch results from the poor selectivity of the nets, trawls, or traps used in fishing. Most of the by-catch is killed, injured, or stunned in the process, so that even though the animals are returned to the ocean, most die or are quickly taken by predators.

Other wasteful fisheries practices exist. **Roe stripping** has been a common practice of fishing fleets serving Japanese and other Asian markets (Alverson 1990). In this practice, roe

(or eggs) are removed from female fish and the rest of the body, together with all male fish, are discarded. About 60,000 tons of pollock taken in the northeast Pacific are stripped of roe and discarded, even though pollock is now marketable in food products such as imitation crab. **Shark finning,** the removal of fins and the discard of the remaining fish, is a growing practice in shark fisheries (Manire and Gruber 1990). Shark fins are considered a delicacy in certain oriental countries.

Worldwide, discarded by-catch of marine fisheries is estimated at 5.4 to 9.1 million tons (Bricklemyer et al. 1989). Discarded by-catch is probably greatest in shrimp trawling (Fig. 16.10), in which the ratio of by-catch to shrimp is commonly 30:1. The otter trawls used in shrimping are wide-mouthed, small-mesh nets pulled across the ocean floor, capturing almost all kinds of swimming animals that live close to the bottom. Much of this by-catch consists of valuable species, and often the by-catch of important sport fish is greater than the catch of these species by sport fishermen. In the Gulf

FIGURE 16.11

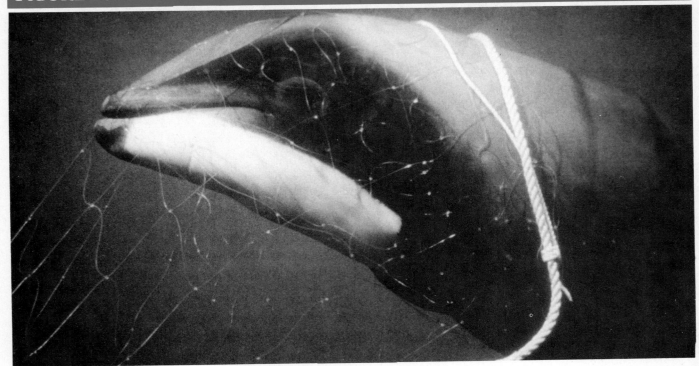

■ Drift gill-netting has been characterized as "strip-mining the sea" because of the large catch of non-commercial fish, seabirds, and marine mammals, such as this Hector's dolphin calf near New Zealand.

of Mexico, shrimp trawling has probably contributed to the decline of several groundfishes, including croaker and red snapper. Furthermore, by depleting the abundance of small fish, predation on the remaining small animal life, including shrimp, by larger fish may be increased.

By-catch of young halibut was probably the major reason for the catastrophic decline of the halibut fishery in the northeast Pacific in the 1970s (International Pacific Halibut Commission 1987). Following the establishment of the International Pacific Halibut Commission in 1923 to regulate United States and Canadian exploitation of this fishery, yields increased from about 20,000 tons in 1931 to about 32,000 tons in 1962. Harvest then declined, coincident with increased fishing by foreign trawlers that captured a heavy by-catch of small halibut. Harvestable halibut yields fell to about 9,000 tons in 1974 and remained below 12,000 tons through 1979. In 1965 the by-catch of halibut was about 12,700 tons, 50 percent or more of which was probably killed in the process. Since establishment of exclusive United States and Canadian fishing zones in this region, trawling has been excluded from major halibut nursery grounds. The by-catch of small halibut has been reduced to less than 4,000 tons, and the harvest of legal fish has risen to over 25,000 tons.

Much fuller information is needed on the types and quantities of by-catch in different fishing operations. Clearly, however, efforts must be made to improve the selectivity of fishing gear. In shrimp trawling, for example, different responses of shrimp and fish to sound, light, or electrical fields might allow fish to be repelled from the trawl mouth (Sternin and Allsopp 1982). Ultimately, much tighter regulation of by-catch must be gained under the United States Fishery Conservation and Management Act and various international conventions.

DRIFT GILL NET FISHERIES

In the 1980s, fishing fleets of several countries began using very large monofilament gill nets for harvesting fish and squid in the open oceans. These nets hang freely from surface floats to a depth of 12 to 18 m, and are up to 50 km long. The nets are deployed in the evening to catch fish or squid that come to the surface to feed at night, and are retrieved the next morning. In the Pacific, about 1,500 fishing boats of Japan, South Korea, and Taiwan are equipped for drift net fishing. The total length of nets for these boats is estimated at 48,000 to 64,000 km.

In the central North Pacific, drift net fishing is ostensibly aimed at squid, but the by-catch of non-target animals is large (Fig. 16.11). Marine mammals and birds are a significant part of this by-catch (See chapter 18). Up to 40 percent of the catch is discarded, and this method of fishing is sometimes characterized as "strip-mining the sea." Particular concern has centered on the catch of salmon in areas where these fish are of North American origin. Squid frequent much warmer water than salmon, and a Squid Regulatory Zone has been established on the basis of sea temperatures. This zone, to which drift net fishing is supposed to be limited, lies south of areas used by salmon, and its boundaries shift north during the summer as sea temperature rises (McCredie 1990). Restricting drift net fishing to this zone is difficult, however. Korean and Taiwanese boats have commonly ignored the zone, and have taken large quantities of salmon. Drift gill netting was probably the main cause of a 70 percent shortfall in the Alaska pink salmon harvest in 1988 (Lenssen 1989).

In the southern Pacific, drift net fishing has also been extended to albacore tuna, and conflicts have arisen with several countries that have historically exploited these stocks. In 1989 the estimated sustainable yield harvest for this fishery was exceeded by about 600 percent (Lenssen 1989).

Because of the highly unselective nature of these nets, their use or possession within the United States exclusive economic zone was prohibited in 1990. The United Nations General Assembly also passed a resolution to impose a moratorium on drift net fishing in the southern hemisphere in July, 1991, and a world-wide moratorium in December, 1992.

MANAGING OFFSHORE MARINE FISHERIES

Ocean fisheries clearly present major management problems because many of them exist in international waters. Since the 1970s, however, most countries have claimed exclusive economic zones extending to a distance of 320 km from shore. This exclusive zone was established by the United States in 1976 under the **Fishery Conservation and Management Act.** In 1982, the **United Nations Convention on the Law of the Sea,** signed by 119 nations, gave formal recognition to this economic zone (Borgese 1983). Since, in fact, the bulk of the world's most productive fisheries lie within this distance of continental coastlines, this development enables nonmigratory fisheries stocks to be managed protectively by the country with jurisdiction.

Major fisheries do exist, however, in regions beyond exclusive economic zones, such as the antarctic waters and the central Pacific. Furthermore, many fish are migratory, and thus move through the economic zones of many countries. Thus, many international commissions have been established to promote the sustained yield management of particular fish or the fisheries of particular regions (Gulland 1980). The membership of these commissions includes the various nations exploiting the fisheries. Examples of commissions concerned with particular types of fish are the **Inter-American Trop-**

ical Tuna Commission, which regulates tuna harvesting in the eastern tropical Pacific, and the **International Pacific Halibut Commission,** which coordinates United States and Canadian harvests of halibut in the exclusive economic zones of these two countries. The **Northeast Atlantic Fishery Commission,** which sets quotas for various stocks and established mesh sizes to control the size of fish taken, is an example of a regional commission.

One of the major shortcomings of marine fisheries management has been the lack of an ecosystem approach. Recently, however, the concept of **Large Marine Ecosystems (LME)** has begun to provide a framework for such management (Sherman 1986). Large Marine Ecosystems are ocean areas with more or less natural boundaries formed by surface currents and submarine topography, and containing species strongly linked into a food web. About 20 such ecosystems have been recognized, primarily in upwelling areas or in partially enclosed ocean basins. The **Antarctic Marine Ecosystem** is one example of an LME (Scully et al. 1986). This ecosystem lies south of the antarctic convergence, where cold antarctic water sinks as it contacts warmer water from farther north. This zone surrounds Antarctica, and corresponds to a sharp biological boundary. The Antarctic Marine Ecosystem is characterized, as described earlier, by the dependence of various fish, squid, birds, and mammals on krill. The **International Commission for the Conservation of Antarctic Living Marine Resources** has undertaken the responsibility for the management of this LME.

The Baltic Sea and the North Sea are other recognized LME's (Sherman and Alexander 1986). These LME's, however, show the difficulties of achieving integrated management, even where natural ecosystem units can be recognized. In the Baltic Sea, for example, scientists of the **International Baltic Sea Fisheries Commission** in 1981 recommended a cod quota of 197,000 tons, but, for political reasons, the commission itself could establish a quota only as low as 272,000 tons, and could not keep actual harvest below 380,000 tons.

KEY MANAGEMENT STRATEGIES FOR OPEN OCEAN ECOSYSTEMS

1. Development of predictive models of productivity and food chain dynamics of both commercial and non-commercial species in exploited marine ecosystems.

2. Establishment of international organizations with effective regulatory powers to manage exploitation of fisheries in international waters.

LITERATURE CITED

Alverson, D. L. 1978. Commercial fishing. Pp. 67–85 *in* H. P. Brokaw (Ed.), *Wildlife in America.* U.S. Government Printing Office, Washington, D.C.

Alverson, D. L. 1990. Roe stripping in the Alaskan pollock fishery. *Fisheries* **15(3):**14–15.

Beddington, J. R. and R. M. May. 1982. The harvesting of interacting species in a natural ecosystem. *Scientific American.* **247(5):**62–69.

Borgese, E. M. 1983. The law of the sea. *Sci. Amer.* **246(3):**42–49.

Bricklemyer, E. C., Jr., S. Iudicello, and H. J. Hartmann. 1989. Discarded catch in U.S. commercial marine fisheries. Pp. 258–295 *in Audubon wildlife report 1989–90.* Academic Press, San Diego, CA.

Brown, L. R. 1985. Maintaining world fisheries. Pp. 73–96 *in* L. Starke (Ed.), *State of the world: 1985.* W. W. Norton Co., New York.

Browning, R. J. 1980. *Fisheries of the North Pacific.* Alaska Northwest Pub. Co., Anchorage.

Cox, G. W. and M. D. Atkins. 1979. *Agricultural ecology.* W. H. Freeman Co., San Francisco, CA.

El-Sayed, S. Z. 1988. The BIOMASS program. *Oceans* **31:**75–79.

Everson, I., J. L. Watkins, D. G. Bone, and K. G. Foote. 1990. Implications of a new acoustic target strength for abundance estimates of Antarctic krill. *Nature* **345:**338–340.

Gulland, J. A. 1980. Open ocean resources. Pp. 347–378 *in* R. T. Lackey and L. A. Nielson (Eds.), *Fisheries management.* John Wiley and Sons, New York.

International Pacific Halibut Commission. 1987. *The Pacific halibut: biology, fishery, and management.* Int. Pac. Halibut Comm. Tech. Rep. No. 22.

Knox, G. A. 1984. The key role of krill in the ecosystem of the southern ocean with special reference to the Convention on the Conservation of Antarctic Marine Living Resources. *Ocean Management* **9:**113–156.

Laws, R. M. 1985. The ecology of the southern ocean. *Amer. Sci.* **73:**26–40.

Lenssen, N. 1989. The ocean blues. *World Watch* **2(4):**26–35.

MacCall, A. D. 1986. Changes in the biomass of the California Current ecosystem. Pp. 33–54 *in* K. Sherman, and L. M. Alexander (Eds.), *Variability and management of large marine ecosystems.* Amer. Assoc. Adv. Sci., Washington, D.C.

Manire, C. A. and S. H. Gruber. 1990. Many sharks may be headed toward extinction. *Cons. Biol.* **4:**10–11.

McCredie, S. 1990. Controversy travels with driftnet fleets. *Sea Frontiers* **36(1):**13–19.

Miller, D. G. M. and I. Hampton. 1989. Biology and ecology of the antarctic krill (*Euphausa superba* Dana): A review. *Biomass* **1:**1–166.

Nicol, S. 1990. The age-old problem of krill longevity. *BioScience* **40:**833–836.

Philander, G. 1989. El Niño and La Niña. *Amer. Sci.* **77:**451–459.

Ryther, J. H. 1969. Photosynthesis and fish production in the sea. *Science* **166:**72–76.

Sahrhage, D. 1989. Antarctic krill fisheries: Potential resources and ecological concerns. Pp. 13–33 *in* J. F. Caddy (Ed.), *Marine invertebrate fisheries: Their assessment and management.* John Wiley and Sons, New York.

Scully, R. T., W. Y. Brown, and B. S. Manheim. 1986. The Convention for the Conservation of Antarctic Marine Living Resources: A model for large marine ecosystem management. Pp. 281–286 *in* K. Sherman and L. M. Alexander (Eds.), *Variability and management of large marine ecosystems.* Amer. Assoc. Adv. Sci., Washington, D.C.

Sherman, K. 1986. Large marine ecosystems as tractable units for measurement and management. Pp. 3–7 *in* K. Sherman and L. M. Alexander (Eds.), *Variability and management of large marine ecosystems.* Amer. Assoc. Adv. Sci., Washington, D.C.

Siegfried, W. R. 1985. Krill: The last unexploited food resource. *Optima* **33:**67–79.

Siniff, D. B. 1991. An overview of the ecology of antarctic seals. *Amer. Zool.* **31:**143–149.

Sternin, V. and W. H. L. Alsopp. 1982. Strategies to avoid by-catch in shrimp trawling. Pp. 61–64 *in Fish by-catch—bonus from the sea.* International Development Research Centre, Ottawa, Canada.

UN FAO. 1987a. *World fisheries situation and outlook.* United Nations Food and Agriculture Organization, Rome.

UN FAO. 1987b. *Review of the state of the world fishery resource.* United Nations Food and Agriculture Organization, Rome.

Weber, M. 1986. Federal marine fisheries management. Pp. 267–344 *in* R. L. Di Silvestro (Ed.), *Audubon wildlife report 1986.* National Audubon Society, New York.

World Resources Institute. 1988. *World resources 1988–89.* Basic Books, Inc., New York.

CHAPTER 17

Coastal Marine Ecosystems

A variety of distinctive marine ecosystems occur along ocean coastlines. These include estuaries and their associated salt marshes, mangrove swamps, and seagrass beds, as well as kelp beds, coral reefs, and other intertidal and shallow water ecosystems. Of all marine ecosystems, these are the most productive, and because of their productivity and accessibility, they are among the ecosystems most impacted by human activities.

Coastal ecosystems have long been exploited by humans. Now, however, intensified exploitation of biotic production is coupled with sewage discharge, chemical pollution, and siltation from growing human populations increasingly crowded into coastal regions. More and more, the shallow marine environment is being probed for oil, phosphates, building materials, and other non-renewable resources. Plans for floating airports and other large facilities are being drawn. Yet the shallow marine environment, if protectively managed, also has enormous potential for mariculture and sea farming.

FIGURE 17.1

■ The open waters of estuaries are often bordered by salt marshes, the result being that primary production is carried out by phytoplankton in open-water areas, surface-living algae on mud flats, and rooted salt grasses and other higher plants in the marsh.

ESTUARIES

An **estuary** is a semi-enclosed body of water occurring where rivers discharge into the ocean and usually has a salinity regime intermediate between fresh waters and the ocean. Estuaries include the regions of river deltas where fresh and ocean waters mix, partially enclosed bays and lagoons that are fed by streams and open to the ocean, and the sounds that lie inside barrier island systems. Salinity grades from nearly fresh in the inner parts of an estuary to almost marine where it opens to the ocean. Frequently, due to seasonal variation in freshwater inflow, the salinity varies.

Estuaries function as partial traps for nutrients, which are carried into them by an upstream flow and confined by barrier islands and barrier beaches, as well as by the tidal prism of high-density salt water, with which the fresh water is often slow to mix. The aquatic portions of an estuary are often in contact with salt marshes that are subject to tidal action (Fig.

17.1). The primary producers of the estuary thus include the salt marsh plants, the algae that live on the surfaces of tidal mud flats, the phytoplankton of the estuary waters, and beds of plants, such as eelgrasses, growing on the estuary bottom. Net primary production of estuaries is thus higher than that of any other aquatic ecosystem.

Estuaries produce large quantities of fish and shellfish that are harvested by man. In North America, for example, most oyster and blue crab production is carried out in estuaries along the Atlantic and Gulf Coasts. Much of the shrimp production in the Gulf of Mexico depends on estuaries that serve as nurseries for the young shrimp. Along the Texas coast, both brown and white shrimp utilize the coastal bays and sounds for their early growth. The adult shrimp spawn in the open ocean, but the young larvae migrate into estuaries to grow. Brown shrimp enter the estuaries in fall, hibernate in the bottom sediments, and complete their growth in spring.

FIGURE 17.2

■ Mangrove ecosystems occupy vast areas of shallow, subtropical and tropical marine waters. Here, red mangroves cover marine shallows around Big Pine Key, Florida.

White shrimp enter as the brown shrimp depart in spring and complete their growth in late summer, at which time they leave the estuaries. A number of fish also use estuaries as nursery areas for the young.

The productivity of estuaries thus depends heavily on the normal inflow of streams. Freshwater inflow carries nutrients into the estuaries and maintains the normal salinity gradient. Water of intermediate salinity is essential to many typical estuarine species by excluding their predators, parasites, and diseases. Oysters, for example, are attacked by predatory oyster drills (a carnivorous snail), parasitic boring sponges that erode the oyster's shell, and certain fungal diseases, none of which can tolerate intermediate salinities. Normal river inflow also functions in certain situations to help maintain the passes through barrier island systems—the water connections through which organisms such as shrimp migrate between the oceans and sounds. How essential this inflow is to the ecology of estuarine species is evidenced by studies that show that the shrimp harvest along the Texas coast is directly related to the rainfall in inland Texas during the two preceding years (Copeland 1966).

In spite of the ecological importance of normal stream inflow, resource policy tends to treat water that enters the ocean as water wasted, particularly in regions such as Texas and California where heavy demands exist for fresh water for urban use and for irrigated farming. Increasingly, rivers that feed into coastal bays and sounds are dammed, and the water diverted for other uses. The nutrients in the flow of these streams thus are lost to coastal waters, and the salinity regimes of the affected estuaries are modified considerably. Sediments that served to maintain beaches and other structural features of coastal ecosystems are also trapped by dams.

Temperate zone estuaries are often bordered by areas of salt marsh. In these marshes, the primary producers are rooted plants, such as the cordgrasses (*Spartina* spp.) and pickleweeds (*Salicornia* spp.), and algae that form mats on tidal mud surfaces. In some cases, these salt marshes play an important role in the dynamics of open estuarine waters because large portions of marshland are flushed daily by tides. In marshes that open broadly to adjacent areas of water, this action transports detritus produced by decomposition of marsh plants into the open waters of the estuary (Odum 1980). In other cases, marshes may import and trap particulate detritus, augmenting detritus food chains and decomposition within the marsh (Zedler 1982). Salt marshes are among the most threatened coastal habitats, often being used as trash dumps, filled to create building sites, or dredged to construct marinas.

In tropical regions, the role of salt marshes is largely taken over by mangrove swamps (Fig. 17.2). "Mangrove" refers to more than 50 species of trees and shrubs, belonging to 12 plant families, that are able to live in shallow seawater or soil periodically flooded by seawater. Some species can invade shallow, permanent water; others are adapted for sites only occasionally inundated. Southeastern Asia possesses a rich flora of mangrove species, but most other areas have only one to four species. Mangroves are unable to tolerate prolonged periods of freezing temperatures, and in North America occur only from Florida and central Baja California southward.

Healthy mangrove stands have a primary productivity higher than any other estuarine habitat (Odum et al. 1982). Insect herbivores and tree crabs sometimes harvest a significant fraction of leaf production of mangroves, but most of the primary production of this ecosystem finds its way into detritus food chains. Substantial export of particulate and dissolved organic materials to the open waters of neighboring estuaries also occurs.

Mangrove forests are extremely valuable ecosystems. They stabilize shorelines and protect them against hurricane damage. They provide habitat for valuable fish and shellfish. They are a source of wood, tannins, and other tree products in many areas. Mangrove forests typically have a distinctive fauna of land birds. In Florida, several West Indian birds reach their northernmost breeding localities in mangrove areas. Mangroves are often the sites of major breeding water bird colonies. The only colony of magnificent frigatebirds in Baja California, for example, breeds in a small mangrove forest area in Magdalena Lagoon. In the Florida Everglades, the American crocodile is also largely confined to this habitat.

Large areas of mangrove forests can be killed by changes in water levels induced by diking, dredging, and impoundment. Oil spills can kill mangroves, as well, by coating the surfaces of the aerial roots that carry on gas exchange. The use of herbicides as forest defoliants in Viet Nam revealed that mangroves are unusually sensitive to these chemicals. More than 1,000 square kilometers of mangrove forests were killed by this spraying.

About 1,900 square kilometers of mangrove swamps occur in Florida, with only a small fraction of the original mangrove having been lost (Odum et al. 1982). Elsewhere in the world, however, large areas of coastal mangrove swamps have been destroyed. Cutting of mangroves for wood is heavy in areas such as Baja California, where wood for fuel and construction is scarce. Mangroves in many parts of southeast Asia are being cut and converted to wood chips for paper pulp production (Fortes 1988). Clearing of mangrove forests to permit coastal development is also a major cause of destruction of this ecosystem. In Ecuador and the Philippines, for example, many mangrove swamps have been cleared for construction of maricultural lagoons. Elsewhere, mangrove areas have been cleared for urban, industrial, and resort development.

Seagrass beds occupy many sheltered temperate and tropical estuaries and coastal waters with sandy or muddy bottoms (Thayer et al. 1984). Seagrasses are true flowering plants

FIGURE 17.3

■ Some of the most extensive beds of eelgrass on the Pacific Coast of North America occur at Padilla Bay in Puget Sound, a major wintering and migratory stopover area for brant.

that are rooted in the bottom, sometimes to a depth of 20 to 30 m in clear water. In temperate regions the dominant seagrass is eelgrass, *Zostera marina* (Fig. 17.3), and in tropical waters *Thalassia testudinum*. Other seagrasses and large algae often occur in seagrass beds, and the surfaces of the seagrass leaves are covered with an epiphytic community of algae and small invertebrates.

The productivity of seagrass beds is high, and the beds are home to large populations of fish and invertebrates (Thayer et al. 1975, 1984). Scallops, in particular, are bivalve molluscs that are specialized for living in seagrass beds. Some waterfowl, such as brant and tundra swans, feed heavily on seagrasses in temperate estuaries. In tropical waters, seagrass beds are home to endangered sea turtles. Much of the productivity of seagrass beds is converted into detritus that feeds into other estuarine food chains.

Dredging and filling often cause high turbidity and sediment deposition that can kill seagrasses. Eutrophication, which reduces light penetration to seagrass beds, can also lead to their demise (see case study). Rakes and dredges that are used to harvest shellfish can also damage seagrass beds by tearing plants loose, as do boat anchors.

The Sacramento and San Joaquin Rivers drain the Central Valley of California and join to discharge into San Francisco Bay, creating the largest, and now the most intensively manipulated, estuary system on the North American Pacific Coast (Nichols et al. 1986, Herbold and Moyle 1989). San Francisco Bay itself is about 100 km from north to south. Farther inland, the Sacramento-San Joaquin delta region is a maze of channels and islands extending from Sacramento over 100 km south to Tracy and 50 km west to San Francisco Bay. Altogether the bay and delta cover an area of nearly 6,000 square kilometers (Fig. 17.4).

About 80 percent or more of the freshwater inflow now comes from the Sacramento River system. Large withdrawals of water for irrigation in the San Joaquin Valley and urban use in southern California are taken from the south end of the delta. In winter, flows in all channels normally move down-channel, toward San Francisco Bay. In summer the withdrawals at the south end of the delta exceed the inflow there, reversing flows in certain channels and causing a large cross-delta movement of water from northern and central portions of the delta. These withdrawals substantially reduce the total inflow into the lower delta and San Francisco Bay.

Originally, this estuary system included about 2,200 square kilometers of tidal marsh, of which only about 5 percent remain today as a result of filling and drainage (Nichols et al. 1986). About 100 species of invertebrates and 20 species of fish have been introduced to the estuary; one of the latter, the striped bass, is now the most important sport fish in the system. Sedimentation from hydraulic gold mining in the 1800s, and pollution from modern urban and agricultural sources have exerted heavy impacts on the estuary. Due to water diversions for urban and agricultural use, freshwater inflow to San Francisco Bay is less than 40 percent of its original level.

FIGURE 17.4

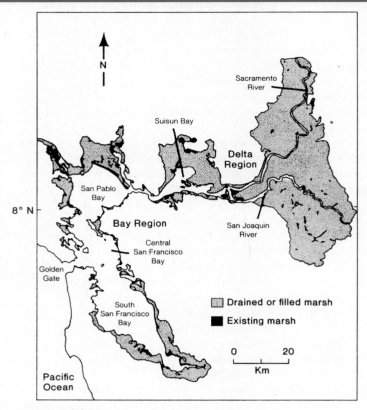

■ San Francisco Bay and the Sacramento-San Joaquin Delta form the largest and most heavily impacted estuary on the Pacific Coast of North America (Modified from Nichols et al. 1986).

The native biota of the estuary has been profoundly changed. Six species of salmon originally passed through the delta; only two, the steelhead and chinook, are still common. Four spawning races of the chinook occur in the Sacramento River, but two of them, the winter and spring runs, are on the verge of extinction. American shad and striped bass, introduced to the delta in the late 1800s, are migratory fish that have fostered major sport fisheries; they too have declined with reduction in freshwater flows. The white sturgeon, a native migratory species, also depends on adequate water flows. The delta proper is also home to two endemic fish, the delta smelt and the Sacramento splittail.

FIGURE 17.5

■ The Sacramento-San Joaquin Delta in California is a major wintering area for tundra swans.

The delta region is home to many mammals and is a major wintering area for swans, geese, and ducks. About 10 percent of the wintering waterfowl in California occur in the delta. These include about 30,000 to 38,000 tundra swans, roughly 85 percent of the California wintering population (Fig. 17.5). More than half a million pintail ducks winter in the delta, and over 30 percent of the statewide hunting harvest of pintails comes from delta counties.

The burgeoning human population of southern California draws much of its water from northern California through the aqueducts of the California Water Project. The growing demand for water in southern California leads to frequent demands for increased diversion from the delta system, and for projects such as a peripheral canal, which would carry water from the Sacramento River around the delta to the head of the aqueduct system leading southward. During the drought years of the late 1980s, plans for protecting the ecological values of the delta ecosystem by retaining freshwater inflows were bitterly criticized by some for ignoring the needs of the southern California population.

FIGURE 17.6

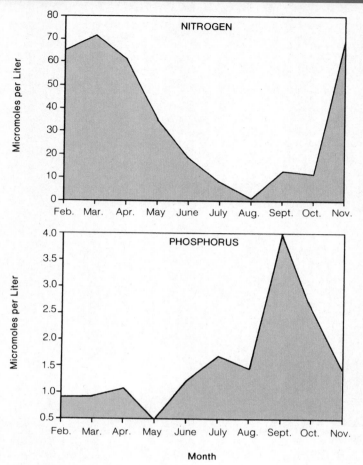

■ In Chesapeake Bay, phosphorus is abundant during the summer production period, making nitrogen the critical limiting factor in primary productivity and eutrophication (Modified from D'Elia 1987).

ESTUARINE EUTROPHICATION

Some estuaries, such as Chesapeake Bay in the eastern United States, are suffering from eutrophication. This bay, the largest estuary in the world, contains some of the most valuable fish, shellfish, and wildlife resources in North America. The symptoms in Chesapeake Bay are similar to those observed in freshwater lakes: reduced transparency of the water, blue-green algal blooms, and oxygen depletion in the deeper waters (D'Elia 1987). Similar conditions afflict the Baltic Sea in Europe (Cederwall and Elmgraen 1990).

The rivers that flow into the bay drain a heavily populated region of about 164,000 square kilometers, and are the ultimate discharge sites for enormous volumes of urban sewage. Most of the change in transparency, however, is not due to particulate matter, such as silt and phytoplankton, washed into the estuary, but is due to growth of phytoplankton within the estuary itself. Large quantities of both phosphorus and nitrogen enter the bay from sewage discharge and runoff from the terrestrial landscape, including fertilized farmland. Studies of seasonal nutrient dynamics show that nitrogen inputs are greatest in winter and early spring, when the freshwater inflow from surface runoff is highest. Because of low temperatures at this season, decomposition is low and large amounts of organic matter accumulate in the sediments. In summer, surface runoff declines, and more of the inflow comes from sewage discharges, which are relatively high in phosphorus. Higher water temperatures favor decomposition, which reduces oxygen levels at the sediment surface. Under these conditions, large quantities of phosphorus are released into the water.

The high concentrations of phosphorus in Chesapeake Bay in summer mean that it is not a limiting nutrient at this season (Fig. 17.6). Thus, summer inputs of nitrogen are critical in increasing algal growth and reducing transparency. Similar observations have been made in Long Island Sound (Ryther and Dunston 1971), and it appears that nitrogen, rather than phosphorus, is usually the key nutrient in eutrophication of coastal estuarine and marine waters.

FIGURE 17.7

■ Kelp forests are a marine habitat important for many coastal fish.

Reduced transparency of the water is evidently responsible for the decline of eelgrass beds in Chesapeake Bay between 1960 and the early 1980s (Orth and Moore 1983). Between 1971 and 1979, at mapped study locations in the lower bay, the percentage of stations lacking eelgrass increased from less than 10 percent to more than 60 percent. Loss of eelgrass is probably a contributing factor to the declines of wintering tundra swans and ducks of several species (Orth and Moore 1983, Terborgh 1989). Deoxygenation of the deeper water of the bay has also become catastrophic at times (Seliger et al. 1985), and has probably contributed to the decline of striped bass, shad, oysters, and other valuable aquatic wildlife.

SHALLOW MARINE WATERS

Shallow marine waters derive a high productivity from nutrient inflows from the adjacent land, but they also receive the detrimental impacts of human activity in coastal areas. Coastal waters also possess several distinctive ecosystems, such as the complexly zoned systems of the intertidal and the kelp forests of cold, shallow waters.

Intertidal and Subtidal Environments

Intertidal and subtidal habitats range from sandy beaches and exposed rocky coastlines to protected muddy shores. Many kinds of invertebrates and fish occupy these habitats, some of them of considerable commercial value. Various species of clams occur in sandy and muddy substrates, and mussels and abalones on rocky substrates. Lobsters, crabs, and sea urchins frequent subtidal areas. Larval fish of many species occur in these shallow waters, as well. Other interesting and ecologically important species also occur in these habitats. Rocky intertidal zones, often dotted by tide pools, are one of the temperate environments of greatest biological diversity.

Exploitation of shellfish in intertidal and shallow subtidal areas has severely depleted many populations of clams, abalones, crabs, and lobsters (Weber 1986). Both commercial and non-commercial fishing, along with pollution and physical disturbance of coastal areas, have contributed to these declines. In California, for example, annual commercial harvest of abalones exceeded 1,800 tons as recently as 1968; now less than 400 tons are taken (Tegner 1989). A similar fate has befallen the California spiny lobster, demonstrating the difficulty of managing species attractive for both commercial and sport fishermen in a heavily populated coastal region.

Kelp Forests

Kelp forests, dominated by the giant *Macrocystis* species, occur in coastal ocean waters along the Pacific coasts of North and South America and around various islands of the southern oceans (Foster and Schiel 1985). Beds of smaller kelps also occur in cool or cold waters at other locations in the northern and southern oceans. Kelp forests are home to a rich variety of marine invertebrates, as well as many fish, some associated primarily with the kelp canopy, others with the bottom (Fig. 17.7).

Kelp forests are of considerable economic value. In California, up to 170,000 tons of kelp are harvested annually to obtain algin, a substance used in foods and pharmaceuticals (Foster and Scheil 1985). Shellfish such as abalone, lobster, and sea urchins are harvested in kelp forests, and in southern California about 70 percent of the coastal catch of sport fish comes from kelp forest areas. The red sea urchin, its roe largely exported to Japan, has recently become a major fishery along the North American Pacific coast.

In some areas, such as the coast of southern California, kelp forests have experienced severe declines. Some of these declines may be due to coastal marine pollution, and others to incursions of warm ocean water to which kelps are intolerant. In California, declines have been attributed in some cases to intense grazing by sea urchins, the populations of which may be sustained at unusually high densities by an ability to assimilate dissolved organic substances introduced into coastal waters by sewage discharges.

Recent studies have revealed a fascinating set of relationships among sea urchins, kelp, and the sea otter, a major sea urchin predator. Sea otters were originally widespread throughout the North Pacific and Bering Sea, south to northern Japan and central Baja California, Mexico (Reidman and Estes 1988). They were harvested intensively in the 18th and 19th centuries by Russian fur hunters. The total population was reduced to perhaps 2,000 individuals (G.R. VanBlaricom, pers. comm.), and the species was completely eliminated from much of its range. Under protection, the sea otter is now increasing in numbers and expanding its range. The total population is now estimated to be about 50,000 animals, mostly in Alaskan waters.

Sea otters reach a weight of about 20 to 23 kilograms, and in favorable coastal waters their density can reach 20 to 30 animals per square kilometer. They are voracious predators, feeding on sea urchins, molluscs, crustaceans, and fish, more or less in that order of preference (Riedman and Estes 1988). Individual sea otters consume about 20 to 23 percent of their body weight per day, so that a population in favorable habitat consumes about 35 tons of marine animal life per square kilometer annually.

Studies of the effects of sea otter predation suggest that in many places the species is a **keystone predator,** an animal that can cause an almost total restructuring of the ecosystem. In the Aleutian Islands, for example, sea otters are absent from some islands and present around others (Estes and Palmisano 1974). Around the island of Shemya, where otters are absent, kelp beds are absent and sea urchins large and abundant, ranging from 400 animals per square meter in shallow water to 45 per square meter at depths of about 18 meters. Other invertebrates, such as chitons, mussels, and barnacles are also abundant. In contrast, around Amchitka Island, where otters have been present for many years, kelp beds are dense and sea urchins small and scarce. Other invertebrates are also scarce. Along the southern Alaska coast, where sea otters have appeared as a result of population growth and range expansion, transformation of the coastal ecosystem has been observed in only a few years (Duggins 1980). When otters appear, sea urchin populations are decimated, and dense, single-species stands of kelps of the genus *Laminaria* appear. Although the keystone predation hypothesis may be too simplistic for conditions throughout the potential sea otter range (Foster and Schiel 1988), sea otters are highly influential predators in the coastal marine environment.

One of the sea otter populations that survived the fur hunters was a group of about 50 to 100 animals off the Monterey Peninsula in central California. This population, too, has been growing and is now estimated at about 1,300 animals. The population has also spread southward, to the dismay of fishermen whose livelihood is lobster, abalone, and sea urchin fishing. Some biologists believe that a recent slowing of the spread of the sea otter has been due to deliberate killing of animals. In spite of the success of the species on its own, under protection from commercial hunting, efforts have been made to introduce the sea otter to San Nicholas Island, one of the California Channel Islands. The purpose of this introduction is to establish a population in an area distant from the present California range, in order to reduce the chance of extinction of the species due to a massive oil spill or other catastrophe. Commercial fishermen operating in the vicinity of San Nicholas Island have opposed this introduction because of the probable severe impact of the sea otter on shellfish populations.

CORAL REEF ECOSYSTEMS

Coral reefs are the tropical rain forests of the sea—marine ecosystems of high productivity and enormous biotic richness. Reefs are also, in a sense, oases of productivity in tropical oceans that are otherwise low in fertility. Their productivity results in large measure because the reef is a flowing-water system. Water from the open ocean is continually flowing across the reef to the lagoon, ultimately to return to the open ocean through deep channels that cut through the reef. As water flows across the reef, however, corals and other reef animals capture and filter living and dead particulate matter from the water, thus harvesting organic food and its contained nutrients from a wide area of ocean.

The nutrients trapped by corals and filter feeders contribute to primary production within the reef ecosystem. The producers themselves are often not obvious. Most reef pri-

FIGURE 17.8

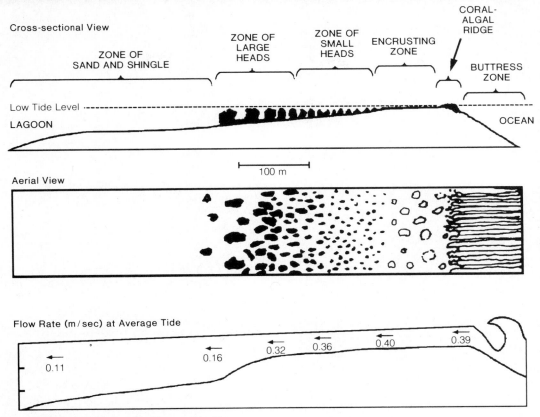

Cross-sectional View

ZONE OF
SAND AND SHINGLE

ZONE OF
LARGE
HEADS

ZONE OF
SMALL
HEADS

ENCRUSTING
ZONE

CORAL-
ALGAL
RIDGE

BUTTRESS
ZONE

Low Tide Level

LAGOON

OCEAN

100 m

Aerial View

Flow Rate (m/sec) at Average Tide

0.11   0.16   0.32   0.36   0.40   0.39

■ Zonation and water flow across a coral reef as illustrated by Japtan Reef, part of Eniwetok Atoll in the Marshall Islands of the western Pacific (Modified from Odum and Odum 1957).

mary production occurs in **algal turfs,** which are low mats of diverse species of algae, some of them encrusting coralline algae that appear more like coral rock than green plant. Other types of algae live in the surface layers of rock, and still others, known as **zooxanthellae,** are single-celled symbiotic organisms that live within the cells of the coral animals. Altogether, these producers contribute to an average net primary production that is higher than that of the oceanic upwelling zones, and only slightly below that of estuaries.

The complex structure of reefs also contributes to the biotic richness of the reef ecosystem. Reefs exhibit a sequence of zones from their ocean face to the backside lagoon (Fig. 17.8). Massive corals form a subtidal buttress zone rising from deep water on the ocean side. These buttresses receive the surge and wave impacts of the open ocean, and by dispersing

this energy protect the inner zones of the reef. Above and behind the buttress zone is often an intertidal coral-alga ridge, which has few living corals and is covered by fragments of coral rock thrown up from the buttress zone by heavy wave action. The coral-alga ridge is home to various encrusting organisms and animals that can hide beneath rock fragments at low tide (Fig. 17.9). Behind the coral-alga ridge are zones of encrusting corals and small coral heads, as the lagoon water deepens. In the deeper waters, massive corals can once again grow, but in this case accompanied by a host of marine organisms that require sheltered waters (Fig. 17.10). Finally, many reefs exhibit sheltered lagoons with a sand and shingle bottom. Each of these zones has a distinctive set of physical conditions and a unique physical structure to which organisms must be adapted. The sequence of diverse microhabitats

FIGURE 17.9

■ The coral-alga ridge of a coral reef is typically covered by fragments of coral rock thrown there by storm waves.

from exposed outer to sheltered inner sides of a reef is analogous to that from the exposed canopy top to the sheltered ground surface in a tropical rain forest.

Fringing or barrier reefs that lie close to continents or islands face growing problems of pollution and physical damage from the activities of human populations. In addition, because of their biotic richness and productivity, reefs are a focus of exploitation, even in remote tropical waters. Worldwide, the most serious threat to reef ecosystems comes from siltation due to soil erosion on nearby land areas. Silt deposition kills most corals and filter-feeding invertebrates by interfering with their systems of food capture and respiration. Locally, sewage discharge has a similar effect. Marine oil pollution can have serious effects on reefs, especially in heavily polluted areas such as the Red Sea and the Persian Gulf (Loya and Rinkevich 1980). Oil appears to stimulate premature

shedding of the planula larvae of corals, retards growth and reproduction, and inhibits colonization of chronically polluted areas. Mining of the coral rock of reefs for construction material and for production of cement, dredging of shipping channels, and the filling of shallow coastal waters to create land for airports and other purposes increasingly take a toll of reef areas in heavily populated coastal areas. On a smaller scale, physical damage to reefs from boat anchors and human trampling also affects many heavily visited reefs.

Exploitation of certain corals and other reef animals has also depleted their populations over large areas. In the Philippine Islands, collecting small reef fish for the aquarium market has become a major form of reef exploitation. One of the common collecting techniques is squirting a solution of sodium cyanide into large corals to stun the fish so they can be captured. Unfortunately, the cyanide often kills much of the coral. Often, the fish themselves are killed or weakened, and many frequently die soon after they are sold. This very wasteful practice has virtually destroyed some reefs.

FIGURE 17.10

■ Sheltered waters on the inner side of a coral reef are home to many kinds of corals, fish, and other animals.

One of the most widespread threats to reefs in the Pacific and Indian Oceans in recent decades has been the crown-of-thorns starfish, *Acanthaster planci* (Fig. 17.11). This starfish reaches a diameter of 0.5 m, and possesses 13 to 16 arms with sharp spines up to 5 cm in length. The crown-of-thorns is normally scarce, with densities of a few individuals per square kilometer. This voracious coral-eating starfish has exhibited devastating outbreaks that have caused the death of most of the large, reef-building *Acropora* corals over vast areas. During outbreaks, densities can reach 100,000 starfish per square kilometer. Full recovery of reefs from such damage requires about 20 to 40 years. In effect, the crown-of-thorns starfish is a keystone species in the coral reef ecosystem. In this case, the role of man as a causal agent is still uncertain.

Major outbreaks of the crown-of-thorns were first noted in 1962 at Green Island Reef, in the Australian Barrier Reef. From here, over the next 12 years, outbreaks spread throughout much of the Pacific and Indian Oceans. In the late 1970s, about when it was thought that the crown-of-

thorns had run its course, a new wave of outbreaks began—one that has continued into the late 1980s. Many reefs were struck again by this second wave, with Green Island being ravaged again in 1981, for example.

Several theories of the cause of crown-of-thorns outbreaks have been offered. Some, such as pesticide pollution that kills predators of small starfish or physical damage to reefs that favors settlement of larval starfish, have received no substantial scientific support. One marine ecologist (Endean 1982) has suggested that widespread depletion of a large carnivorous snail, the triton snail, *Charonia tritonis,* and perhaps other such predators, has freed the crown-of-thorns from natural biological control. The triton can kill and consume crown-of-thorns starfish, but many marine ecologists doubt that this predator was originally abundant enough to have been a strong control on the starfish.

FIGURE 17.11

■ The crown-of-thorns starfish, a voracious predator on stony corals, has shown long-term outbreaks in the Pacific and Indian Oceans in recent decades.

Another theory is that unusually high survival of larval starfish during their planktonic life permits the settlement of juvenile starfish on reefs in abnormally large numbers. Birkeland (1982) postulated that heavy rainfall events might wash large quantities of nutrients into coastal waters, fostering a bloom of plankton that provide a rich food source for the crown-of-thorns larvae. He noted that outbreaks in several instances occurred three years after extreme rainfall events in Samoa, Guam, and Palau, and that outbreaks were also more frequent around islands with nutrient-rich volcanic soils than around atolls formed of nutrient-poor coral sand and shingle. Other workers, however, have noted that outbreaks do occur on isolated reefs that are not near any such source of nutrients. Recent experiments (Olson 1987) also question whether crown-of-thorns larvae are food-limited during their planktonic life. However, the pattern of spread of outbreaks suggests that some factor has favored unusual survival of larvae of this species (Moran 1987).

Estuarine and Marine Preserves

The marine environment is the last of the major divisions of the biosphere to receive formal protection through the designation of parks, preserves, and sanctuaries. The coastal marine environment has now begun to receive such protection. The United States has initiated systems of national estuarine and marine sanctuaries. Many other countries have also begun to designate marine parks and preserves.

The United States National Marine Sanctuary and National Estuarine Research Reserve Programs were established in 1972. By 1987, seven National Marine Sanctuaries had been designated (Foster and Archer 1988). One of these, the sunken wreck of the U.S.S. Monitor, a Civil War ship, is a historical monument. Four of the remaining sites preserve coral reefs, three in the southeastern states and one in American Samoa. The remaining two sites, near the California Channel Islands and surrounding the Farallones Islands off San Francisco, preserve coastal water areas used heavily by marine birds and mammals. The National Estuarine Research Reserves emphasize long-term monitoring and ecological research. Seventeen of these reserves have been designated at representative locations along the Atlantic, Gulf, and Pacific coasts, and in Hawaii and Puerto Rico. Additional sites are being considered for addition to this system.

Many other countries have begun to designate protected coastal and marine areas of various types. By the late 1980s, about 430 marine preserves had been created in 69 countries (World Resources Institute 1988). Australia, for example, has established the Great Barrier Reef Marine Park, covering an area of 350,000 square kilometers of ocean and extending along 2000 km of coastline (Morris 1983). This park is zoned in a comprehensive way, with activities such as fishing being permitted in some areas but not in others. In many cases, however, designated preserves lack effective protection and active management. In other cases, the primary purpose of the preserve is tourism, and heavy use by visitors leads to physical damage and overfishing.

Key Management Strategies for Estuaries, Coastal Waters, and Reefs

1. *Regulation of the quantity and quality of freshwater discharges into coastal marine environments to maintain favorable fertility and salinity relationships and avoid detrimental eutrophication and siltation effects.*

2. *Management of keystone species to keep their influence on the biotic structure of coastal ecosystems within defined limits.*

Literature Cited

Birkeland, C. 1982. Terrestrial runoff as a cause of outbreaks of *Acanthaster planci* (Echinodermata: Asteroidea). *Mar. Biol.* **69:**59–64.

Cederwall, H. and R. Elmgraen. 1990. Biological effects of eutrophication in the Baltic Sea, particularly the coastal zone. *Ambio* **19:**109–112.

Copeland, B. J. 1966. Effects of decreased river flow on estuarine ecology. J. Water Poll. *Control Fed.* **38:**1831–1839.

D'Elia, C. F. 1987. Nutrient enrichment of the Chesapeake Bay. *Environment* **29:**6–11, 30–33.

Duggins, D. O. 1980. Kelp beds and sea otters: An experimental approach. *Ecology* **61:**447–453.

Endean, R. 1982. Crown-of-thorns starfish on the Great Barrier Reef. *Endeavour* **6:**10–14.

Estes, J. A. and J. F. Palmisano. 1974. Sea otters: Their role in structuring nearshore communities. *Science* **185:**1058–1060.

Fortes, M. D. 1988. Mangrove and seagrass beds of East Asia: Habitats under stress. *Ambio* **17:**207–213.

Foster, M. S. and D. R. Schiel. 1985. The ecology of giant kelp forests in California: A community profile. *U.S. Fish Wildl. Serv. Biol. Rep.* 85(7.2).

Foster, M. S. and D. R. Schiel. 1988. Kelp communities and sea otters: Keystone species or just another brick in the wall? Pp. 92–115 in G. R. VanBlaricom and J. A. Estes (Eds.), *The community ecology of sea otters*. Springer-Verlag, Berlin.

Foster, N. M. and J. H. Archer. 1988. The National Marine Sanctuary Program—policy, education, and research. *Oceanus* **31(1):**4–12.

Herbold, B. and P. B. Moyle. 1989. The ecology of the Sacramento-San Joaquin Delta: A community profile. *U.S. Fish Wildl. Serv. Biol. Rep.* 85(7.22).

Loya, Y. and B. Rinkevich. 1980. Effects of oil pollution on coral reef communities. *Mar. Ecol. Prog. Ser.* **3:**167–180.

Moran, P. 1987. The *Acanthaster* phenomenon. *Oceanogr. Mar. Biol. Annu. Rev.* **24:**379–480.

Morris, G. C. 1983. The Great Barrier Reef Marine Park: A unique management concept. *Parks* **8(3):**1–4.

Nichols, F. H., J. E. Cloern, S. N. Luoma, and D. H. Peterson. 1986. The modification of an estuary. *Science* **231:**567–573.

Odum, E. P. 1980. The status of three ecosystem-level hypotheses regarding salt marsh estuaries: Tidal subsidy, outwelling, and detritus-based food chains. Pp. 485–496 in V. S. Kennedy (Ed.), *Estuarine perspectives*. Academic Press, New York.

Odum, H. T. and E. P. Odum. 1955. Trophic structure and productivity of a windward coral reef community on Eniwetok Atoll. *Ecol. Monogr.* **25:**291–320.

Odum, W. T., C. C. McIvor, and T. J. Smith, III. 1982. *The ecology of the mangroves of south Florida: A community profile.* U.S. Fish Wildl. Serv., Office of Biol. Services, Washington, D.C. FWS/OBS-81/24.

Olson, R. R. 1987. In situ culturing as a test of the larval starvation hypothesis for the crown-of-thorns starfish, *Acanthaster planci. Limnol. Oceanogr.* **32:**895–904.

Orth, R. J. and K. A. Moore. 1983. Chesapeake Bay: An unprecedented decline in submerged aquatic vegetation. *Science* **222**:51–53

Reidman, M. L. and J. A. Estes. 1988. A review of the history, distribution and foraging ecology of sea otters. Pp. 4–21 *in* G. R. VanBlaricom and J. A. Estes (Eds.), *The community ecology of sea otters.* Springer-Verlag, Berlin.

Ryther, J. H. and W. M. Dunstan. 1971. Nitrogen, phosphorus, and eutrophication in the coastal marine environment. *Science* **171**:1008–1013.

Seliger, H. H., J. A. Boggs, and W. H. Biggley. 1985. Catastrophic anoxia in the Chesapeake Bay in 1984. *Science* **228**:70–73.

Tegner, M. J. 1989. The California abalone fishery: Production, ecological interactions, and prospects for the future. Pp. 401–420 *in* J. F. Caddy (Ed.), *Marine invertebrate fisheries: Their assessment and management.* John Wiley and Sons, New York.

Terborgh, J. 1989. *Where have all the birds gone?* Princeton Univ. Press, Princeton, NJ.

Thayer, G. W., D. A. Wolfe, and R. B. Williams. 1975. The impact of man on seagrass systems. *Amer. Sci.* **63**:288–296.

Thayer, G. W., W. J. Kenworthy, and M. S. Fonseca. 1984. *The ecology of eelgrass meadows of the Atlantic coast: A community profile.* U.S. Fish Wildl. Serv. FWS/OBS-84/02.

Weber, M. 1986. Federal marine fisheries management. Pp. 267–344 *in* R. L. Di Silvestro (Ed.), *Audubon wildlife report 1986.* National Audubon Society, New York.

World Resources Institute. 1988. *World resources 1988–89.* Basic Books, Inc., New York.

Zedler, J. B. 1982. *The ecology of southern California coastal salt marshes: A community profile.* U.S. Fish Wildl. Serv. Biol. Serv. Program, Washington, D.C. FWS/OBS-81/54.

Special Problems of Aquatic Ecosystems

Like terrestrial ecosystems, the aquatic environment exhibits a number of special problems. Marine mammals and sea birds, for example, have been among the most carelessly and severely exploited animals on earth. Efforts to protect and manage them have likewise become among the most controversial issues of international conservation. Like terrestrial ecosystems, the aquatic portion of the biosphere has experienced general impacts that affect many specific types of ecosystems. Foremost among these are pollution by various chemicals and synthetic materials such as plastics, and deposition of massive amounts of acidic compounds released by combustion processes and other human activities.

CHAPTER 18

Marine Mammals and Birds

Few groups of animals have held greater attraction for humans than marine mammals. Aristotle and Pliny the Elder wrote accounts of the natural history of dolphins and whales in ancient times. The greatest attraction to marine mammals, however, soon became economic: oil, baleen, pelts, and meat. The lure of gain from whaling and sealing drew men in sailing ships to the wildest regions of the arctic and antarctic oceans—many to their deaths—and sustained commercial whaling until near the end of this century. More recently, the seeming intelligence and friendliness of certain marine mammals have gained them increasing prominence in literature, art, and folklore. These characteristics have also helped to give the group as a whole a special prominence in the efforts of animal protectionists and conservation ecologists. Some marine mammals have, in fact, become rallying standards for popular conservation efforts.

Marine mammals belong to four orders (Table 18.1). The Cetacea, or cetaceans, include 78 species of whales, dolphins, and porpoises (Gaskin 1982, Perrin 1989). The Pinnipedia, or pinnipeds, comprise 33 species of seals, sea lions, and the walrus (Riedman 1990,

Reeves et al. 1991). The Sirenia, so-named because of their fancied relation to the sirens of Greek mythology, include four living species of manatees and dugongs. Two species, the polar bear and sea otter, belong to the Carnivora, an order containing the major families of flesh-eating land mammals. All of these groups contain species that are or were seriously endangered.

Seabirds range widely, visiting the most distant parts of the world oceans, and the impressive nesting colonies that they form have also attracted human interest and exploitation. They likewise belong to several orders, most importantly the Sphenisciformes (penguins), Procellariiformes (albatrosses, shearwaters, and petrels), Pelecaniformes (pelicans, boobies, gannets, cormorants, and frigate birds), and Charadriiformes (auks, puffins, gulls, and terns).

We shall consider first the history of whaling and the current status of efforts to protect and manage populations of the great whales. Then we shall examine conservation issues for other groups of marine mammals and seabirds.

WHALES AND WHALING

The great whales, the principal objects of commercial whaling, are derived from an early evolutionary line allied with ungulates. Whales first appeared about 70 million years ago, following the disappearance of giant marine reptiles. By 30 million years ago, the two modern groups of whales—toothed and baleen whales—had appeared. Of the modern great whales, one, the sperm whale, is a toothed whale and the remaining ten are baleen whales (Table 18.2).

The sperm whale, a medium-sized species, is a toothed whale that feeds on larger fish and squid (Fig. 18.1). The largest males rarely exceed 16 m in length, and few females exceed 11 m. Sperm whales largely confine their activities to the open temperate and tropical oceans. They are deep divers, and can remain submerged for up to two hours. Adult females travel in breeding herds of 10 to 15 animals, which are visited for short periods by individual bulls for mating. Nonbreeding males travel in bachelor herds. The present population is thought to be of the order of a million animals, but this estimate is highly speculative. Two related species, the dwarf and pygmy sperm whales, are poorly known and probably rare forms.

The baleen whales lack true teeth, and feed mostly on krill, small fish, and other small marine animals, which they capture by use of the specialized **baleen apparatus** suspended from their upper jaws. This structure, made of fibrous plates, acts as a filter. Several of the baleen whales have "pleated" throats that are enormously expandable, enabling them to take a large volume of water, containing food organisms, into their mouth cavity. The water is then expelled through the baleen plates, so that the food organisms are retained and swallowed. Others, such as the gray and right whales, take food-rich materials into the baleen cavity by sucking or skimming. Baleen whales seem to have a promiscuous breeding system and live either alone or in herds containing varying numbers of individuals of one or both sexes.

Baleen whales are varied in size and distribution (Fig. 18.2). They frequent coastal waters and upwelling zones where their food is abundant. The largest is the blue whale, which reaches 30 m in length and 160 tons in weight—the largest animal ever to live on earth. Whaling has reduced this species to less than a tenth of its original number. It is also one of the most endangered of the great whales, with its total number probably being 5,000 to 10,000. Like all of the baleen whales, it is migratory, moving to higher latitudes in the summer. The fin, northern and southern right, bowhead, sei, humpback, Bryde's, and gray whales are progressively smaller species, all of which were hunted heavily and are greatly reduced in abundance compared to their original numbers. The

TABLE 18.1

Major Groups of Living Marine Mammals

GROUP	NUMBER OF SPECIES	NUMBER ENDANGERED U.S. LISTING	IUCN
Order Cetacea.			
Whales, Dolphins, and Porpoises.			
Baleen Whales	11	7	4
Sperm Whales	3	1	
Beaked and Bottlenose Whales	19		
Beluga and Narwhal	2		
Pilot and Killer Whales	6		
River Dolphins	5	1	2
Porpoises	6		
Marine Dolphins	26		
Order Pinnipedia.			
True Seals, Sea Lions, Fur Seals, and Walrus.			
True Seals	18	3	
Sea Lions and Fur Seals	14		
Walrus	1		
Order Sirenia.			
Sirenians.			
Manatees	3	2	
Dugong	1	1	
Order Carnivora.			
Carnivores.			
Sea Otter	1	1 (partial)	
Polar Bear	1		

FIGURE 18.1

A sperm whale is butchered at a whaling station in Taiji, Japan. The sperm whale, largest of the toothed whales, is an open-ocean, deep-diving animal that feeds primarily on squid.

smallest of the commercial baleen whales is the minke whale, which is just over 9 m in maximum length and about 10 tons in maximum weight. Minke whales, the least persecuted and most abundant of the baleen whales, are believed to number about 725,000, and may be about twice as abundant now as before commercial whaling began. The smallest baleen whale, the pygmy right whale, is a rare, poorly known species that reaches a length of about 6 m.

The objectives of whaling have been to obtain meat, oil, baleen, and other materials (Allen 1980). Aboriginal peoples have long sought whales for subsistence harvests of meat, blubber, and oil, and this type of whaling is still practiced by a few groups, such as the Inuit of Alaska, Canada, and Greenland. Early commercial whaling concentrated on oil for use in lamps and candles and for lubrication. The sperm whale yielded oil that was prized for use in oil lamps and as a high-quality industrial lubricant. **Spermaceti,** a waxy material from the oil deposit in the whale's head, was used in making the finest candles. A ton or more of spermaceti could be obtained from a large male sperm whale. Another substance, **ambergris,** formed in the intestine of the sperm whale, was used as a chemical stabilizer in perfumes. In modern time, sperm whale oil was used as an industrial lubricant, but the meat was used mostly as animal feed. Baleen whales were sought originally for their oil, used in lamps, and for baleen. A large right whale could yield over 300 barrels of oil. The cartilaginous baleen apparatus was also used to manufacture corset stays, buggy whips, and other products. In recent years, the oil of baleen whales was used in margarine and cosmetics. The meat

TABLE 18.2

Distribution and Estimated Original and Present Populations of Great Whales

SPECIES	RANGE	MAXIMUM LENGTH (M)	ORIGINAL POPULATION	PRESENT POPULATION
Baleen Whales				
Gray	N. Pacific	14	(24,000)	21,000
Minke	Cosmopolitan	10	(350,000)	(725,000)
Sei	Cosmopolitan	18	(250,000)	(50,000)
Bryde's	Cosmopolitan	15	(92,000)	(90,000)
Blue	Cosmopolitan	30	(200,000)	(10,000)
Fin	Cosmopolitan	26	(500,000)	(150,000)
Humpback	Cosmopolitan	15	(125,000)	(10,000)
Bowhead	Arctic Ocean	18	(65,000)	9,000
N. Right	N. Temperate and Polar Oceans	18	(50,000)	1,000
S. Right	S. Temperate and Polar Oceans	18	(100,000)	3,000
Sperm Whales				
Sperm	Temperate and Tropical Oceans	18	(2,000,000)	(1,000,000)

Source: Based on various sources. Population figures in parentheses represent rough estimates for present populations, and no more than educated guesses for original populations.

FIGURE 18.2

■ A California gray whale, one of the baleen whales, "spy-hopping" in San Ignacio Lagoon, Baja California, Mexico. The function of this behavior is uncertain. Gray whales migrate to coastal lagoons and bays in this region in winter to calve.

of some baleen and toothed whales has become a popular food in Japan.

Commercial whaling began in the 12th century, when Basque fishermen began hunting the right whales in the Bay of Biscay, off the northern coast of Spain (Gambell 1976). Whaling spread to several other North Atlantic locations in the 17th and 18th centuries, and to the rest of the world oceans in the 19th century. The greatest destruction of whales, however, has occurred during the modern era. Modern whaling dates from the development of steam-powered ships and the harpoon gun in the 19th century. These advances enabled whalers to hunt whales effectively in the most remote areas of the oceans. Modern harpoon guns can fire either a "cold" harpoon, lacking an explosive charge, or a harpoon with an explosive charge that kills or stuns the animal, facilitating its capture. The explosive harpoon is regarded as the more humane version, since it kills the animal quickly. In recent years, use of the explosive harpoon was required for all species except the small minke whale, this exception reflecting the fact that an explosive charge damaged such a high percentage of the flesh of this species.

Whaling harvests increased until the mid-1900s. The harvest tonnage peaked in 1930–1931, when 3.6 million tons were taken. The numbers of whales taken continued to increase until the early 1960s, when harvests reached about 90,000 animals per year. This pattern reflected the progressive overexploitation of species from large to small. Beginning in the 1930s, the average size of the animals taken began to decline, reflecting the overexploitation of the largest species and individuals. In 1903, the average whale taken weighed 66 tons, in 1970 only 23 tons.

Concern over the rapid decline of whale stocks, beginning in the 1930s, led to the formation of the **International Whaling Commission (IWC)** in 1946 (Gambell 1990). The members of this commission are primarily nations with territorial waters containing whales, and include almost all whaling nations (except Portugal), as well as many nonwhaling countries. In 1979, the commission designated a sanctuary zone in the Indian Ocean, giving partial protection to several species, and set quotas on the remaining species. In the 1970s and early 1980s, quotas were reduced rapidly, as well as being shifted more and more onto the minke whale, which had not been heavily exploited because of its small size. This reduction culminated in adoption of zero quotas for a 5-year period following the 1984–1985 whaling season. Russian and Japanese whalers continued to operate until the 1986–1987 whaling season, however, and Japan, Norway, and Iceland continued to take small numbers of several species in so-called "scientific" harvests (Fig. 18.3). A boycott of Icelandic fish products finally forced Iceland to abandon

FIGURE 18.3

■ Iceland continued to take a scientific harvest of small numbers of fin whales and other species until 1989.

scientific harvests in 1989. In 1990, Japan, Norway, and Iceland petitioned the IWC for quotas on certain species, and the IWC voted against non-zero quotas until the data needed to devise sound management plans for the more abundant species become available. Iceland has threatened to withdraw from the IWC if quotas are not granted in 1991.

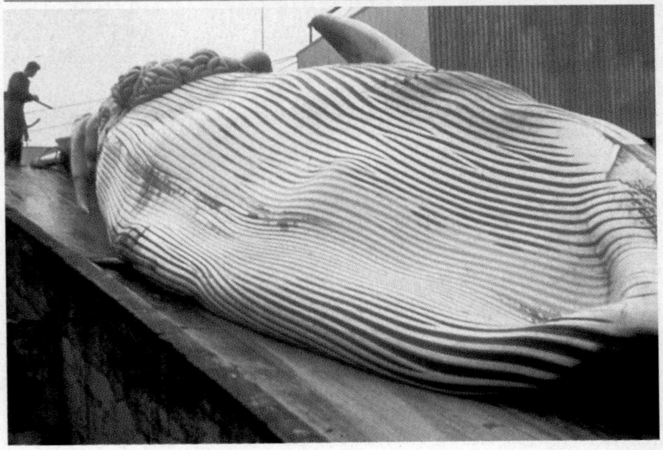

Due to the great difficulties of estimating the numbers of diving marine animals, the present populations of the great whales are only roughly known, and their original numbers can only be guessed at (Table 18.2). Even so, the present numbers of all species except the northern right whale appear to be at least a few thousand, and their recovery appears to be possible if the moratorium on harvest is indeed respected. Some economists believe that once terminated, whaling will be uneconomical to resume due to the high costs of outfitting vessels for the enterprise and the fact that alternate sources of oil and meat will have become entrenched in the marketplace. Continued interest in whaling by several countries, however, indicates that this may not be true.

ABORIGINAL WHALING

Aboriginal whaling is still practiced for several species of the great whales. In Alaska, hunting of bowhead whales is carried out in eight villages near and above the Bering Strait (Braham 1989). Whaling by Inuit people predates western contact, and was carried out perhaps as long as 4,000 years ago. The original technique was to pursue whales in sealskin boats and strike them with stone-bladed spears to which skin ropes and sealskin floats were attached.

FIGURE 18.4

■ In spring, Alaskan Eskimos capture small numbers of bowhead whales in open-water "leads," using sealskin boats and explosive harpoons thrown by hand or shot from shoulder guns.

Whaling by the Alaskan Inuit uses a mix of modern, 19th century, and traditional techniques (Fig. 18.4). The bowhead whales are captured with a small explosive harpoon that is thrown by hand or shot from a shoulder gun. Whaling teams operate during the spring, when the offshore ice begins to shift and retreat. At this time, movements of the ice open up long gaps, or "leads," through which the bowheads migrate. Whaling teams haul their boats, usually made of sealskins sewn over a wooden frame, out to the edge of these leads, which are up to several miles from shore. When a whale appears, the boats are launched and driven by paddle or sail to a point from which the harpoon can be thrown or shot. If the whale is taken, it is hauled onto the ice and butchered. Some whaling is also pursued in fall, during the return migration period.

During the late 1970s, concern over the decline in whale populations coincided with a sharp increase in whaling activity by the Alaskan Inuit. This increase reflected the prestige of whaling and the fact that more money was flowing into the villages from governmental programs and activities such as petroleum development. Increased affluence enabled many individuals to acquire equipment and funds to support a whaling crew, in spite of a lack of whaling experience. One consequence of increase in whaling activity was an increase of the total number of whales taken from 10 to 15 per year to 25 to 50 per year. A second was an even greater increase in the number of animals struck with explosive harpoons, but not recovered. In 1977, for example, 79 animals were struck and lost. As a result, in the late 1980s, Inuit whaling was restricted by quotas both on the number of animals taken and on the number struck. The most recent quota (1988–1989) allows a maximum annual harvest of 41 animals or a maximum of 46 animals struck.

The broader concern over Alaskan Inuit whaling centers on the fact that the bowhead may still be one of the most depleted of great whales. Populations in four other northern ocean areas probably amount to only a few hundred individuals each (Braham 1989). That of Alaskan waters, estimated at 7,800 individuals, is the only healthy population.

Aboriginal whaling is also conducted for minke and humpback whales along the western coast of Greenland and for sperm and some baleen whales in Indonesia. A similar form of traditional whaling was practiced for humpback whales near the Lesser Antillean island of Bequia, with an annual quota of three animals, until 1990.

SMALL CETACEANS

Several species of small cetaceans have now become more endangered than most species of great whales, and once important populations of many species have been extirpated or threatened with extirpation (Brownell et al. 1989, Wursig 1989). Annually, more than 150,000—perhaps as many as 500,000—small cetaceans are killed, deliberately or incidental to other fishery activities. Others are captured alive for research or display (Meith 1984). Populations of several species have suffered from depletion of their foods or disturbance of their habitat. The International Whaling Commission has generally ignored the conservation issues of small cetaceans, except for certain populations of bottlenose and killer whales that have been taken by commercial whalers.

Hunting some species of small cetaceans for meat has been carried out for centuries or millennia. North American and Greenland Inuit hunt the beluga (white) whale and the narwhal (Hertz and Kapel 1986). In recent decades, commercial whaling vessels equipped with harpoon guns also

FIGURE 18.5

■ Pantropical spotted dolphins are one of the species that swim in association with schools of yellowfin tuna in the eastern tropical Pacific.

shifted their attention to some of the smaller whales, including the killer whale in 1979–1980 and the Baird's beaked whale more recently. Major harvests of several species of dolphins and porpoises are taken by Japan in the western Pacific. In the late 1980s, after the moratorium in whaling was implemented, Japanese fishermen severely overexploited Dall's porpoises over a three-year period. Along the coast of Chile, killing of dolphins and seals to obtain meat for baiting crab traps has nearly eliminated the Commerson's dolphin, one of the least common delphinids, from this region.

Hunting harvests, however, are second in total numbers to the mortality of several other species incidental to tuna, squid, and salmon fishing. In the eastern Pacific Ocean, schools of spinner, striped, and spotted dolphins often swim above schools of tuna (Fig. 18.5). Tuna boats equipped with purse seines use these schools of dolphins to locate tuna, and set their nets around both tuna and dolphins. Many dolphins are trapped and drowned when the net is closed. In 1972, purse seining killed over 420,000 dolphins (Hofman 1990). Although techniques have been developed to permit the release of dolphins over the edge of the net before it is finally hauled aboard (Fig. 18.6), losses still remain high. The Marine Mammal Protection Act of 1972 placed a quota of 20,500 kills by United States tuna boats. United States boats have usually managed to keep within this limit, although it was reached both in 1976 and 1986. Many animals are killed in this fashion by tuna boats of other countries, however, and in 1986 and 1987 the total dolphin mortalities incidental to tuna fishing were 125,000 and 112,000 animals, respectively

(Hofman 1989). In 1990, several tuna packing companies announced that they would not market tuna captured by setting purse seines around herds of dolphins, and that their products would be labelled "dolphin safe." The impact of this action on dolphin mortality remains to be seen.

Drift gill-netting for tuna, squid, and salmon in the Pacific is similarly estimated to kill tens of thousands of small cetaceans (mainly right whale dolphins and Dall's porpoises) annually. In this case, the animals simply become entangled in the nets and drown. Efforts to restrict or eliminate drift gill-netting (See chapter 16) may reduce these kills.

In the northwestern Atlantic, harbor porpoises have fallen victim to gill nets in the Bay of Fundy (Read and Gaskin 1988). These nets are anchored at depths of 35 to 100 m to capture groundfish. The estimated annual mortality in these nets is about 7.5 percent of the minimum population, which is close to the maximum rate of potential population growth for animals of this type. Thus, net mortality could be a major contributor to population decline. Gill-netting along the California coast has caused mortality of several marine mammals, including some whales. In the Gulf of California, the endemic vaquita, a close relative of the harbor porpoise, is now on the verge of extinction due to heavy mortality in gill nets used to catch totoaba, a small tuna (Brownell et al. 1989).

Several other cetaceans have become endangered as a result of fishing mortality or disturbance of their river, estuary, or mangrove swamp habitats (Meith 1984, Brownell et al. 1989). In Asia, the endemic river dolphins in the Indus and Yangtze Rivers have been reduced to a few hundred

FIGURE 18.6

■ Mortality of dolphins in purse seining for tuna can be reduced by a "backing down" procedure in which the surface edge of the net is drawn down to allow them to escape before the net is brought on board.

individuals or fewer, and those of the Ganges and Brahamaputra to a few thousand. Both species of humpbacked dolphins, which inhabit estuaries and mangrove swamps have been severely affected by habitat destruction and disturbance. In the lower St. Lawrence River, Canada, the beluga whale may be succumbing to the effects of chemical pollution from urban-industrial sources (See chapter 19).

PINNIPEDS

Almost all species of pinnipeds have been hunted for meat, oil, fur, or ivory, although only three species of monk seals (one, the Caribbean monk seal, is probably extinct) are classified as endangered by the United States Fish and Wildlife Service (Fig. 18.7). The Mediterranean monk seal, with a declining population of about 300 individuals, is severely threatened (Reeves et al. 1991).

Two species of pinnipeds have been exploited on a commercial basis in the Northern Hemisphere in recent decades: the northern fur seal and the harp seal. Northern fur seals, which breed almost entirely in the Pribilof Islands of the Bering Sea, were first exploited in the 18th century by Russian fur hunters. During the 18th and 19th centuries most fur seals were hunted at sea, where both males and females were taken. The result of heavy exploitation was serious depletion of fur seal numbers. In 1911, under a treaty between the United States, the USSR, Japan, and Canada, pelagic sealing was halted, and an annual harvest of young male fur seals was scheduled to be taken from the Pribilof Islands. The proceeds from sale of pelts was divided among the treaty nations, with the meat being utilized by the Aleut residents of the islands, who also provided the work force for the hunts. The annual harvest of fur seals under this arrangement was 20,000 to 60,000 animals annually, from a population of 2.0 to 2.5 million animals.

For many years, the harvest of fur seals was considered to be one of the most successful programs of sustained yield management of a wild animal population. However, in the early 1970s the numbers of fur seals unexpectedly began to decline, falling to about 900,000 animals in the late 1980s (D. DeMaster, pers. comm.). Between 1973 and the early 1980s the decline in young fur seals born in the Pribilof colonies averaged 6 percent annually. This decline, together with decreasing profitability of the harvest and pressure from environmentalists, led to a cessation of the harvest in 1983. The exact cause of the population decline is still uncertain. The population appears to be below carrying capacity, the birthrate of young is high, parental feeding seems to be normal, and the animals are healthy (Loughlin et al. 1987). The most likely cause is high mortality of fur seals in drift gill nets and by entanglement in discarded fragments of trawl nets used for other fish in the Bering Sea and North Pacific (Fig. 18.8). Estimates of possible net mortality are as high as 30,000 animals per year (Alexander 1986). Disease, chemical pollutants, predation, depletion of seal foods by commercial fishing, and the lingering effects of an experimental harvest of female seals from 1956 to 1968 have not been ruled out as partial causes, however (Scheffer and Kenyon 1989).

In the Bering Sea, populations of Steller's sea lions (Stirrup 1990) and harbor seals (Pitcher 1990) have also declined. Net entanglement is believed to be less common for these species than for fur seals. Greatly increased harvests of finfish, particularly pollock, in the feeding grounds of these species may be the prime contributor to their decline.

Harp seals, which occur in the North Atlantic, have also been harvested in large numbers. This species is quite abundant, with an estimated total population of 2.2 to 3.0 million (Ronald and Dougan 1982). In this case, commercial harvest has largely been centered on the newborn pups, because of their dense, white fur. Young seals are taken in early spring on the offshore ice pack, where they are born. Until 1983, about 50,000 animals were harvested in Norwegian and

FIGURE 18.7

■ The population of the Hawaiian monk seal, classified by the United States as endangered, is about 2,000 individuals or fewer.

FIGURE 18.8

■ Entanglement in active fishing nets and discarded net fragments may be a major cause of recent declines in northern fur seal numbers in the Pribilof Islands.

FIGURE 18.9

■ A walrus breeding colony.

Russian waters and about 170,000 taken in Canadian waters. An additional 10,000 or so harp seals were also taken by Canadian Inuit, so that the overall harvest of harp seals was about 230,000. Strong pressure from environmental and animal rights groups led the European Economic Council to suspend the import of white harp seal pelts in 1983. The Canadian harvest of harp seal pups has continued at a lower level, and increasing numbers of older seals, which have lost the white coat, are now being taken. In 1990, about 70,000 harp seals were taken in Canada (K. Ronald, pers. comm.).

The walrus, limited in distribution to the Arctic Ocean, has been harvested by Inuit and other northern peoples for meat, skin, and ivory (Fig. 18.9). The world population of the species is estimated at 235,000 to 245,000, and the annual harvest in Asia and North America at about 12,000 (MacKay 1987). The worldwide increase in value of ivory has also affected walrus ivory, which, in 1989, brought about $90 per kilogram. Reports in 1989 indicated that an average pair of walrus tusks was being sold for $500 to $700 in Alaska.

OTHER MARINE MAMMALS

Manatees and dugongs, which occupy fresh and saltwater areas along tropical and subtropical coastlines (Fig. 18.10), have been greatly reduced in numbers, largely by habitat distur-

bance and hunting for their meat. In the 1980s, the Florida population of the West Indian manatee had been reduced to about 1,000 animals, compared to an estimated 10,000 animals in the 1940s (St. Aubin and Lounsbury 1990). In addition to drainage and conversion of its habitat, Florida manatees now suffer heavy mortality from collisions with motorboats (Fig. 18.11). Another member of this group, the Steller's sea cow, occurred in the North Pacific but, in the late 1700s, was hunted to extinction for its oil.

Other marine mammals, the polar bear and sea otter, were discussed in detail in chapters 10 and 17.

MARINE MAMMALS AND FISHERY CONFLICTS

The **Marine Mammal Protection Act** of 1972 established a moratorium on the harvest or intentional killing of pinnipeds, sea otters, and other marine mammals in United States waters, along with the importation of such animals or their products (Hofman 1989). Exceptions were made for limited "unintentional" killing of porpoises in tuna fishing, the Inuit harvest of certain species, harp seal harvests in Canada, and captures for research and public display. The act has since been amended to cover other unintentional harvests. The act further specifies that the populations of these animals are to be

FIGURE 18.10

■ Only about 1,000 West Indian manatees remain in their Florida range.

FIGURE 18.11

■ Injuries from boat propellers are a major cause of mortality of manatees in Florida.

FIGURE 18.12

■ Populations of northern elephant seals, such as these animals on San Miguel Island, and several other pinnipeds have increased greatly off the California coast since passage of the Marine Mammal Protection Act in 1972.

managed to maintain "optimal sustainable populations," a level defined as lying between the maximum supportable population and that at which annual growth is greatest. Not surprisingly, under this protection, marine mammal populations have increased substantially in many locations. In California, for example, populations of California sea lions, harbor seals, northern fur seals, and elephant seals have grown rapidly (Fig. 18.12). California sea lions have increased from about 3,000 to 4,000 animals in 1930 to about 90,000 animals in the late 1980s (D. DeMaster, pers. comm.). Elephant seals, reduced to about 100 animals on Guadalupe Island, Mexico, in the early 1900s, have increased in numbers and spread north into California (Cooper and Stewart 1983). The California population of elephant seals had grown to about 120,000 animals in the late 1980s. The Guadalupe fur seal is now increasing in numbers, and could move into California in the near future.

The growing numbers of these species, together with the activities of sea otters and several small cetaceans, are leading to conflicts between marine mammals and fishermen. In effect, the Marine Mammal Protection Act gave priority for protection to the top carnivores of the marine ecosystem (Manning 1989). In 1984, pinnipeds along the California coast consumed 953,000 tons of sea life, nearly five times the 202,000 tons harvested by commercial fishermen in the same area. A similar situation prevailed in the Bering Sea in the 1970s, where pinnipeds alone consumed nearly 2.8 million tons of finfish (Lowry and Frost 1985). In addition to the consumption of finfish and shellfish, the larger mammals often rob or damage nets and other fishing gear. Along the California coast, such losses amounted to about $600,000 in 1980 (DeMaster et al. 1985). Some of the mammals that breed in California also migrate north into Oregon, Washington, and British Columbia, where they prey on salmon and other fish.

Similar problems of competition between marine mammals and human fishing interests, both commercial and sport, have appeared in Maine, Scotland, Norway, and Chile (Manning 1989). Ultimately, it appears that active control of some marine mammal populations will be needed to balance the interests of commercial and sport fishing with those of wildlife conservation.

FIGURE 19.1

■ The oil tanker *Exxon Valdez* went aground on a reef in Prince
William Sound, Alaska, in March, 1989, spilling 41,000 tons of crude
oil, polluting waters and shorelines over 25,000 square kilometers.

Several areas of the world oceans show very heavy oil
pollution. These include enclosed marine basins such as the
Mediterranean Sea, Gulf of Mexico, Caribbean Sea, Red Sea,
and, of course, the Persian Gulf. The spills associated with
production and transport in these enclosed basins are less easily
dispersed and are more likely to affect coastal ecosystems than
elsewhere. Major tanker routes also experience heavy oil pollution. Thus, both the east and west coasts of Africa receive
heavy pollution from tanker traffic between the Middle East
and Europe. The route from the Middle East across the Indian
Ocean and through the South China Sea to far eastern ports
is another area of heavy pollution.

EFFECTS OF OIL ON MARINE ECOSYSTEMS

The fate and effects of oil pollution vary greatly in different
marine environments (Teal and Howarth 1984, Neff 1990).
Oil discharged into the ocean forms slicks, which may range
from a fraction of a millimeter to 20 mm in thickness. Much
of the lightest components evaporate within a few days, re-

ducing the total amount of petroleum by 60 percent or more.
The lighter, non-tar components rapidly dissolve or disperse,
as well, and their concentrations usually fall to background
levels within two to six months. Heavier components may
be churned into a viscous emulsion, known as **chocolate
mousse,** that is about 75 percent water. Mousse forms readily
in cold, icy waters, and may persist for a long time. Over time,
photochemical and microbial processes break down petroleum compounds. Oil components are gradually precipitated
from surface waters by mixing with mineral sediment, adsorption on particulate matter, incorporation in zooplankton
fecal pellets, or uptake by phytoplankton, all of which associate the petroleum with particles that eventually sink. The
heaviest components eventually form tar balls, which float
indefinitely but are relatively inert chemically.

In the open ocean, oil slicks may cause mortality of
plankton and fish, but severe effects on these organisms usually last only for days to a few weeks. Sea birds and mammals
may be fouled by these slicks, as examined later in detail.

When oil reaches coastal areas, longer-term impacts
occur. On sandy beaches, the effects of oil pollution may last

FIGURE 19.2

■ Efforts were made to clean rocky shores contaminated with oil in Prince William Sound after the *Exxon Valdez* spill, but these may have been more harmful than beneficial to marine invertebrates.

for weeks to months. Oil may be buried for periods by sand deposition, and exposed later by normal beach erosion. Ultimately, the reworking of beach sands leads to chemical and mechanical dispersal of the petroleum residues. When oil pollution is frequent, this removal of sand detracts from recreational use of the beaches.

Rocky coastlines experience still longer impacts; here the effects may last for months to years. Deposition of oil on rocky intertidal surfaces kills almost all organisms, and coats the surfaces with tars that weather very slowly, inhibiting recolonization (Fig. 19.2). The inhabitants of rocky intertidal areas, however, are adapted to colonization of disturbed sites, and in time recovery occurs without serious damage.

The most prolonged impacts of coastal oil pollution occur in estuary and salt marsh areas, where they may last from years to decades. This long-term effect occurs because large quantities of oil may become buried in anoxic sediments where little or no chemical degradation occurs. After months or years, this oil may be re-exposed and exert toxic effects anew.

A striking example of the long-term impact of pollution on an estuary ecosystem followed the breakup of the barge *Florida* near Buzzard's Bay, Massachusetts in September, 1969 (Sanders, et al. 1980). This spill was small, involving only 630 tons of fuel oil. The immediate effects were severe, and included extensive mortality of the *Spartina* marsh grass and an estimated 95 percent kill of bottom-living vertebrates and invertebrates. Shellfish harvests were closed because of contamination by oil.

Much of the oil was buried in estuary sediments, to be re-exposed and re-buried time after time by tidal action. As a result, the area affected slowly increased, and one year later the area closed to shellfish harvests had to be extended. In 1982, some 12 years later, the estuary and marsh were judged to have shown 95 percent recovery, but significant amounts of oil were still present (Teal and Howarth 1984).

Unfortunately, some of the techniques used to disperse oil slicks or in attempts to clean oiled coastal areas have done more harm than good (Kerr 1991). Solvents used to disperse oil from the *Torrey Canyon* spill on the coast of England killed

FIGURE 19.3

■ Diving birds such as western grebes can become completely oiled by repeated passage through the ocean surface in their feeding activities.

FIGURE 19.4

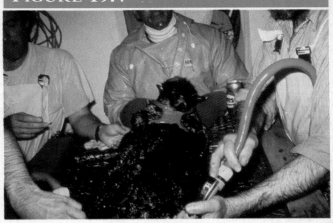

■ In spite of efforts to clean and rehabilitate sea otters that were oiled by the 1989 *Exxon Valdez* spill, most heavily oiled animals died.

most marine life. The hot-water washing of rocky coasts oiled by the Exxon Valdez spill killed or buried many invertebrates and carried the oil to lower beach levels that often had not been affected severely. As a result, treated coastlines were more heavily damaged than untreated areas.

IMPACTS OF OIL ON SEABIRDS AND MAMMALS

The most serious overall impacts of marine oil pollution are on seabirds (Ohlendorf et al. 1978). These impacts are greatest for four groups of seabirds: alcids, cormorants, sea ducks and eiders, and penguins. Alcids, or members of the auk family, include puffins, murres, auklets, murrelets, and the razorbill. The penguin most affected by oil pollution is the jackass penguin, which inhabits the coast of South Africa, on the main tanker route from the Middle East to Europe and North America. Members of these four families are diving birds that feed on fish or molluscs. In areas with oil slicks, their diving carries them back and forth through the oil-covered surface, causing heavy oiling of the plumage. Other diving birds, such as western grebes, are similarly affected (Fig. 19.3). Birds such as gulls, which rest on the surface but do not dive, are much less affected by surface slicks.

Oiling of the plumage affects birds in several ways. Matting of the plumage leads to loss of flight ability and buoyancy. Birds unable to remain afloat drown or are forced ashore, where they are unable to feed. The matted plumage is also less effective as insulation, leading to more rapid loss of body heat and an increased rate of metabolic consumption of body fat reserves. Preening of the oiled plumage leads to

ingestion of oil, which may cause toxic effects on internal organs such as the pancreas, liver, and kidneys. Irritation of the intestinal lining by ingested oil reduces water absorption, promoting dehydration. Sea otters fouled by oil also lose the insulative value of their pelage, and the ingestion of oil in efforts to clean it leads to toxic effects on internal organs (Geraci and Williams 1990). Most heavily oiled birds and sea otters die, even when efforts are made to clean and rehabilitate them (Fig. 19.4).

Birds that experience chronic oil pollution of plumage and food also experience a number of reproductive effects, including reduced egg production, mortality of eggs contaminated with oil by incubating adults, and reduced growth of young birds whose foods are contaminated with oil.

The most obvious seabird and mammal impacts are those associated with major tanker and offshore platform accidents. In February, 1952, collision of the tankers *Ft. Mercer* and *Pendleton* off the coast of Massachusetts led to the death of about 350,000 eider ducks. The grounding of the *Gerd Maersk* near the mouth of the Elbe River in Germany in January, 1955 killed an estimated 500,000 birds of 19 species. The 1989 Exxon Valdez accident probably killed about 350,000 to 390,000 seabirds, mostly common and thick-billed murres (Piatt et al. 1990), 200 or more bald eagles, 3500 to 5500 sea otters, and at least 200 harbor seals. Breeding by seabirds in colonies affected by the spill was also reduced.

The grounding and breakup of the *Torrey Canyon* off the French coast in March, 1967 remains one of the best-documented tanker accidents from the standpoint of effects on seabirds and marine life in general. After this accident, some 30,000 birds or their carcasses were recovered on the coast of Brittany, leading to an overall estimate of 40,000 to 100,000 bird deaths. Of 7,851 oiled birds recovered alive, only 6 percent survived rehabilitation efforts, and many of these apparently died after release.

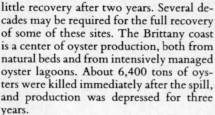

In March, 1978, the tanker *Amoco Cadiz* lost power and, in stormy weather and heavy seas, drifted onto a submerged reef close to the coast of Brittany, France. All 220,000 tons of cargo and fuel oil spilled, the largest tanker accident on record. About 30 percent of this oil came ashore, spreading along about 400 km of the Brittany coastline.

Oiling of the coastline produced massive mortality of coastal marine life, especially intertidal and shallow subtidal organisms (National Academy of Sciences 1985). Most invertebrates on oiled sandy beaches and mud flats were killed. Large amounts of oil were stored in buried layers in beaches, to be exposed by seasonal beach erosion and deposition processes over the next several years. Salt marsh vegetation, as well as animal life, was killed in many locations, and heavily oiled sites showed little recovery after two years. Several de-

little recovery after two years. Several decades may be required for the full recovery of some of these sites. The Brittany coast is a center of oyster production, both from natural beds and from intensively managed oyster lagoons. About 6,400 tons of oysters were killed immediately after the spill, and production was depressed for three years.

Offshore, short-term decreases in phytoplankton and zooplankton production were noted. Kelp production and harvest was also reduced. Oil reached the bottom sediments over a large area of the western English Channel, killing many of the bottom invertebrates. Although the immediate fish kill was small, the growth of several commercial fish that depended on estuary habitats was reduced in the year following the spill. Flatfish also showed an abnormally high incidence of fin rot in the months after the spill.

More than 4,500 dead birds of 33 species, but mostly auks, cormorants, and loons, were recovered in the month fol-

lowing the spill. Considering the chances of killed birds coming ashore and being recovered, the total estimated bird kill was 19,000 to 37,000.

The spill caused various other damages and costs. Some farm crops near the coast were damaged by oily mists carried inland by winds. Major expenditures were made in cleanup activities, and the region experienced a significant loss of tourist income. An evaluation of the total social costs of this spill (National Oceanic and Atmospheric Administration 1983) yielded an estimate of $190 to 290 million. In 1990, 12 years after the disaster, French plaintiffs were awarded damages of $120 million in a suit against the Amoco Company. This award, probably to be appealed, was the outcome of an appeal by Amoco of an earlier judgement for $85.2 million in damages.

On the longer term, effects of the spill were evident in the decline of populations of the shag (a cormorant) and three species of alcids at bird colonies on the Sept Isles, off the coast of Brittany (Table 19.2). These observations suggest that one of the major causes of decline of several species of alcids and other seabirds in the eastern North Atlantic has been chronic oil pollution over recent decades. In the late 1800s, moreover, 12,000 pairs of puffins were estimated to breed on the Sept Isles, compared to the 2,500 pairs just before the *Torrey Canyon* accident.

SYNTHETIC ORGANIC POLLUTANTS

Several other types of chemical pollution now exist on a global scale. **Polychlorinated biphenyls (PCBs),** a group of industrial chemicals used as insulation fluids in transformers and as plasticizers and fire retardants in various materials, have become widespread in aquatic environments through leakage and trash disposal (Waid 1986, Tanabe 1988). PCBs are close relatives of chlorinated hydrocarbon pesticides, and share their features of stability and high solubility in lipids. Thus, they show food chain magnification, sometimes reaching concentrations of hundreds to thousands of parts per million in birds and mammals at the ends of aquatic food chains (Fig. 19.5). Their mobility seems to be less than that of some other chlorinated hydrocarbons, such as DDT, and they are most serious as pollutants in semienclosed water areas such as the

TABLE 19.2

Bird Populations on the Sept Isles, Off the Coast of Brittany, France, Before and After the Torrey Canyon Grounding

	NUMBER OF BREEDING PAIRS	
	BEFORE TORREY CANYON	AFTER TORREY CANYON
Fulmar	20	20
Gannet	2,600	2,600
Shag (Cormorant)	250	190
Gulls (Four Species)	7,692	7,482
Common Murre (Auk)	270	50
Razorbill (Auk)	450	50
Puffin (Auk)	2,500	400

Source: Data from W. R. P. Bourne, "Special Review—After the 'Torrey Canyon' Disaster" in Ibis, 112:120–125, 1970, Blackwell Scientific Publications, Oxford, England.

FIGURE 19.5

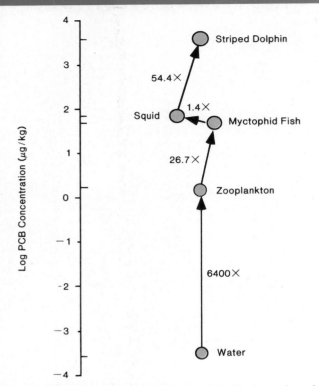

■ Data from a food chain in the western North Pacific Ocean show the strong food chain concentration and magnification pattern of PCBs, similar to that of chlorinated hydrocarbon pesticides. Source: Data from S. Tanabe, et al., "Polychlorophenyls, DDT, and Hexachlorocyclohexane Isomers in the Western North Pacific Ecosystem" in *Archives of Environmental Contamination and Toxicology*, 13:731–738, 1984.

Baltic Sea, Long Island Sound, and San Francisco Bay. PCBs have been implicated in reproductive impairments of a number of birds and seals (Kubiak et al. 1989, Reijnders 1986). Because of the widespread occurrence of PCBs in the marine environment and their tendency to show food chain magnification, marine mammals are likely to be the animals most affected (Tanabe 1988). More thorough studies are needed to determine the seriousness of this threat (Addison 1989).

Still other petroleum derivatives, the **polynuclear aromatic hydrocarbons (PAHs),** have recently been recognized as major environmental pollutants. PAHs are stable molecules consisting of two to several fused benzene rings. Some are apparently natural components of petroleum, others are formed by combustion of fossil fuels. They apparently enter aquatic environments through urban runoff and atmospheric precipitation. Many of these compounds are highly carcinogenic, and are thought to be responsible for liver cancers and fin deterioration of fish in polluted areas of Puget Sound.

HEAVY METALS AND CYANIDE

Mercury and other heavy metals such as copper, cadmium, chromium, lead, nickel, and zinc all occur naturally in fresh and marine waters. Use of these elements by man, however, has substantially increased their rates of entry into active pools in natural ecosystems. For example, human activities have roughly doubled the rate of entry of mercury into active circulation in the biosphere. Mercury also shows food chain magnification, and on several occasions has reached concentrations in marine fish, such as tuna, that exceed those considered safe for human consumption. In Sweden, contamination of the landscape by mercury from industrial air pollution is widespread. Acidification of lakes and streams (See chapter 20) is now bringing increased amounts of this mercury into biologically active form, leading to increased mercury levels in many fish (Hakanson 1988). About 250 lakes have been closed to fishing because of this, and recent reports suggest that levels in about 9,400 additional lakes justify similar closing. Failure of egg hatching in white-tailed sea eagles, a fish-eating species, has also been attributed in Sweden to mercury contamination. Cadmium is also a heavy metal of increasing concern. Ohlendorf and Fleming (1988) identified it as a pollutant of particular concern for birds in both San Francisco and Chesapeake Bays.

Mercury and cyanide have also become serious pollutants in gold mining areas. In the Amazon Basin, mercury is being used in large quantities to separate gold from lighter minerals. Since 1958, about 1,200 tons of waste mercury have been discharged into the Tapajos River in the southern portion of the basin, and fish from the river are heavily contaminated. In the western United States, cyanide is being used to leach gold from old mine tailings and low-grade ore. Large numbers of birds and mammals have died after feeding or drinking at ponds of cyanide-containing water used in leaching.

Tributyltin (TBT) is an organic tin-containing compound that has proven to be a highly effective active ingredient for antifouling paints for marine vessels (Champ and Lowenstein 1987). Paints based on TBT protect hulls against growth of attached invertebrates such as barnacles for five to seven years, compared to about one year for copper-based antifouling paints. Fouling of ship hulls, by increasing friction between hull and water, greatly increases the fuel costs of operating vessels, so that an effective antifouling agent is extremely valuable. The United States Navy, for example, has estimated that use of TBT would save about $150 million annually. TBT has been shown to be highly toxic to marine organisms in concentrations of less than 1 part per billion. In France, TBT was shown to cause mortality and deformation of oysters in bays where it was used in antifouling paints on fishing boats. Concentrations of TBT in the range of potentially serious impact have been detected in bays and harbors in several locations in North America (Goldberg 1986). France has banned the use of TBT on small vessels, and its use has been restricted by several other European countries. In 1988 the United States Environmental Protection Agency proposed restrictions that included a ban on use of TBT paints on most boats under 20 m in length, and restrictions on the rate of release of TBT from the treated surfaces of larger vessels (Weiss 1988).

FIGURE 19.6

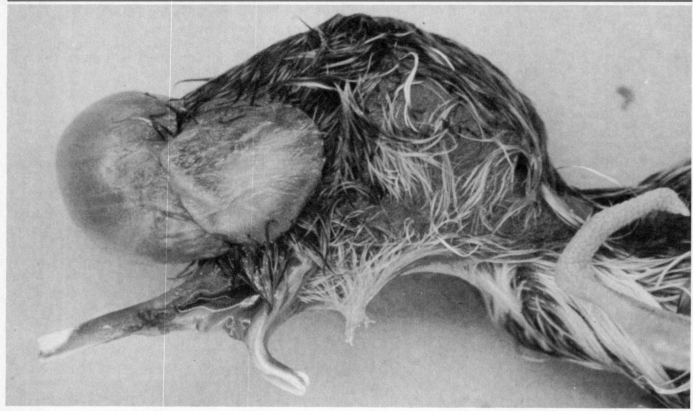

■ A deformed black-necked stilt chick from Kesterson Reservoir in California is an example of the effects of selenium pollution in several locations in western North America.

SELENIUM

Selenium is an element required by most vertebrates in trace amounts, but toxic to them in high concentrations (Ohlendorf 1989). Selenium is low in concentration in most soils, because it is easily removed by leaching, but is abundant in some desert soils. Irrigated farming in desert regions and the leaching of selenium by drainage waters has created pollution problems in several wildlife areas in western North America. Selenium is concentrated by organisms, compared to its levels in the environment, and appears to show some degree of food chain magnification.

Kesterson Reservoir, in Kesterson National Wildlife Refuge, California, received irrigation runoff water from large areas of irrigated land in the San Joaquin Valley. In 1983, the water entering Kesterson Reservoir contained selenium in concentrations about 200 times normal (Ohlendorf 1989). All plants and animals sampled showed high concentrations of selenium. American avocets and black-necked stilts experienced nearly complete reproductive failure at Kesterson Reservoir in 1984 and 1985, whereas success was normal at a nearby, unpolluted area (Williams et al. 1989). Eared grebes and American coots nesting at Kesterson in 1983 also expe-

rienced low nesting success, and these species, together with black-necked stilts (Fig 19.6) showed frequent embryonic abnormalities (Ohlendorf et al. 1989). Many adult water birds that died at the reservoir also showed toxic selenium levels. As a result of these observations, the reservoir has been closed until the selenium contamination problem is overcome.

Selenium contamination has also been detected in other parts of the San Joaquin Valley, as well as in San Francisco Bay. In the Tulare Basin of the southwestern San Joaquin Valley, a high incidence of bird deformities has also been reported. From 1982 to 1989, selenium levels in certain bottom-feeding ducks in San Francisco Bay were equal to or greater than those in aquatic birds at Kesterson Reservoir that had died or had reproductive problems. Selenium pollution has also become a problem at wildlife areas associated with irrigation projects in other locations in Nevada, Utah, and Wyoming.

RADIOISOTOPES

With the worldwide proliferation of nuclear technology has come the threat of environmental pollution by illegal disposal of radioactive wastes and release of radioisotopes by plant ac-

FIGURE 19.7

■ Plastic articles and their fragments have become a global form of marine pollution.

cidents. The materials released include radioisotopes of cesium, strontium, and phosphorus, which have moderate to long half-lives and which also show biological concentration and food chain magnification. In many cases, the occurrence of such releases has been kept from public knowledge.

In the United States, disposal of radioactive wastes from Department of Energy facilities at Savannah River, South Carolina, Oak Ridge, Tennessee, and Hanford, Washington has contaminated wildlife, both terrestrial and aquatic (Holloway 1990). At the Hanford facility, radioisotopes entered the Columbia River by waste discharge and the return of contaminated reactor cooling water to the river from 1945 through the early 1960s (Stenehjem 1990). The entire food chain of the river was contaminated to the extent that human health impacts became a serious concern. Even so, information about the situation was withheld from the public.

The 1986 disaster at Chernobyl in the Soviet Union has also led to contamination of marine life in the northern part of the Black Sea. In this case, surface runoff carried radioisotopes into the Black Sea from regions receiving fallout from the plant explosion. The extent of this contamination is yet to be evaluated.

PLASTICS

Plastics are still another group of aquatic pollutants. Plastics make up more than half of the man-made debris littering the ocean (O'Hara 1988). The plastics found at sea include plastic products and their fragments, **polyethylene spherules,** and **polystyrene spherules** (Fig. 19.7). Over time, plastic products entering the ocean by trash disposal become brittle and break into small, irregular pieces. These fragments remain buoyant and float until they are washed up on beaches or become trapped in other materials that eventually sink. Polyethylene spherules, termed "nibs" in the plastics industry, are beads two to five mm in diameter that are the raw material for the production of molded plastic products. Nibs are transported by ships designed to carry bulk products of various sorts. Quantities of these spherules enter sea waters when the holds are cleaned or by ballasting practices like those used by oil tankers. Polyethylene spherules remain buoyant indefinitely, and are now found worldwide in oceans and on ocean beaches. Plastic fragments and nibs also become concentrated in the centers of ocean gyres, such as the Sargasso Sea. Polystyrene spherules, 0.1 to 3 mm in size, are the raw

FIGURE 19.8

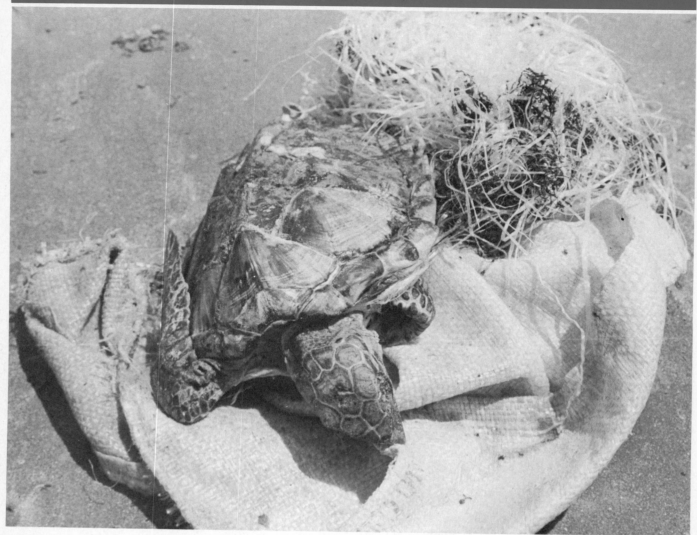

■ Sea turtles, such as this hawksbill, often die after they have become entangled in discarded vegetable bags, or after ingesting plastic bags, sheets, or strips that they apparently mistake for food organisms.

material from which styrofoam products are made. Like nibs, they are shipped by bulk carriers, and enter ocean waters in similar fashion. Polystyrene spherules ultimately become waterlogged and sink, so that they are not as widespread as polyethylene spherules.

The consequences of plastic pollution for marine wildlife are uncertain, although these items are known to be ingested by many kinds of invertebrates, fish, and birds. Sea turtles, in particular, are known to ingest plastic bags, sheets, or other bulky items, apparently reacting to them as if they were jellyfish or other food items (Fig. 19.8). Laysan albatrosses, nesting on one of the most remote islands in the Pacific Ocean, feed large numbers of plastic items to their chicks, and most dead chicks of this species have such items in their stomachs (O'Hara 1988).

INTERACTIONS OF POLLUTANTS

In most fresh and marine waters, wildlife is exposed to several pollutants, each potentially serious by itself, but having potential additive or synergistic effects. In the lower St. Lawrence River, for example, beluga whales have been dying in large numbers in recent years. The tissues of these animals contain high levels of chlorinated hydrocarbon pesticides, PCBs, heavy metals, and an industrial chemical, benzo-a-pyrene, that might have come from an aluminum plant on a tributary river. Many of the animals appeared to have died of infectious diseases, suggesting that their immune systems had failed, possibly as a result of pollutant effects.

Major die-offs of aquatic mammals have occurred in several other locations in recent years. These have involved bottlenose dolphins along the eastern seaboard of the United States, gray and ringed seals in the Baltic Sea, harbour seals

FIGURE 19.9

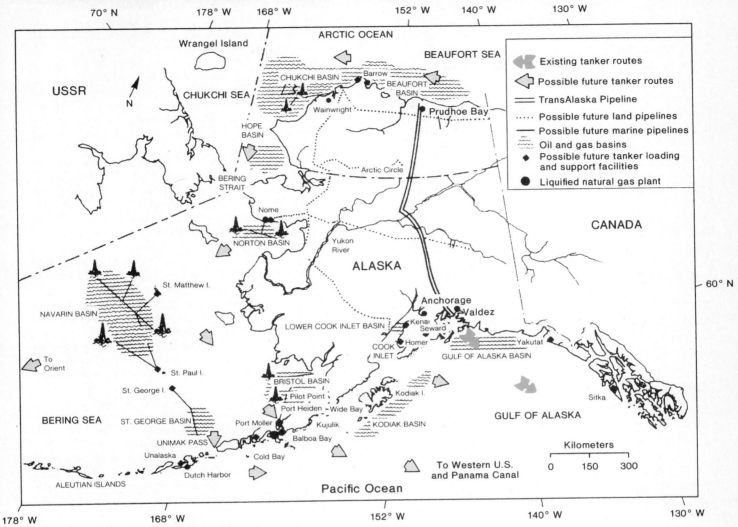

■ Offshore oil and gas basins in Alaskan waters, and possible future pipelines, processing and loading facilities, and tanker routes (Modified from Bolze and Lee 1989).

in the Dutch Waddensee, and the Lake Baikal seal in the Soviet Union. These die-offs are the apparent result of disease epidemics or, in the case of bottlenose dolphins, toxins produced by red tide organisms. In all cases, these animals were exposed to substantial levels of several chemical pollutants. It is far from certain that pollutant-induced failure of the immune system is involved in these episodes, but it is a possibility that cannot be ignored.

CONTROLLING CHEMICAL POLLUTION

Reducing marine oil pollution is one of the major challenges for the last decades of the petroleum age. Most of the oil and gas basins remaining to be tapped lie offshore, and many of these are in high latitudes (Fig. 19.9). In Alaska, for example,

petroleum basins occur in the Gulf of Alaska, along the Aleutian Peninsula, in the Bering Sea, in Norton Sound, and along the Arctic Ocean (Bolze and Lee 1989). Other potential basins lie along the coasts of California, North Carolina, Florida, and other states. Developing the technology for safe exploitation of these reserves is essential.

Since the Exxon Valdez accident, efforts to reduce the risk of tanker accidents have centered on a requirement that oil tankers be constructed with double hulls to reduce the chance of oil spillage in groundings. In 1990, the United States passed legislation requiring double hulls on new tankers and the retrofitting of double hulls for existing tankers.

Several national and international efforts have been made to control the discharge or dumping of hazardous materials in lakes and oceans. The broadest international agreement is the **London Dumping Convention,** drafted in 1972, and ratified by 64 countries, including the United States, as of 1989 (Duedall 1990). This agreement bans the deliberate discharge of organohalogens (chlorinated hydrocarbons, PCBs, and related chemicals), certain heavy metals, plastics, petroleum substances, high-level radioactive wastes, and some other materials. Many other materials can be dumped only by permit, which essentially serves to control the location of dumping.

The **International Convention for the Prevention of Pollution from Ships,** commonly known as the **MARPOL Convention,** prohibits the dumping of plastics at sea. The United States signed this agreement, effective in December, 1988. The MARPOL Convention has been signed by 39 other nations.

Neither the London Dumping Convention nor the MARPOL convention covers discharge or dumping of sewage or sewage sludge, nor of wastes from offshore mineral exploration or mining. The United States, in 1988, passed legislation requiring that all ocean dumping of sewage sludge and industrial waste cease by 31 December 1991, however. This legislation is likely to encourage similar legislation in many developed nations.

KEY STRATEGIES FOR CONTROLLING AQUATIC POLLUTION

1. Application of improved technologies for handling tanker ballast, ship bilge waste, and bulk carrier cargo residues.

2. Application of waste recycling and disposal technologies that reduce direct disposal of toxic materials to aquatic environments.

3. Strengthening and extending international conventions relating to discharge and dumping of materials into aquatic environments.

LITERATURE CITED

Addison, R. F. 1989. Organochlorines and marine mammal reproduction. *Can. J. Fish. Aq. Sci.* **46**:360–368.

Bolze, D. A. and M. B. Lee. 1989. Offshore oil and gas development: Implications for wildlife in Alaska. *Marine Policy* **13**:231–248.

Champ, M. A. and F. L. Lowenstein. 1987. TBT: The dilemma of high-technology antifouling paints. *Oceanus* **30(3)**:69–77.

Duedall, I. W. 1990. A brief history of ocean disposal. *Oceanus* **33(2)**:29–38.

Geraci, J. R. and T. D. Williams. 1990. Physiologic and toxic effects on sea otters. Pp. 211–221 *in* J. R. Geraci and D. J. St. Aubin (Eds.), *Sea mammals and oil: Confronting the risks.* Academic Press, San Diego, CA.

Goldberg, E. D. 1986. TBT: An environmental dilemma. *Environment* **28(8)**:17–20, 42–44.

Hakanson, L. 1988. Mercury in fish in Swedish lakes. *Env. Pollution* **49**:145–162.

Holloway, M. 1990. Hot geese: Contaminated wildlife roam nuclear reservations. *Sci. Amer.* **263**:22.

Kerr, R. A. 1991. A lesson learned, again, at Valdez. *Science* **252**:371.

Kubiak, T. J., H. J. Harris, L. M. Smith, T. R. Schwartz, D. L. Stalling, J. A. Trick, L. Sileo, D. E. Docherty, and T. C. Erdman. 1989. Microcontaminants and reproductive impairment of the Forster's tern on Green Bay, Lake Michigan—1983. *Arch. Env. Contam. Toxicol.* **18**:706–727.

National Academy of Sciences. 1985. *Oil in the sea: Inputs, fates, and effects.* National Academy Press, Washington, D.C. 601 pp.

National Oceanic and Atmospheric Administration. 1983. *Assessing the social costs of oil spills: The Amoco Cadiz case study.* NOAA, Washington, D.C.

Neff, J. M. 1990. Composition and fate of petroleum and spill-related agents in the marine environment, Pp. 1–33 *in* J. R. Geraci and D. J. St. Aubin (Eds.), *Sea mammals and oil: Confronting the risks.* Academic Press, San Diego, CA.

O'Hara, K. J. 1988. Plastic debris and its effects on marine wildlife. Pp. 395–434 *in* W. J. Chandler (Ed.), *Audubon wildlife report 1988/89.* Academic Press, San Diego, CA.

Ohlendorf, H. M. 1989. Bioaccumulation and effects of selenium in wildlife. Pp. 133–177 *in* Selenium in agriculture and the environment. Soil Science Society of America, Madison, WI.

Ohlendorf, H. M. and W. J. Fleming. 1988. Birds and environmental contaminants in San Francisco and Chesapeake Bays. *Marine Pollution Bull.* **19**:487–495.

Ohlendorf, H. M., R. L. Hothem, and D. Welsh. 1989. Nest success, cause-specific nest failure, and hatchability of aquatic birds at selenium-contaminated Kesterson Reservoir and a reference site. *Condor* **91**:787–796.

Ohlendorf, H. M., R. W. Risebrough, and K. Vermeer. 1978. Exposure of marine birds to environmental pollutants. *U.S. Fish and Wildl. Service Wildl. Res. Report No. 9,* 40 pp.

Piatt, J. F., C. J. Lensink, W. Butler, M. Kendziorek, and D. R. Nysewander. 1990. Immediate impact of the "Exxon Valdez" oil spill on marine birds. *Auk* **107**:387–397.

Reijnders, P. J. H. 1986. Reproductive failure in common seals feeding on fish from polluted coastal waters. *Nature (Lond.)* **324**:456–457.

Sanders, H. L., J. F. Grassle, G. R. Hampson, L. S. Morse, S. Garner-Price, and C. C. Jones. 1980. Anatomy of an oil spill: Long-term effects of the grounding of the barge *Florida* off West Falmouth, Massachusetts. *J. Mar. Res.* **38**:265–380.

Stenehjem, M. 1990. Indecent exposure. *Nat. Hist.* **9/90**:6–21.

Tanabe, S. 1988. PCB problems in the future: Foresight from current knowledge. *Env. Pollution* **50**:5–28.

Tanabe, S., H. Tanaka, and R. Tatsukawa. 1984. Polychlorophenyls, DDT, and hexachlorocyclohexane isomers in the western North Pacific ecosystem. Arch. Environ. *Contam. Toxicol.* **13**:731–738.

Teal, J. M. and R. W. Howarth. 1984. Oil spill studies: A review of ecological effects. *Env. Management* **8**:27–44.

Waid, J. (Ed.) 1986. *PCBs and the environment.* Vol. I–III. CRC Press, Boca Raton, FL.

Weiss, J. S. 1988. Action on antifouling paints. *BioScience* **38**:90.

Williams, M. L., R. L. Hotham, and H. M. Ohlendorf. 1989. Recruitment failure in American avocets and black-necked stilts at Kesterson Reservoir, California, 1984–1985. *Condor* **91**:797–802.

Wilson, R. D., P. H. Monaghan, A. Osanik, L. C. Price, and M. A. Rogers. 1974. Natural marine oil seepage. *Science* **184**:857–865.

CHAPTER 20

Acid Deposition

Pollution of the atmosphere by the waste products of human activity, mainly those of combustion, has long been a serious environmental problem in and near large urban areas. In recent decades, however, these problems have evolved into major regional air pollution syndromes that are impacting natural ecosystems over large continental areas. Urban air pollution still remains a major human health problem, but regional air pollution has now become a serious problem for the health of plant and animal life in streams, lakes, and forests.

Acid deposition, the increased input of acids to aquatic and terrestrial ecosystems through rain, fog, snow, dry particulate fallout, and capture of gaseous compounds is now the most general of these regional problems. Acid deposition has directly and indirectly influenced the structure and function of many ecosystems. The most serious effects of acid deposition have been on aquatic ecosystems, but increasing evidence exists that forests are also damaged by this phenomenon.

THE NATURE OF ACID DEPOSITION

Acid deposition results from the release of sulfur (S) and nitrogen (N) compounds into the atmosphere, in forms that may ultimately react with water and oxidants to form sulfuric and nitric acids. This reaction may occur in the atmosphere, yielding **acid precipitation**—rain and snow that are unusu-

ally acidic in nature. The fallout of ash or dust particles containing S and N compounds and the direct adsorption of gaseous forms of these elements, or **dry deposition,** also leads to the formation of acids in the soil, in the water of streams and lakes, or in the tissues of organisms. The S and N oxides are derived mostly from combustion processes (Fig. 20.1). Oxides of S result largely from the combustion of fossil fuels and the smelting of ores, with electrical power producing sta-

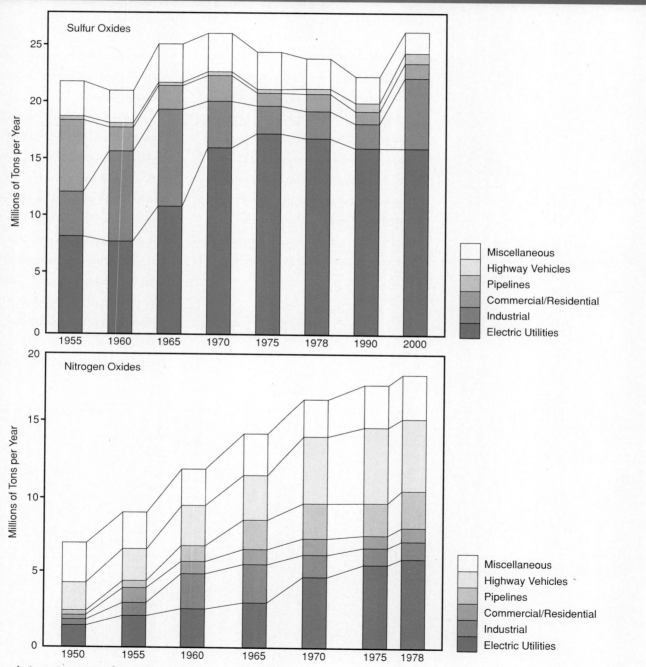

■ Sulfur and nitrogen emissions from various sources in North America (Modified from MacDonald 1985).

tions being the greatest contributor in the eastern United States (MacDonald 1985). Oxides of N result largely from the burning of organic matter, the union of N and oxygen (O) under the high pressures and temperatures of vehicle engines, and the conversion of N in fertilizers to N oxides by certain groups of soil bacteria. In some areas of the tropics, substantial quantities of nitrogen and sulfur oxides enter the atmosphere from burning of savannas or newly cleared forest lands (Rodhe and Herrera 1988). Where industry and space heating activities predominate, as in the industrial belt of eastern North America, perhaps two-thirds of the oxides in acid deposition are those of S; where vehicle emissions predominate the N oxides are relatively more important. Dry deposition, which can be as much as five to eight times the wet precipitation of acidic materials, tends to be of greater importance in localities close to major combustion sources, such as smelters or power plants.

Unpolluted rain and snow show low levels of acidity. On the **pH scale** (Fig. 20.2) from 0 to 14, acidity increases as the scale value decreases. Each unit change on this scale represents a ten-fold change in the concentration of acid ions (H^+). A pH value of 7.0 is neutral, meaning that equal numbers of acid and basic (OH^-) ions are present. On this scale, pure water has a pH of about 5.6. This slight acidity results from carbon dioxide (CO_2) in the air. When it dissolves in droplets of rain, CO_2 reacts with water to form carbonic acid (H_2CO_3), which partially ionizes to release H^+ ions. In nature, furthermore, the pH of precipitation in remote areas that are unaffected by human activities and free of calcareous dust is usually about 5.0, due to the presence of other acidic substances of natural origin (Schindler 1988). Thus, acid precipitation as an unnatural phenomenon is rain or snow with a pH less than about 5.0.

Acid deposition is greatest in areas downwind from urban and industrial regions where large S and N emissions occur. In North America, the northeastern United States and the eastern maritime provinces of Canada are most seriously affected, but most of the area east of the Mississippi River experiences heavy acid deposition (Fig. 20.3). Sulfate concentration in precipitation is about ten times higher in areas of eastern North America than in areas remote from pollution sources, with 90 percent being of human origin (Galloway et al 1984, Schindler 1988). In Europe, the Scandinavian countries and the portions of eastern Europe downwind from the heavily industrialized parts of the United Kingdom and central Europe receive the heaviest levels of acid deposition (Fig. 20.3). A smaller area of acid deposition is centered on Korea and Japan.

Regions affected by acid deposition experience rain, snow, or fog with pH values down to 3.0 or lower. Over the entire year, the pH of rain and snow averages 4.0 to 4.4 in many locations in eastern North America and Europe (Park 1987). Annual mean values as low as 3.8 have been recorded in parts of the Netherlands. In the White Mountains of New Hampshire, rainstorms have been recorded with pH values as low as 2.85; in Pennsylvania a storm value of 2.70 was noted. In Scotland rain during a 1974 storm registered a pH of 2.4. Acid fogs with pH values as low as 2.2 sometimes occur in

FIGURE 20.2

The pH scale

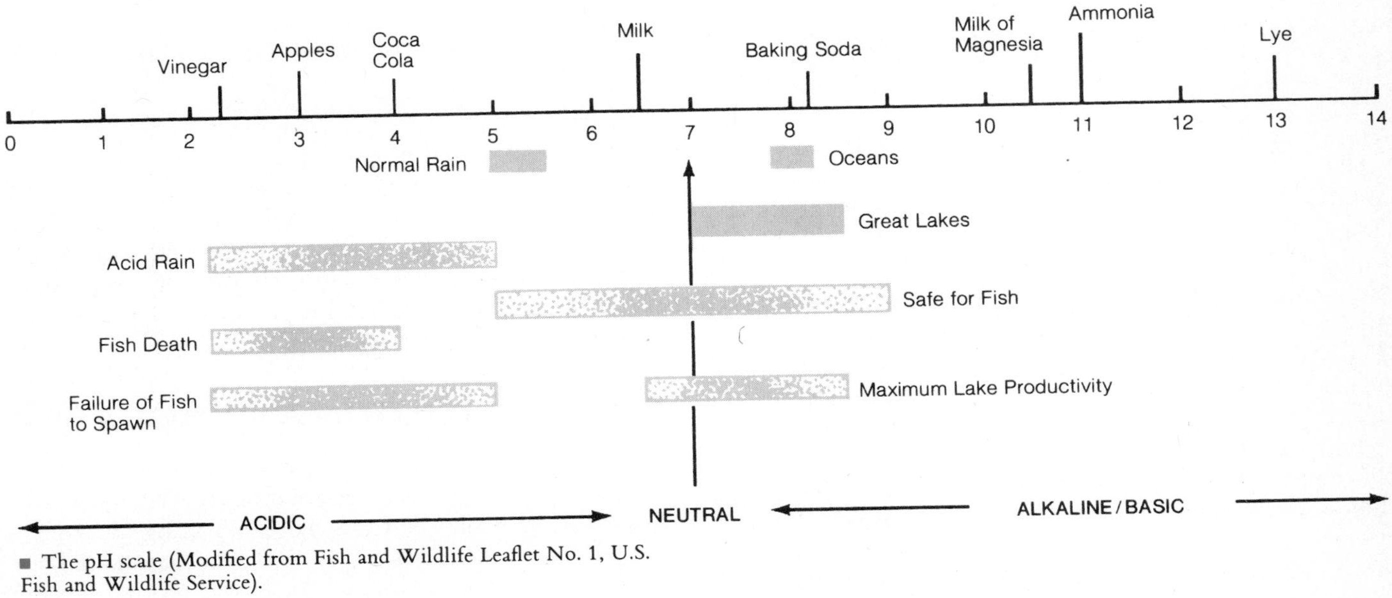

■ The pH scale (Modified from Fish and Wildlife Leaflet No. 1, U.S. Fish and Wildlife Service).

FIGURE 20.3

■ Regions of eastern North America and northern Europe affected by acid precipitation. Source: Data from L. A. Barrie and J. M. Hales, "The Spatial Distributions of Precipitation Acidity and Major Ion Wed Deposition in North America during 1980" in *Tellus,* 368:333–355, 1984.

coastal urban areas of southern California (Waldman et al. 1982). The moisture in clouds bathing the spruce-fir forests on Mount Mitchell, North Carolina, has registered a pH as low as 2.1.

Why has acid deposition only recently become a serious environmental problem, considering that the activities that release S and N oxides have been occurring for centuries? Part of the reason is simply the continued growth of the manufacturing, heating, transportation, and farming activities that release large quantities of S and N oxides. Increases in the release of pollutants that catalyze the oxygenation of S and N compounds has also contributed (Schindler 1988). In addition, major industrial operations, such as smelters, power plants, and steel mills, have attempted to relieve intense local air pollution problems by building taller smokestacks. Dis-

charge of combustion wastes at higher levels in the atmosphere results in their dispersal by winds over a much broader area. Many industries and power plants now have stacks over 300 m high, and in a few cases over 400 m high (Fig. 20.4). These taller stacks have been a major factor in the conversion of many intense local air pollution problems into the general regional problem of acid deposition.

ACID DEPOSITION AND AQUATIC ECOSYSTEMS

Lakes and streams are sensitive to acidification in regions where watersheds are weakly buffered against pH change due to increased rates of input of acids. In fact, many aquatic ecosystems in such regions are naturally quite acid, because of basic

FIGURE 20.4

■ Tall stacks like these in Sudbury, Ontario, have helped transform severe local acid pollution problems into regional acid deposition syndromes. The stack of the Inco smelter (background) is over 400 m tall, and in 1989 released about 637,000 tons of sulfur oxides.

climatic, topographic, and ecological conditions (Patrick et al. 1981). Acid bogs and bog lakes, which contain many interesting plants and animals, are common in much of the coniferous forest region of the Northern Hemisphere, for example. Many lakes and streams that were much less acid, or nearly neutral, in pH are sensitive to the effects of increased acid deposition, however. In general, sensitive aquatic ecosystems are those of regions underlain by granitic or other igneous bedrock, or by sandstones, quartzites, or other strata low in carbonate minerals. Resistant ecosystems occur in regions with underlying limestones, shales, or other formations rich in carbonates. Watersheds of larger size, or having deeper soil or glacial till layers, are also more resistant to acidification. In North America, mountain regions such as the Cascades, Sierra Nevadas, Rocky Mountains, Appalachians, and Adirondacks—all areas underlain in large measure by granitic rocks—tend to be sensitive to acid deposition. Most of eastern Canada, underlain by the granites of the Canadian Shield, is also sensitive. This region, furthermore, is one of thousands of lakes and streams. Much of the eastern and southeastern lowlands of the United States are also moderately sensitive to acid precipitation due to shortage of carbonate minerals in their waters. In Europe, large areas of Scandinavia are likewise sensitive because of their underlying granitic bedrock.

The buffering effect of carbonate minerals results from the equilibrium that is formed between carbonate ions, bicarbonate ions, un-ionized carbonic acid, and carbon dioxide in water. When acids are added to such a system, the hydrogen ions react with bicarbonate ions to form carbonic acid, and this, in turn decomposes to release carbon dioxide and water. The bicarbonate used up in this reaction is replaced by conversion of carbonate to bicarbonate. Thus, until the bi-

carbonates and carbonates in the watershed buffering system are exhausted, addition of acids produces no change in pH. In eastern North America, the buffering capacity of watersheds now averages about 40 percent less than normal, and for many lakes this capacity is completely exhausted (Schindler 1988). Depletion of this buffering capacity has probably been the product of decades of acid deposition, another reason why acidification of ecosystems has not been a serious problem until recently.

The seasonal input of acids to aquatic ecosystems is also an important consideration. In regions with cold winters, large quantities of acid materials may accumulate in winter snow. In spring, rapid melting of snow may flush these acids into streams and lakes, creating a heavy pulse of acidity. The extremes of pH that result may cause serious effects, even though the average pH of the water over the year is not greatly lowered.

The effects of acidification result not only from the acid ion itself, but also by increased leaching or chemical activation of aluminum and perhaps other heavy metals (Haines 1981). Heavy metals, in general, are enzyme poisons, and even slight changes in concentration of their biologically active forms can cause major effects. Aluminum is almost always present in soils in large quantities, and in its ionic form it is toxic to many plants and animals (Schindler 1988).

The response of aquatic ecosystems to acidification involves all major groups of organisms: producers, decomposers, and consumers (Stenson and Eriksson 1989). Among the producers, filamentous algae tend to increase in abundance, often forming mats on the bottom and on submerged surfaces. Increase in filamentous algae is largely due to the disappearance of snails, which graze on them (see below). Increase in filamentous algae is often at the expense of phytoplankton, reflecting competition for nutrients between these two groups (Hendrey et al. 1976). Rooted higher plants, which utilize bicarbonates as their source of carbon for photosynthesis, also tend to decline in acid waters, and sphagnum mosses, which are more efficient at using dissolved carbon dioxide, increase in abundance. Moderate acidification does not appear to reduce overall primary productivity (Schindler et al. 1985). When acidification becomes severe enough to slow nutrient recycling, the rate of primary production may be reduced, however.

Decomposition processes often are not affected by moderate acidification (Schindler et al. 1985), but in other cases, rates of decomposition appear to be reduced, largely because of reductions in the abundance and activity of detritus-feeding invertebrates, termed "shredders." Shredders serve to fragment leaves and other bulky units of dead tissue into smaller bits that are more easily attacked by bacteria. When the rate of decomposition declines, undecomposed material accumulates, and nutrient regeneration is slowed. Accumulation of undecomposed organic matter may seal the surface of the bottom sediments, also reducing the regeneration of nutrients. Ultimately, this results in the creation of a bog system, where peat deposits accumulate.

FIGURE 20.5

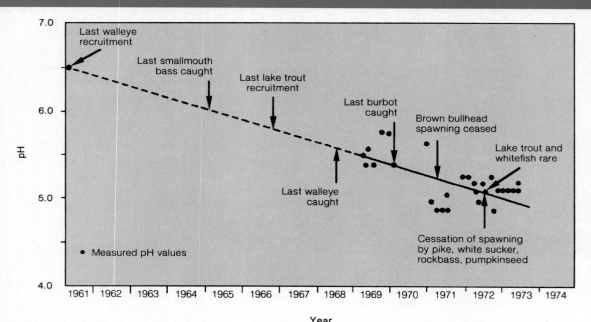

■ Progressive disappearance of fish species from Lumsden Lake, Ontario, with increasing acidity. Source: Data from R. J. Beamish, et al., "Long-Term Acidification of a Lake and the Resulting Effect on Fishes" in *Ambio* 4:98–102, 1975, Royal Swedish Academy of Sciences, Stockholm, Sweden.

Acid conditions lead to the decline in abundance and diversity of both zooplankton and benthic invertebrates (Hendry et al. 1976). The number of species of zooplankton declines rapidly as pH drops. In Ontario, for example, lakes with pH values above 5 possessed 9 to 16 species, three to four of which were of high abundance. Below pH 5, only one to seven species were present, of which only one to two were abundant. Among larger invertebrates, snails are highly sensitive to acid conditions because of their calcium carbonate shells. Below a pH of 6.6 the diversity of snails begins to decline, and below pH 5.2, snails are absent. Insect larvae such as those of mayflies and stoneflies show reduced diversity as pH drops, but also exhibit considerable replacement of acid-sensitive species by acid-resistant species (Hall and Ide 1987). Disappearance of fish also permits the dominance of certain groups of insect predators, such as certain kinds of dragonfly larvae, backswimmers, and midge larvae (Evans 1989, Stenson and Eriksson 1989). These invertebrates, in turn, modify the zooplankton community, reducing the abundance of some cladocerans and favoring dominance of certain copepods.

Loss of fish populations from acidified lakes and streams has attracted the greatest concern. Few comprehensive sets of data on changing conditions of lake acidity and fish populations over long periods exist, however. Nevertheless, it is clear that in several areas of eastern North America, such as the Adirondack Mountains and Nova Scotia, progressive acidification has led to loss of fish from lakes and streams (Haines 1981, 1986). Different species of fish show differential sensitivity to acid conditions, but salmonid fish such as lake whitefish, lake trout, rainbow trout, and Atlantic salmon are highly sensitive. At certain thresholds, acid conditions result in the disturbance of reproductive physiology in adults and in survival of eggs and young (Schofield 1976). Acid-related changes in calcium, aluminum, and organic matter content of the water influence these processes, too (Henrikson et al. 1989). As pH declines, fish recruitment typically fails first, but populations of older and larger fish survive for a while. At pHs of 4.5 or lower, most species are unable to reproduce, and lakes and streams with such pH values usually lack fish.

Loss of fish from some lakes can be rapid or slow. Lumsden Lake, located 65 km southwest of Sudbury, Ontario in the granitic La Cloche Mountains, showed a decline in pH from 6.8 in 1961 to 4.4 in 1971 (Beamish and Harvey 1972). This change was accompanied by disappearance of six of the seven species present, leaving only the lake chub, which was declining in abundance (Fig. 20.5). A similar pattern of rapid, progressive loss of species was noted in George Lake, Ontario (Beamish et al. 1975). In other cases, loss of fish may be a gradual process, requiring decades. In Norway, for example, adult fish may survive for 10 to 20 years after all reproduction has ceased (Henricksen et al. 1989).

Stream fish communities are also seriously affected by acidification (Lacroix 1987). In Nova Scotia, for example, Atlantic Salmon and several other fish had disappeared by the early 1980s from Halfway Brook, which was strongly acidified (Table 20.1). In nearby Westfield River, which was slightly less acidified, and North River, which was much less acidified, these species were still abundant. As river conditions

TABLE 20.1

Percent Composition and Total Biomass of Fish in Three River Areas in Eastern Nova Scotia in Relation to Acidity of Water.

SPECIES	NORTH RIVER (pH 5.6 TO 6.3)	WESTFIELD RIVER (pH 4.7 TO 5.4)	HALFWAY BROOK (pH 4.5 TO 5.0)
Atlantic Salmon	36%	7%	0%
White Sucker	28%	32%	3%
Chubs and Shiners	19%	1%	0%
American Eel	15%	52%	92%
Yellow Perch	2%	<1%	4%
Brook Trout	<1%	3%	<1%
Other	0%	5%	1%
BIOMASS (G/100 M²)	292	398	99

Source: Data from G. L. Lacroix, "Fish Community Structure in Relation to Acidity in Three Nova Scotia Rivers" in *Canadian Journal of Zoology,* 65:2908–2915, 1987, National Research Council of Canada, Ottawa, Ontario, Canada.

became more acid, both density and biomass of fish declined. In Westfield River, the decline of Atlantic salmon and several chubs and shiners was partially offset by increased abundance of white suckers and American eels. The only species favored by the strongly acid conditions of Halfway Brook was the American eel, which spawns in the ocean and lives as an adult in fresh water. In Halfway Brook, more than 90 percent of the overall fish biomass was contributed by this species.

Amphibians are also sensitive to acidification of waters, especially during their embryonic or larval stages (Freda 1986). Critical pH values for embryonic survival of frogs, toads, and salamanders range from 3.8 to 5.0, being lowest for species that inhabit naturally acidic habitats. Temporary ponds, in which many amphibians breed, are especially vulnerable to acidification. The effects of acid deposition might be a contributor to the widespread declines of amphibians noted in recent years (Milstein 1990).

More recently, secondary effects of acidification have been noted. The declines of aquatic invertebrates and fish have been implicated in declines of populations of several water birds (Mitchell 1989). Populations of fish-eating birds such as loons and mergansers have decreased in eastern North America in recent decades. Some studies show reduced reproductive success of these species in lakes without fish. In other studies, in lakes where amphibians and invertebrates remain abundant, no such decreases were noted. Many species of waterfowl feed primarily on aquatic invertebrates, such as midge fly larvae, during the breeding season. Declines of some of these species, such as the black duck (Fig. 20.6), might be related in part to decreased food availability due to acidification of aquatic systems. Songbirds, such as swallows and flycatchers, that nest near water and feed heavily on insects with aquatic larvae have been reported to show thin-shelled eggs and reduced reproductive success in areas with acidified waters. Recent reports from the Netherlands also suggest that eggshell thinning is occurring in woodland insectivorous birds, possibly due to low calcium availability caused by acidifica-

FIGURE 20.6

■ Acid deposition may be contributing to the decline of the black duck in eastern North America by decreasing the abundance of its aquatic invertebrate foods.

tion of the forest ecosystem. Clear evidence that these effects on birds are due to acid deposition, however, has not been obtained (Huckabee et al. 1989).

EXTENT OF ACIDIFICATION OF AQUATIC ECOSYSTEMS

The potential impact of acid deposition on aquatic ecosystems in sensitive regions downwind from major urban and industrial regions is enormous. Large numbers of lakes and streams in eastern Canada and the northeastern United States are at risk. About 350,000 lakes in eastern Canada are sensitive to acidification. About 14,000 of these lakes have become severely acidified, and 150,000 have experienced some ecological effects (French 1990). In Ontario, alone, an estimated 48,000 lakes are experiencing effects of acid deposition

(Loucks 1980). In the United States, about 1,000 lakes have been severely affected and 3,000 or so are somewhat affected (French 1990). The most concentrated impact has been in the Adirondack Mountains of New York State, where about 50 percent of the lakes above 600 m in elevation have pH's less than 5.0 and lack fish; in the 1930s only a few of these same lakes lacked fish.

Acidification exists on a similar scale in Scandinavia. In Norway, 200,000 lakes are sensitive to acidification, and nearly 78 percent of the lakes in the southernmost counties are now barren of fish (Henricksen et al. 1989). Between the early 1970s and 1986, the number of barren lakes in southern Norway doubled. In neighboring Sweden 90,000 sensitive lakes exist, about half of which have experienced significant acidification. About 90,000 km of streams in Sweden have also become acidified (World Resources Institute 1986).

OTHER EFFECTS OF ACID DEPOSITION ON AQUATIC ECOSYSTEMS

Increased deposition of acidic materials in watersheds resistant to acidification may have quite a different effect. In this situation, acid deposition may increase the transport of carbonates from watershed sources into streams and lakes, increasing their alkalinity and pH (Kilham 1982). Weber Lake, in the lower peninsula of Michigan, for example, has increased in pH from a range of 6.2 to 6.6 in the early 1950s to 7.2 to 8.2 in 1980. The total concentration of carbonates and bicarbonates in the lake water roughly doubled over this period. One of the effects of this change is to increase the availability of phosphorus, the limiting nutrient to productivity. Thus, acid deposition may contribute to eutrophication of lakes in areas that have watersheds rich in carbonates.

In other situations, such as estuarine and coastal marine waters that are well buffered against pH change, the nitrogen in acid deposition may act as a nutrient input, contributing to eutrophication. A case in point may be Chesapeake Bay, where eutrophication, with attendant decrease in water clarity, has been implicated in the decline of eel grass beds. About 25 percent of the N entering this estuary comes as atmospheric inputs, largely due to air pollution (French 1990).

EFFECTS OF ACID DEPOSITION ON FOREST ECOSYSTEMS

Regional acid deposition is an extension of local air pollution problems to broad areas of the terrestrial environment (Bormann 1985). Severe impacts of sulfur dioxide pollution on vegetation near power plants, smelters, and industries using large amounts of fossil fuels have long been recognized. Likewise, effects of photochemical smog, due largely to nitrogen oxides released by vehicles, on forest vegetation downwind from urban areas such as Los Angeles are well known. Not surprisingly, regional acid deposition may be contributing to general impacts on terrestrial vegetation. Increased mortality

FIGURE 20.7

■ Acid precipitation may be a major contributor to mortality of red spruce on Camels Hump Mountain, Vermont.

and reduced growth, collectively termed **forest decline,** have been observed over large areas of western Europe and in parts of eastern North America.

Forest decline was first noted for spruce and fir in the Black Forest of West Germany in the 1960s. In the 1970s and 1980s decline spread to other countries and began to affect pines, beech, and other hardwoods (Park 1987). By 1988, forest decline in Europe had affected about 35 percent of all forest area (French 1990). In Czechoslovakia, over 71 percent of forests were affected, and in West Germany 52 percent were damaged. The estimated loss in timber production for 1990 amounted to over $30 billion for Europe as a whole. In Sweden, where soil pH declined about 0.8 pH units over a period of 35 years, substantial change in composition of the herbaceous stratum of forests has also occurred (Falkengren-Grerup 1986).

In North America, forest decline has affected red spruce and Fraser fir at high elevations in the Appalachian Mountains (Fig. 20.7) and sugar maple in New England and eastern Canada (Vogelmann et al. 1985, Pitelka and Raynal 1989). Recently, it has been suggested that widespread mortality of dogwood in eastern deciduous forests is linked to acid precipitation that increases its vulnerability to fungal disease.

Four major hypotheses about how acid deposition might contribute to forest decline have been offered. These involve 1) direct effects on the foliage, 2) leaching of nutrient cations from the soil, 3) aluminum toxicity in the soil, and 4) increased frost vulnerability due to excess N availability (Foster 1989). Good evidence now exists that acidic cloud moisture increases the sensitivity of red spruce foliage to winter cold, and that acid deposition contributes to soil acidification and nutrient leaching (L. F. Pitelka, pers. comm.). Many of the stands showing decline, however, are old, overstocked, or affected by insects or drought. It is likely that forest decline is not a single acid-induced phenomenon, but the result of multiple stresses involving the effects of acid deposition, other air pol-

FIGURE 20.8

■ Wet (left) and dry (right) precipitation collectors on Whiteface Mountain, New York, are one of many sets of collectors in operation to measure the input of acids and other pollutants to forests in eastern North America.

FIGURE 20.9

■ Fraser fir trees in greenhouses at the Boyce Thompson Research Institute at Cornell University are exposed to pollutant levels comparable to those on Whiteface Mountain, New York, to determine the role of pollutants in forest decline.

lutants, weather conditions, insects, diseases, stand age, and repeated harvesting without fertilization (Chevone and Linzon 1988, Klein and Perkins 1988, Garner et al. 1989). The chemistry of precipitation in these forest areas is monitored continuously to determine the quantitative input of pollutants (Fig. 20.8). Controlled greenhouse studies are also being conducted to determine the role of specific pollutants in forest decline (Fig. 20.9).

RECOVERY OF ACIDIFIED LAKES

Reduction of acid deposition can permit the recovery of acidified lakes and streams. In Ontario, for example, sulfate and heavy metal emissions have been reduced to about one-third of early 1970s levels (Schindler 1988). This reduction has been accompanied by increase in pH and decrease in toxic metal levels of many lakes. Whitepine Lake, 90 km north of Sudbury, is one of the lakes that has shown both chemical and biological recovery (Gunn and Keller 1990). In 1980, it had a pH of 5.4, and several acid-sensitive fish, including lake trout, were not reproducing. By 1988, lake pH had increased to 5.9 and aluminum content of the water had declined 42 percent. Lake trout had resumed breeding, and recovery of invertebrate communities typical of non-acidified lakes had begun. Ontario has required major smelters, steel mills, and power plants to reduce sulfur oxide emissions 43 percent between 1989 and 1994 (APIOS 1991).

Estimates of the reductions that must be made in acid deposition to permit recovery of lakes in sensitive areas suggest that about 9 to 14 kg of sulfate per ha is the maximum annual deposition level allowable in areas with sensitive lakes (Schindler 1988). This compares to levels of 20 to 50 kg ha^{-1} yr^{-1} that are being received in most areas of eastern North America and Europe. Even given reductions of acid deposition, recovery may be slow where the buffering capacity of lake and stream watersheds has been exhausted by decades of acid leaching. Even given the chemical recovery of aquatic ecosystems, biological recovery may require a long process of recolonization and ecological interaction by acid sensitive species, before food webs typical of non-acidified ecosystems are restored.

On the short term, local treatments to mitigate the effects of acid precipitation on a small scale are possible for lake ecosystems. Where fish reproduction is declining due to moderate acidification, stocking of young fish that are past the most acid-sensitive age can be carried out. Species with greater acid tolerance can also be introduced as the pH of the water declines, although at the risk of a permanent change in the biotic structure of the lake ecosystem (Porcella 1989).

FIGURE 20.10

■ Experimental liming of lakes in the Adirondack Mountains as part of the Lake Acidification Mitigation Project suggests that the costs of maintaining normal acidity in this way are about $1,000 to $3,000 per hectare over a 10–year period.

Liming of lakes or their watersheds to restore some measure of buffering capacity can also be carried out (Fig. 20.10). In Sweden, several thousand lakes and over a hundred streams have been treated with ground limestone in efforts to counteract acidification. Experimental studies of the effects of liming lakes have been carried out in Ontario and New York (Porcella 1989, Keller et al. 1990). In small lakes in which the water volume is replaced within a short time by inflow and outflow, the effects of liming are transitory. For most larger lakes, liming mitigates acidification for only three to six years. Still, this is probably too expensive to be applied except in special circumstances. Assuming the application of five tons of calcite per ha, at a cost of about $200 per ton, for example, Huckabee et al. (1989) calculated that the cost of treatment of the watersheds of 1,419 lakes identified as needing such treatment in the northeastern United States would total $128 million over the five-year period of treatment effectiveness.

CONTROL OF ACID EMISSIONS AND LAKE RECOVERY

Over the long term, however, acid precipitation is a problem that can only be resolved by reducing regional air pollution, particularly by nitrogen and sulfur oxides. This resolution must be achieved on a broad regional, or even international, basis, as illustrated by the acid precipitation problem involving Canada and the United States in eastern North America and several countries in western Europe. The international aspects of acid deposition and its relation to basic industrial and urban economics have made progress toward resolving this problem very difficult. Effective international controls have yet to be devised, either in Europe or North America.

In Europe, efforts have been coordinated through the United Nations Economic Commission for Europe (ECE). The ECE presented a plan for a 30 percent reduction of S oxide releases over 1980 levels, to be achieved by 1993. This plan also calls for N oxide emissions to be held at 1987 levels. Most western European countries except the United Kingdom have agreed to this plan. The United Kingdom, which has the highest S emissions of any western European country, "exports" over 70 percent of these via weather systems that carry them eastward. At the same time, the United Kingdom receives less inflow of S from the west than any of the European mainland countries. Thus, acid deposition effects have been less serious in the United Kingdom than in Scandinavia or central Europe.

In North America, a long-standing argument has existed between Canada and the United States over the seriousness of acid deposition, and of the need for reduction in oxide emissions. John R. Roberts, the Canadian Minister of the Environment, has termed acid deposition "the most serious environmental threat ever to face the North American continent" (Whetstone and Rosencranz 1983). Emissions of S oxides in the United States are over five times those in Canada, and the transport of these oxides across the Canadian border means that about 50 percent of the S deposition in Canada is of United States origin (Schindler 1988). In New England, however, some of the S deposition is of Canadian origin. The Canadian government has pressured the United States to reduce oxide emissions, but the United States has generally responded that further study must be conducted before regulatory actions are taken. Within the United States, similar controversy has flared between the New England states, where release of oxides is low but acid deposition heavy, and the industrial midwestern states where most oxides originate. In 1990, reauthorization of the **Clean Air Act** included amendments to reduce S oxide emissions from power plants to 8.9 million tons by the year 2000, a reduction of 10 million tons over the 1980 level, and to reduce N oxide emissions to 2.0 million tons below the 1980 level by 1997. These reductions should go far to reduce acid deposition to levels that do not affect acid-sensitive ecosystems.

KEY STRATEGIES FOR REDUCING THE IMPACTS OF ACID DEPOSITION

1. Integrated international efforts to reduce S and N oxide discharges from industrial, vehicular, and agricultural sources.

2. Development of techniques to restore or enhance the buffering capacity of lakes and streams.

LITERATURE CITED

APIOS (Acidic Precipitation in Ontario Study). 1991. *Annual program report 1989/1990.* Ontario Ministry of the Environment, Toronto.

Beamish, R. J. and H. H. Harvey. 1972. Acidification of the La Cloche mountain lakes, Ontario, and resulting fish mortalities. *J. Fish. Res. Bd. Canada* **29**:1131–1143.

Beamish, R. J. and H. H. Harvey. 1972. Acidification of the La Cloche mountain lakes, Ontario, and resulting fish mortalities. *J. Fish. Res. Bd. Canada* **29**:1131–1143.

Beamish, R., W. Lockhart, J. Van Loon, and H. Harvey. 1975. Long-term acidification of a lake and the resulting effect on fishes. *Ambio* **4**:98–102.

Bormann, F. H. 1985. Air pollution and forests: An ecosystem perspective. *BioScience* **35**:434–441.

Chevone, B. I. and S. N. Linzon. 1988. Tree decline in North America. *Env. Poll.* **50**:87–99.

Evans, R. A. 1989. Response of limnetic insect populations of two acidic, fishless lakes to liming and brook trout (*Salvelinus fontinalis*). *Can. J. Fish. Aq. Sci.* **46**:342–351.

Falkengren-Grerup, U. 1986. Soil acidification and vegetation changes in deciduous forests in southern Sweden. *Oecologia* **70**:339–347.

Freda, J. 1986. The influence of acidic pond water on amphibians. *Water, Air and Soil Pollution* **30**:439–450.

Foster, N. W. 1989. Acidic deposition: What is fact, what is speculation, what is needed? *Water, Air, and Soil Pollution* **48**:299–306.

French, H. F. 1990. Clearing the air. Pp. 98–118 *in* L. Starke (Ed.), *State of the world 1990.* W. W. Norton, New York.

Galloway, J. N., G. E. Likens, and M. E. Hawley. 1984. Acid precipitation: Natural versus anthropogenic components. *Science* **226**:829–831.

Garner, J. H. B., T. Pagano, and E. B. Cowling. 1989. *An evaluation of the role of ozone, acid deposition, and other airborne pollutants in the forests of eastern North America.* USDA Forest Service Gen. Tech Pep. SE-59.

Gunn, J. M. and W. Keller. 1990. Biological recovery of an acid lake after reductions in industrial emissions of sulphur. *Nature* **345**:431–433.

Haines, T. A. 1981. Acidic precipitation and its consequences for aquatic ecosystems: A review. *Trans. Amer. Fish. Soc.* **110**:669–707.

Haines, T. A. 1986. Fish population trends in response to surface water acidification. Pp. 300–334 *in* National Academy of Sciences. *Acid deposition: Long-term trends.* National Academy Press, Washington, D.C.

Hall, R. J. and F. P. Ide. 1987. Evidence of acidification effects on stream insect communities in central Ontario between 1937 and 1985. *Can. J. Fish. Aq. Sci.* **44**:1652–1657.

Hendrey, G. R., K. Baalsrud, T. S. Traaen, M. Laake, and G.R. Raddum. 1976. Acid precipitation: Some hydrobiological changes. *Ambio* **5**:224–227.

Henricksen, A., L. Lien, B. O. Rosseland, T. S. Traaen, and I. S. Sevaldrud. 1989. Lake acidification in Norway: Present and predicted fish status. *Ambio* **18**:314–321.

Huckabee, J. W., J. S. Mattice, L. F. Pitelka, D. B. Porcella, and R. A. Goldstein. 1989. An assessment of the ecological effects of acidic deposition. *Arch. Environ. Contam. Toxicol.* **18**:3–27.

Keller, W., D. P. Dodge, and G. M. Booth. 1990. Experimental lake neutralization program: overview of neutralization studies in Ontario. *Can. J. Fish. Aq. Sci.* **47**:410–411.

Kilham, P. 1982. Acid precipitation: Its role in alkalization of a lake in Michigan. *Limnol. Oceanogr.* **27**:856–867.

Klein, R. M. and T. D. Perkins. 1988. Primary and secondary causes and consequences of contemporary forest decline. *Bot. Rev.* **54**:1–43.

Lacroix, G. L. 1987. Fish community structure in relation to acidity in three Nova Scotia rivers. *Can. J. Zool.* **65**:2908–2915.

Loucks, O. L. 1980. Acid rain: Living resource implications and management needs. *Trans. N.A. Wildl. Nat. Res. Conf.* **45**:24–37.

MacDonald, G. J. 1985. *Climate change and acid rain.* MITRE Corporation, McLean, VA.

Milstein, M. 1990. Unlikely harbingers. *National Parks* **64(7–8)**:19–24.

Mitchell, B. A. 1989. Acid rain and birds: How much proof is needed? *American Birds* **43**:234–241.

National Research Council. 1986. *Acid deposition: Long-term trends.* Nat. Acad. Press, Washington, D.C.

Park, C. C. 1987. *Acid rain: Rhetoric and reality.* Methuen, London.

Patrick, R., V. P. Binetti, and S. G. Halterman. 1981. Acid lakes from natural and anthropogenic causes. *Science* **211**:446–448.

Pitelka, L. F. and D. J. Raynal. 1989. Forest decline and acidic deposition. *Ecology* **70**:2–10.

Porcella, D. B. 1989. Lake acidification mitigation project (LAMP): An overview of an ecosystem perturbation experiment. *Can. J. Fish. Aq. Sci.* **46**:246–248.

Rodhe, H. and R. Herrera (Eds.). 1988. *Acidification in tropical countries.* John Wiley and Sons, New York.

Schindler, D. W. 1988. Effects of acid rain on freshwater ecosystems. *Science* **239**:149–156.

Schindler, D. W., K. H. Mills, D. F. Malley, D. L. Findlay, J. A. Shearer, I. J. Davies, M. A. Turner, G. A. Lindsay, and D. R. Cruikshank. 1985. Long-term ecosystem stress: The effects of years of experimental acidification on a small lake. *Science* **228**:1395–1401.

Schofield, C. L. 1976. Acid precipitation: Effects on fish. *Ambio* **5**:228–230.

Stenson, J. A. E. and M. O. G. Eriksson. 1989. Ecological mechanisms important for the biotic changes in acidified lakes in Scandinavia. *Arch. Environ. Contam. Toxicol.* **18**:201–206.

Vogelmann, H. W., G. J. Badger, M. Bliss, and R. M. Klein. 1985. Forest decline on Camels Hump, Vermont. *Bull. Torrey Bot. Club* **112**:274–287.

Waldman, J. M., J. W. Munger, D. J. Jacob, R. C. Flagen, J. J. Morgan, and M. R. Hoffmann. 1982. Chemical composition of acid fog. *Science* **218**:677–680.

Whetstone, G. S. and A. Rosencranz. 1983. *Acid rain in Europe and North America: National response to an international problem.* Env. Law Inst., Washington, D.C.

World Resources Institute. 1986. *World resources 1986.* Basic Books, Inc., New York.

Conservation Theory and Practice

To conserve the earth's biological diversity, we must go beyond the diagnosis of problems, difficult as this can often be. We must define strategies of managing nature that are both ecologically sound and acceptable to society. In this section we shall consider some of the major areas in which the development of conservation strategy is centered. These range from very applied topics such as sustained yield management, economic valuation of ecological resources, and practical ways to control exotic species, to theoretical topics such as how to maintain the genetic health of populations. At an even more general level, strategies must be developed to aid species that have declined to the brink of extinction and to develop comprehensive systems of preserves to protect as much of the earth's biodiversity as possible.

CHAPTER 21

Sustained Yield Exploitation

Harvesting populations of wild plants and animals is still a major human activity. Some human populations, such as the North American Inuit, still rely heavily on harvests of wild plants and animals for food. Aside from a few intensive aquacultural operations, most fishery harvests are taken from wild populations of fish, shellfish, and plants such as kelps. Likewise, although modern forestry increasingly emphasizes the production of timber on intensive "tree farms," timber harvests worldwide still mostly involve the cutting of natural forest stands. Can such harvests be made on a sound, continuing basis? And if so, how?

In this chapter we shall consider some of the theory on which the harvesting of natural populations is based. Following this, we shall evaluate the success of the theory in its application to particular species. Finally, we shall consider some special problems relating to the harvest of wild species of plants and animals: clearcutting and game cropping.

HARVESTING TERMINOLOGY

Resource exploitation theory has attempted to define patterns of sustainable yield for wild populations. **Sustained yield** is any level of harvest that can be taken from a population indefinitely. **Maximum sustained yield (MSY)** is the greatest harvest that can be taken indefinitely. MSY does not take into account factors such as the cost of taking such a harvest relative to the various kinds of benefits that may result. Nor does it consider the various benefits that may result from leaving some or all of the MSY unharvested, such as recreational benefits for people who come to view and photograph the unharvested plants or animals. **Maximum sustained revenue (MSR)** is the greatest net profit—difference between the value of the harvest and the costs to take it—that can be obtained on a continuing basis. The yield that provides the best compromise of all of the economic and societal considerations is termed the **optimum sustained yield.**

In its most general terms, harvesting theory suggests that up to some threshold, animals that are removed are partially or completely replaced by the increased reproduction or survival of those remaining (Caughley 1984). If the threshold is exceeded, the population declines toward extinction.

EARLY HARVEST MODELS

One of the earliest sustained yield formulations was suggested by Paul Errington (1946), based largely on his studies of bobwhite quail populations in the midwestern United States, and popularized by Durward Allen (1954). Errington visualized midwestern farmland as having a certain **threshold of protection** for quail during the winter: a certain average capacity, based on availability of food and cover, to carry birds through to the next breeding season (Fig. 21.1). This threshold was regarded as being essentially constant from year to year. If the fall population was below this threshold, over-winter survival would be very high, whereas if the fall population was above this level, the excess would certainly die from factors such as starvation or predation. Thus, the portion of the fall population in excess of the threshold of protection constituted a harvestable surplus—birds that would otherwise die anyway due to lack of winter food and cover.

The threshold-of-protection concept assumes that hunting mortality and natural winter mortality are almost perfectly compensatory. When this is true, and if a population follows strong seasonal changes in resources such as food and cover occur, the concept may provide a good guide to harvesting, especially for a species like the bobwhite quail with a high reproductive potential. For less seasonal environments and for species whose populations fluctuate less from season to season, the threshold-of-protection idea fails to recognize the great adjustment that most populations can make to mortality, and thus greatly underestimates the harvest that can be taken (McCullough 1979).

CARRYING CAPACITY AND POPULATION REGULATION

Modern harvesting models are based on the concepts of carrying capacity of the habitat and of how populations grow relative to this carrying capacity. The **carrying capacity** is the maximum population that can exist in equilibrium with average conditions of resource availability—the supplies of food, nest sites, cover, and similar environmental features that are used by organisms. For most plants, resources comprise water, nutrients, and sunlight, with occupation of an area of substrate often providing the guarantee of these resources in the case of rooted or attached plants. For animals, the supply of appropriate foods is often the primary determinant of carrying capacity, but the availability of nest sites or sites that provide protection from predators or bad weather may force carrying capacity to be less than that possible on the basis of food.

Although carrying capacity can be defined as an average level of resource availability, it is, in fact, a variable. Actual resource conditions change seasonally and from year to year. Often, as well, long-term trends in resource conditions occur due to biotic succession, climatic shifts, or human impacts on the environment.

Many factors affect the growth of populations. The population in an area of habitat grows when reproduction and immigration of individuals exceeds mortality and emigration of individuals to other areas. When populations are low, relative to carrying capacity, competition for resources is likely to be low, and individuals find it easy to protect themselves against predators or normal fluctuations of weather. Reproduction is active, mortality infrequent, individuals from neighboring areas may be attracted to the site, and the population growth rate is high. When populations are close to the carrying capacity, however, competition for food and other resources is strong; individuals often may be unable to find cover against predators or bad weather. Reproduction is likely to be low, mortality high, and emigration to less densely populated areas common. Thus, some of the factors that affect population growth increase the intensity of their action as the population approaches the carrying capacity. These are termed **density dependent factors,** because the intensity of their effects is determined by the density of the population relative to carrying capacity.

Other factors may act quite differently. Diseases, catastrophic weather events, episodes of pollution, and other factors may reduce reproduction, kill individuals, or trigger the movement of individuals in ways quite unrelated to population density. These are termed **density independent factors** because the intensity of their effects is not related to the size of the population relative to its carrying capacity.

Obviously, density dependent and independent factors are the ends of a continuum; one factor might be strongly dependent on population density, another weakly related to

FIGURE 21.1

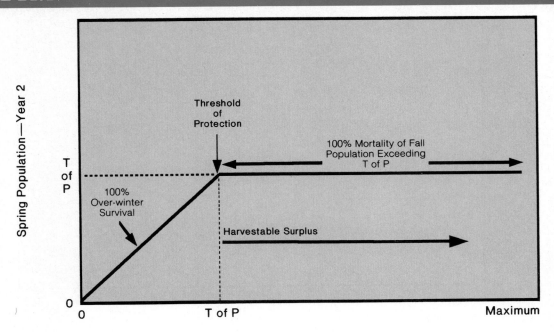

Threshold of Protection

100% Mortality of Fall Population Exceeding T of P

100% Over-winter Survival

Harvestable Surplus

Spring Population—Year 2

T of P

0

0   T of P   Maximum

Fall Population—Year 1

■ The threshold of protection (T of P) concept assumes that the habitat can support a certain density of animals over an unfavorable season, such as winter in the midwestern United States, and that the excess population entering that season is a harvestable surplus.

density. Generally, biotic factors such as competition and predation tend to show strong density dependence and abiotic factors such as severe weather and chemical pollution density independence. However, the action of a disease agent may be determined by the genetic composition of a population, rather than its density. The action of a pollutant might also be more severe in a dense than in a sparse population if individuals are weakened because of competition for food. Populations may thus be subject to influences that are primarily density dependent over certain periods, and largely density independent over other periods (McCullough 1990).

In any case, the size of any population at a given instant is the result of the action of both density dependent and density independent factors. For large, long-lived organisms such as most vertebrates, density dependent factors are usually assumed to be the strongest determinants. Small, short-lived organisms such as most insects and other invertebrates are assumed to be much more strongly influenced by density independent factors.

LOGISTIC HARVESTING MODELS

Much sustained yield theory has been based on the **logistic equation,** which describes the S-shaped curve of growth of a population in a habitat with a certain carrying capacity, and a pattern of growth determined primarily by density dependent factors (Fig. 21.2). The logistic equation is given below:

$$\frac{dN}{dt} = rN \left(\frac{K - N}{K} \right)$$

In this equation, N is the number of individuals in the population at a given time, K is the average carrying capacity, r is the per capita potential population growth rate (under nonlimiting conditions), and t is time. The symbol d means "change in," so that the left side of the equation reads, "change in number of individuals divided by change in time," or growth per unit time. The right side of the equation states that this growth is equal to rN multiplied by (K − N)/K.

FIGURE 21.2

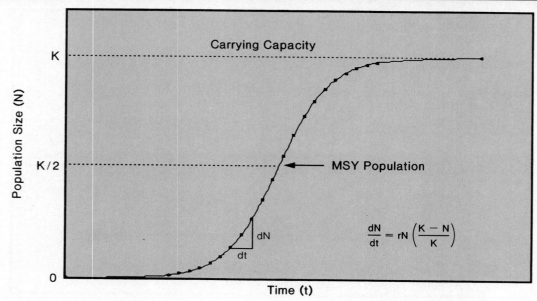

■ The logistic growth curve assumes a population whose growth toward a certain carrying capacity (K) is continuously and gradually slowed by various restraints. According to this equation, the point of maximum sustained yield (MSY) is halfway between zero and the carrying capacity.

The term rN is the potential growth that would occur if resources were not limiting, based on the number, N, of individuals present and the growth potential, r, associated with each. The remaining term applies a "degree of realization" factor to potential growth. This term, in effect, expresses the increasing intensity of density dependent factors as the population approaches carrying capacity. When N is very small relative to K, this expression yields a numerical value near 1.0; most of the potential growth is thus realized. When N is near K, the value of the expression drops to near zero; most of the potential growth is not realized. Thus, the logistic equation can be stated:

Realized growth = Potential growth × Degree of realization.

If we graph realized growth against N, as it increases from 0 to K, we see a dome-shaped curve with a peak halfway between 0 and K (Fig. 21.3a). In other words, population growth is greatest when N equals $K/2$. If exactly this number is harvested, the population will not increase, and during the next time interval it will once again produce this maximum growth (which could once again be harvested). This suggests that MSY is possible when a population is about half its carrying capacity. How many individuals does this represent? When N is half of K, only half the potential growth is realized; this equals a quarter of the potential growth at K, or $rK/4$. What harvesting effort (fraction of the $K/2$ population) does this involve? If realized growth is only 50 percent of potential growth at $K/2$, this equals $r/2$ times N; harvest effort thus equals $r/2$. In summary:

■ MSY should be possible at a
  population of $K/2$
■ The MSY, in numbers of individuals,
  equals $rK/4$
■ For MSY, the fraction of the
  population to harvest is $r/2$

A curve of this sort was constructed for the Peruvian anchoveta fishery (See chapter 16) using data from harvests and fishing effort in the 1960s (Fig. 21.3b). The MSY predicted from this relation was about ten million tons annually, a figure exceeded by about three million tons in 1970. This predicted harvest has not been sustained, however. Some possible reasons for this will be noted later in the chapter.

Having determined the maximum sustained yield from a population, how should this yield be taken to maintain a population at its maximum production level? Two possibilities exist, the removal of a fixed quota per unit time, or the application of a fixed harvesting effort per unit time. In a hunted population, a fixed quota is illustrated by the removal of a certain number of animals, regardless of the effort involved, and fixed effort by control over the total number of hunter days, regardless of the number of animals removed.

The fixed quota option, illustrated in Fig. 21.4a, assumes the removal of the same number of individuals regardless of the population size. If the population is somewhat above the $K/2$ level, this means that the harvest will exceed the actual growth of the population, and the population will thus decline toward the $K/2$ level of highest growth rate. However, if the population has fallen somewhat below the

FIGURE 21.3

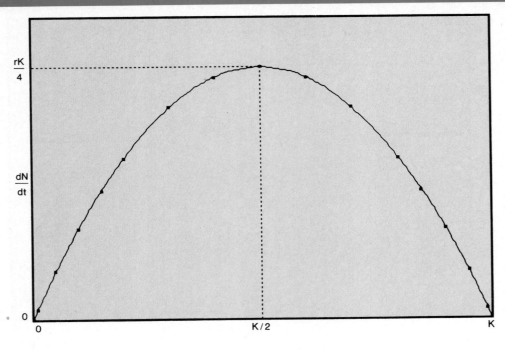

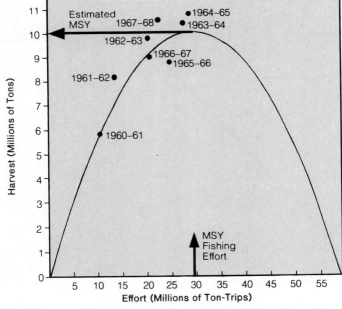

a. Population growth per unit time (dN/dt) in relation to population size (N), according to the logistic equation, indicating the point of maximum sustained yield (K/2). b. Logistic population growth in relation to fishing effort, fitted to data for the Peruvian anchoveta fishery, with an estimate of maximum sustained yield (Modified from L. K. Boerma and J. A. Gulland. 1973. *J. Fish. Res. Bd. Canada* **30**:2226–2235).

K/2 level, harvest will also exceed growth, and the population will therefore decline further. As long as the same quota is applied in this situation, the population will be forced downward, ultimately to extinction.

Application of a fixed harvest effort (Fig. 21.4b), however, acts in a different fashion. If the population is above the K/2 level, harvest exceeds growth, and the population declines toward the level of greatest growth rate. If the population falls below the K/2 level, however, the harvest resulting

FIGURE 21.4

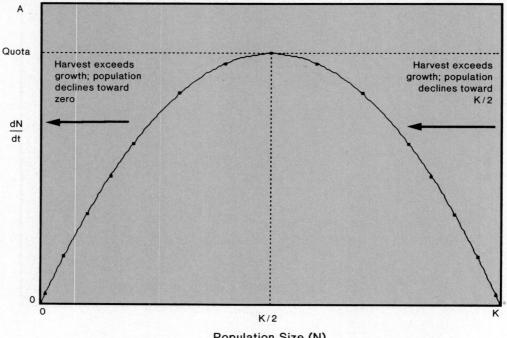

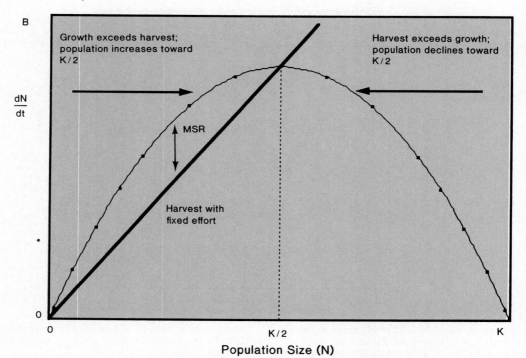

■ a. The effect of fixed quota harvests on population size when populations stray from the MSY level. The MSY level is unstable since deviations below this level initiate a sequence leading to extinction. b. The effect of fixed harvest effort on population size when populations stray from the MSY level. The MSY level is stable since deviations both upward and downward return the population toward this level. If the harvest effort line is equivalent to the cost of harvesting, maximum sustained revenue occurs at the point where the difference between effort and harvest curves is greatest.

FIGURE 21.5

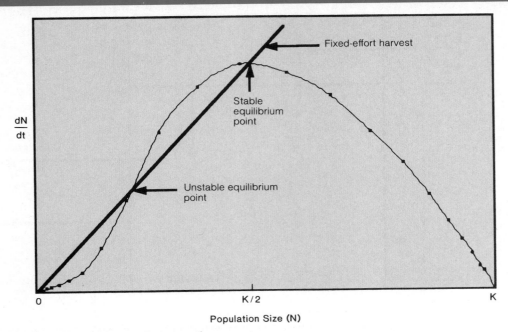

■ Curves of population growth per unit time may have more than one equilibrium point, the lower of which is unstable, as indicated in Fig. 21.4a.

from the same effort becomes less than the actual population growth, and the population increases toward the level of maximum growth. Thus, a fixed effort strategy of harvest should have the effect of regulating the population at the desired level of K/2. A fixed effort harvesting technique is therefore, in principle, a sounder approach than a harvest by quota.

In some cases the curve of population growth per unit time may have more than one intersection point with a line representing fixed harvest effort (Fig. 21.5). In this case, if the population falls below the lower point of intersection, the tendency of continued harvesting is to drive the population toward zero. The continued depressed state of the Peruvian anchoveta harvest following its decimation by heavy exploitation in combination with a severe El Niño may be an example of this phenomenon.

How do wildlife managers determine where a population stands relative to the MSY level, given the difficulties of determining the carrying capacities of wild populations? At any sustained yield level, recruitment into the population must equal mortality. For animals in which yearlings can be distinguished from older adults, the maximum ratio of yearlings (the component just recruited into the adult population) to total adults is thus an estimate of the fraction of the adult population that can be harvested (Fig. 21.6). This practice assumes that at the MSY population density most of the unrealized growth of the population involves poor reproduction and pre-yearling survival, and that hunting harvest and natural mortality of older adults are nearly fully compensatory.

Sustained yield theory based on the simple logistic equation is usually too oversimplified to be applied in most cases, however. The attractive simplicity of the logistic equation hides several unrealistic assumptions about the members of a population. For example, representation of a population by N assumes that all individuals are the same; each counts "one." Real populations of 100 individuals might in one instance consist of young, highly reproductive animals, and in another of old, largely post-reproductive individuals, however. The potential population growth rates, r, for these two groups would be quite different. The model also assumes that there is an exact, 1:1 trade-off among different causes of mortality in harvested populations, e.g., complete compensation of hunting mortality by reduced natural mortality.

The populations of most higher plants and animals violate several assumptions of the simple logistic equation. If reproduction is concentrated in certain age classes, or if harvesting increases the average fecundity of individuals, the MSY population tends to lie between half and three-quarters of the carrying capacity, K (Fig. 21.7). **Age-structured populations,** or populations in which reproductive potential differs for individuals of different age, are common. Many populations have a carrying capacity set by shortages of food or other resources, and show substantial change in mean fecundity of individuals in the population range close to the carrying capacity. Thus, the MSY population is very frequently greater than K/2 (Goodman 1980).

FIGURE 21.6

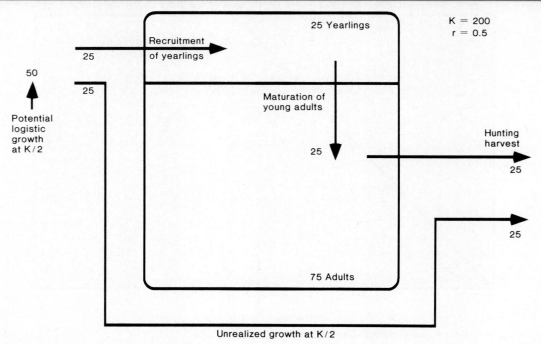

■ The maximum ratio of yearlings to total adults in a population (25/100 in this case) is an indication of the fraction of the adult population that can be harvested at MSY (rK/4, or 25 in this case).

FIGURE 21.7

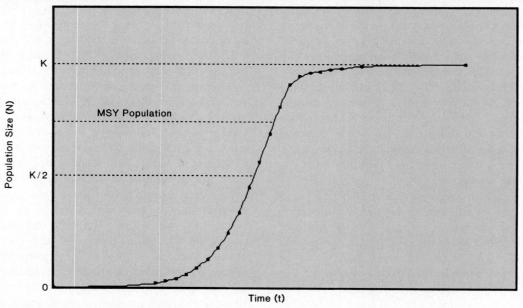

■ The shape of population growth curves of many species causes the MSY population to lie closed to K than predicted by the simple logistic relation.

FIGURE 21.8

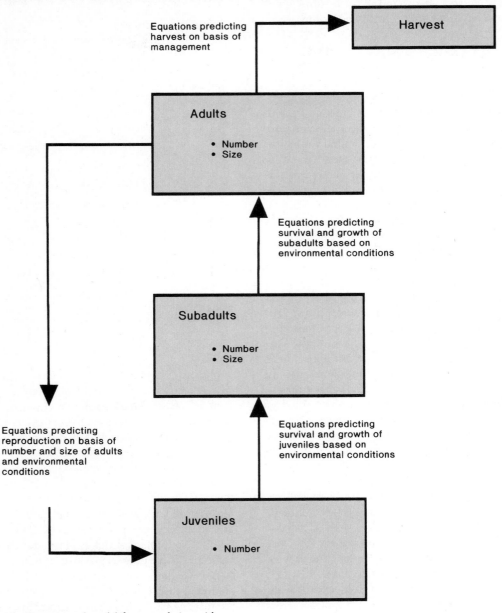

Equations predicting
harvest on basis of
management

Harvest

Adults

• Number
• Size

Equations predicting
survival and growth of
subadults based on
environmental conditions

Subadults

• Number
• Size

Equations predicting
reproduction on basis of
number and size of adults
and environmental
conditions

Equations predicting
survival and growth of
juveniles based on
environmental conditions

Juveniles

• Number

■ A simple example of a dynamic pool model for a population with three age classes.

DYNAMIC POOL MODELS

Simple logistic models do not explicitly take the age structure of the population into account, nor do they explicitly consider such relationships as recruitment rate, body growth rate, natural mortality rate, and harvest rate of individuals of various ages or sizes. **Dynamic pool models** (Pitcher and Hart 1982, Getz and Haight 1989) describe the numbers or biomass of different age (or size) classes as a function of these different processes, each of which is described by one or more equations.

Dynamic pool models can become very complicated, but are structured in the general fashion indicated in figure 21.8. The population is divided into several age or size classes. Recruitment into the youngest or smallest class is calculated as a function of the population of adult organisms (older or larger size classes) and other important environmental variables. The growth of individuals is described mathematically; together with mortality factors, this determines the number and biomass of individuals in older or larger classes. Natural mortality is related mathematically to factors such as predation, physical conditions, and degree of compensation with harvesting mortality. Mortality due to harvesting is determined by management practices, such as regulation of net size, bag limit, length of harvest season, and so forth. Thus, a dynamic pool model consists of a set of many equations.

HARVEST MANAGEMENT IN PRACTICE

How effective is regulation of harvest in promoting high productivity in exploited populations? Does reduction of populations to some level below the carrying capacity increase reproduction, improve health, and reduce natural mortality? From a conservation standpoint, does controlled harvesting promote healthier populations and greater ecosystem stability? How effective are short-term adjustments in harvest as management tools? Unfortunately, although many examples can be cited of how populations have behaved under particular management schemes, almost no carefully designed experiments address these questions (Caughley 1984).

For large mammals, many observational studies show that populations can reach, or even exceed, the maximum level that the habitat can support, leading to poor reproduction and high mortality. Often, such populations develop in the absence of natural predators, as well as in the absence of hunting by humans. These populations may severely disturb the ecosystem. On St. Matthew Island, Alaska, for example, a herd of 29 introduced reindeer grew to about 6,000 animals in only 19 years (Klein 1968). By this time, the tundra vegetation had become severely overgrazed, and in three years starvation had reduced the herd to only 42 individuals. The vegetation following the crash showed severe damage, with the lichen component being reduced to about 10 percent of normal.

White-tailed deer on the George Reserve of the University of Michigan (Fig. 21.9) provide one of the best documented examples of these facts, and of how harvesting affects population dynamics and ecosystem conditions (McCullough 1979). The reserve, completely fenced, is a 4.64 square kilometer area that contains a mixture of forest, grassland, and wetland habitats. In 1928, two male and four female deer were introduced to the area, at that time a private estate. In 1933, the first effort to estimate the number of deer was made by counting the animals passing a line of 30 people walking across the reserve. A count of 162 animals was obtained, a value later corrected to an estimate of 181 deer. This, itself, is one of the most remarkable examples of exponential growth by a large animal population. During the late 1930s, the population was estimated to have exceeded 220, and severe browsing effects became evident. Substantial culling of the herd did not begin until the late 1930s, by which time considerable natural mortality must have begun to occur.

Systematic study of the George Reserve deer herd began in 1952, after the reserve had become a research area of the University of Michigan, and has continued until present. These studies have led to an estimate of carrying capacity of about 176 deer, and of 99 deer in the post-harvest population at which maximum sustained yield, about 49 animals per year,

FIGURE 21.9

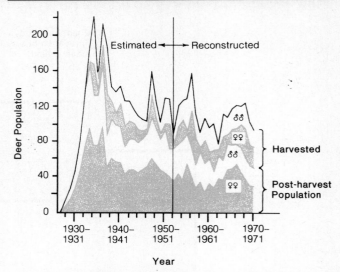

■ The white-tailed deer population on the University of Michigan's George Reserve has been subject to carefully controlled harvesting since the early 1930s in an effort to determine maximum sustained yield and to determine the relationship between hunting harvest and recruitment of deer into the adult population (Modified from McCullough 1979).

is obtained. Under this management regime, virtually the entire mortality for the population is from hunting. Based on a model of the dynamics of the population, a harvest level that permitted the population to remain at 176 animals would result in about 18 adult and 26 pre-adult deaths from natural causes (e.g., malnutrition, disease).

Thus, strong compensatory responses in mortality and recruitment—close to those predicted by basic harvesting theory—indeed are shown by the George Reserve deer. Members of George Reserve population are also healthier than deer in unharvested populations (McCullough 1984). Storch (1989) also found that this held true for chamois in a heavily harvested population in Bavaria, in which animals of all age classes weighed more than those in a lightly harvested population in similar habitat. Thus, for some ungulates, the basic premises of harvesting theory appear to be sound.

How effective are short-term adjustments in harvest or bag limits to maintaining populations at MSY levels? Because hunting harvest of waterfowl is such a large fraction—about half—of total winter mortality, adjustments in bag limits have been considered to be essential means of management to assure that an adequate breeding population survives to produce the next season's crop of waterfowl. This practice assumes that lower hunting harvests in years of low populations lead to increased over-winter survival, and thus to larger breeding populations in the following summers.

Recent studies of the influence of variable bag limits for mallard ducks (Fig. 21.10), and of the relation of bag limits to hunting harvest, have raised questions about the efficacy of this approach (Anderson and Burnham 1978). In years of

FIGURE 21.10

■ Analyses of the influence of liberal and restrictive bag limits for mallard ducks suggest that a close trade-off exists between winter mortality due to hunting harvest and natural causes.

TABLE 21.1

Over-Winter Survivorship of Mallard Ducks in North America in Years of Low and High Abundance and Differing Bag Limits.

	1962	1970
Total Population	7.6 million	11.6 million
Bag Limit	2	4–6
Hunting Harvest	2.2 million	4.1 million
Percent of Population	28.9%	35.3%
Over-Winter Survival	56%	56%

Source: Data from D. R. Anderson and K. P. Burnham, "Effect of Restrictive and Liberal Hunting Regulations on Annual Survival Rates of the Mallard in North America" in *Transactions of North American Wildlife and Natural Resources Conference,* 43:181–186, 1978, Wildlife Management Institute, Washington, DC.

low numbers, lower bag limits did, in fact, correlate with reduced percentages of the total population taken by hunters. However, the reduced hunting mortality did not translate into increased over-winter survival (Table 21.1). Nichols et al. (1984) reached similar conclusions after a review of studies of various species of ducks. Hunting harvest and other causes of mortality appeared to be strongly compensatory; an increase in one was balanced by a decrease in the other.

These observations led to a test of the effect of stabilized duck hunting regulations for North American mallards from 1979 through 1985 (Sparrowe and Patterson 1987). During this period, hunting season length and bag limit were held constant. A complicating factor during this study period was a series of drought years and the decline of populations of several duck species, including mallards, to low levels. Nevertheless, analyses of hunting mortality in relation to overall mortality supported the conclusion that hunting mortality was largely compensatory with natural mortality. The implication of this conclusion is that changes in bag limits from year to year have little management value.

Recent studies of population fluctuations of the pintail also tend to support the conclusion that yearly adjustments of bag limits are of little value (Raveling and Heitmeyer 1989). Analysis of population and harvest data from 1961 to 1985 shows that changes in populations are most closely related to spring habitat conditions. The number of pintails harvested was correlated most closely with the total size of the population. In turn, these populations varied from one extreme to the other, even during years of large bag limits, suggesting that habitat conditions rather than changes in bag limit were the major control on populations.

Thus, for waterfowl, carrying capacity can change greatly from year to year. Except in extreme cases, efforts to increase the next year's breeding population by reducing the present year's harvest by a restrictive bag limit result mainly in greater natural mortality over the winter. The concept of stabilized hunting regulations has been adopted as part of the North American Waterfowl Management Plan for the period 1986 to 2000 (See chapter 11).

SHORTCOMINGS OF POPULATION MODELS

A major shortcoming of most models is that they usually focus on the population dynamics of single exploited species, and do not consider the ecosystem as a whole. Few models have attempted, however, to consider the role of an exploited species in relation to other species in its food chain, or to species that are competitors or have close symbiotic or commensal ties. Harvests that reduce the population of one herbivore to 50 percent of its carrying capacity, for example, may free food resources for use by other herbivores. Increase in the populations of these species may ultimately utilize the initial excess of resources, in effect establishing a new, lower carrying capacity for the exploited species. Or, as in the case of the Antarctic krill fishery, exploitation of a species at one trophic level may reduce populations of other species of higher trophic levels, some of which may be of equal or even greater value in the long run (Beddington and May 1980).

On top of this, populations of some species are subject to exploitation in different locations. The Pacific whiting, *Merluccius productus,* a migratory fish of the Pacific coast of North America, illustrates the complexity of managing migratory fish (Getz and Haight 1989). Whiting spawn in waters off southern California and Mexico. As young fish, they are prey to various other fish, but as they grow they become predators on young of some of these same species. As they migrate north in summer, they are harvested by fisheries based at several ports in the United States and Canada. Thus, harvesting strategy must consider a complicated set of ecological, economic, and political relationships.

FIGURE 21.11

■ Clearcutting has been the primary technique of forest harvest in the Pacific Northwest.

HOW SHOULD FORESTS BE MANAGED?

Two major alternative systems of management of forests for timber harvest exist: even-aged and uneven-aged stand management. Even-aged management emphasizes plantations of single tree species of the same age, which are managed to maximize growth to some pre-determined size and harvested at about the same time. The most common pattern of harvesting is **clearcutting,** in which all trees are cut and all harvestable timber removed at one time (Fig. 21.11). Regeneration may occur from the natural seed bank, from seed broadcast onto the site after cutting, or from seedlings planted after cutting. About two-thirds of United States timber is harvested in this fashion.

Modified forms of clearcutting include **shelterwood cutting** and **seed-tree cutting.** In shelterwood cutting the mature trees are removed in several loggings over several years. This retains a partial canopy that favors germination and early growth of a new generation of trees, and is used for species that require shade during their early life. All mature trees are nevertheless eventually cut, and a young, even-aged stand replaces them. In seed-tree cutting, all trees are cut except for a scattering of mature trees that are left to provide the seed for forest regeneration.

Uneven-aged management involves harvesting of individual mature trees at frequent intervals (Fig. 21.12). **Selective harvesting** is most suitable for tree species whose seedlings are tolerant of shaded conditions. For species whose seedlings require strong sunlight, groups of trees—in effect, small clear cuts—must be done. Uneven-aged management has long been the basic technique of forest management in Europe.

It is often argued that clearcutting represents the most efficient and profitable technique of forest management for timber production. Haight (1987), however, showed that this does not necessarily apply to management initiated in forests of mixed composition and age structure. In a specific example involving management of California white and red fir forests,

FIGURE 21.12

■ In the Panhandle National Forest, Idaho, selective logging is carried out with the aid of a helicopter that lifts logs from steep slopes to waiting trucks.

Getz and Haight (1989) found that the profitability of uneven-aged management was greater than that resulting from conversion of the stands to even-aged management by initial clearcutting.

GAME CROPPING AND WILDLIFE CONSERVATION

It is often suggested that sustained yield harvesting of wildlife, or **game cropping,** in ecosystems such as the East African savannas can produce greater yields of meat and other animal products than can livestock ranching in the same areas. The logic of this suggestion is that a diverse fauna of wildlife species specialized for feeding on different plants and at different heights can exploit plant production more fully than can a single domestic species such as cattle. Furthermore, native wildlife are often resistant to endemic diseases, such as trypanosomiasis in Africa, to which domestic animals are susceptible. The advantage of game cropping for conservation, according to this argument, is that it gives strong economic justification for retention of native ecosystems rather than replacing them by ranching operations.

Several efforts have been made to test the economic viability of game cropping. However, few long-term, economically viable game cropping operations have yet been established. **Game farming,** a ranching operation that mixes livestock with selected species of wildlife, has proved successful in several areas (Pollock 1969). The game animals are usually species for which trophy hunters are willing to pay large fees, or animals that provide meat and skins for tourist restaurants or markets. In South Africa, however, several species of medium to large ungulates are commonly raised for meat production on private farms (Skinner 1985), and some of these operations appear to be equally profitable to beef ranching.

Despite the logic about harvest of primary production by a diverse fauna of native species, a larger sustained yield by game cropping than by ranching has never been demonstrated. This is largely due to the fact that a ranching operation is able to control the age and sex composition of a herd of domestic animals very closely, maintaining the herd in a highly productive state with regard to the desired product (Caughley 1976). Thus, the MSY of a herd of domestic animals seems to be higher than that possible from a wild fauna harvested in any practical fashion. According to this view, the economic arguments for wildlife conservation relate to tourism and sport hunting, rather than to game cropping.

On the other hand, productive livestock ranching depends on the maintenance of vegetation that domestic animals can utilize fully (McCullough 1979). In areas that cannot be maintained as open grasslands, cattle are unable to harvest plant production as effectively as can a diverse community of wild species. In many places livestock ranching may also be difficult because of shortage of water or presence of poisonous plants (Skinner 1985). Efforts to extend ranching activities into such regions are likely to destroy wildlife communities, but create inefficient ranching operations.

With recognition that preservation of wildlife ecosystems depends heavily on the creation of economic benefits for local human populations, conservationists are viewing subsistence harvest of game with renewed interest. In Zimbabwe, for example, efforts are being made to promote sustained use of wildlife resources on communal lands that were created from the tribal reserves of colonial days (Anon. 1989). These efforts emphasize local development and management of activities such as wildlife-centered tourism, trophy hunting, commercial game and fish harvests, and subsistence harvests. Under this arrangement, profits and benefits from wildlife resources would accrue to the local population, encouraging the sustained management of wildlife and minimizing poaching. Some of these projects are being studied to see if mixed-species systems of domestic animals and wildlife are economically viable and ecologically stable. Thus, game ranching and game farming may yet find an important niche in the conservation scheme.

KEY PRINCIPLES OF SUSTAINED YIELD MANAGEMENT

1. Determination of harvest level for a population must consider the carrying capacity of the habitat and the effects of harvesting on the age structure and fecundity of the population.

2. Determination of harvest levels for one or more species should consider the relationships of all species in the ecosystem.

3. Harvesting policy should consider benefits to society of both harvested and unharvested portions of the population.

LITERATURE CITED

Allen, D. L. 1954. *Our wildlife legacy.* Funk and Wagnalls, New York.

Anderson, D. R. and K. P. Burnham. 1978. Effect of restrictive and liberal hunting regulations on annual survival rates of the mallard in North America. *Trans. N. Am. Wildl. Nat. Res. Conf.* **43**:181–186.

Anonymous. 1989. *Wildlife utilization in Zimbabwe's communal lands.* Unpublished report of the Centre for Applied Social Sciences, World Wide Fund for Nature, and Zimbabwe Trust, Harare, Zimbabwe.

Beddington, J. R. and R. M. May. 1980. Maximum sustainable yields in systems subject to harvesting at more than one trophic level. *Mathematical Biosciences* **51**:261–281.

Caughley, G. 1976. Wildlife management and the dynamics of ungulate populations. Pp. 183–245 *in* T. H. Coker (Ed.), *Applied Biology,* Vol. I. Academic Press, New York.

Caughley, G. H. 1984. Harvesting of wildlife: Past, present, and future. Pp. 3–14 *in* S. L. Beasom and S. F. Roberson (Eds.), *Game harvest management.* Texas A & I University, Kingsville.

Errington, P. L. 1946. Predation and vertebrate populations. *Quart. Rev. Biol.* **21**:144–177, 221–245.

Getz, W. M. and R. G. Haight. 1989. *Population harvesting: Demographic models of fish, forest, and animal resources.* Princeton Univ. Press, Princeton, NJ.

Goodman, D. 1980. The maximum yield problem: Distortion of the yield curve due to age structure. *Theor. Pop. Biol.* **18**:160–174.

Haight, R. G. 1987. Evaluating the efficiency of even-aged and uneven-aged stand management. *Forest Sci.* **33**:116–134.

Klein, D. R. 1968. The introduction, increase, and crash of reindeer on St. Matthew Island. *J. Wildl. Manag.* **32**:350–367.

McCullough, D. R. 1979. *The George Reserve deer herd: Population ecology of a K-selected species.* Univ. Mich. Press, Ann Arbor.

McCullough, D. R. 1984. Lessons from the George Reserve, Michigan. Pp. 211–242 *in* L. K. Halls (Ed.), *White-tailed deer: Ecology and management.* Stackpole Books, Harrisburg, PA.

McCullough, D. R. 1990. Detecting density dependence: Filtering the baby from the bathwater. *Trans. N. A. Wildl. Nat. Res. Conf.* **55**:534–543.

Nichols, J. D., M. J. Conroy, D. R. Anderson, and K. P. Burnham. 1984. Compensatory mortality in waterfowl populations: A review of the evidence and implications for research and management. *Trans. N. A. Wildl. Nat. Res. Conf.* **49**:535–554.

Pitcher, T. J. and P. J. B. Hart. 1982. *Fisheries ecology.* Croom Helm, London, England.

Pollock, N. C. 1969. Some observations on game ranching in southern Africa. *Biol. Conserv.* **2**:18–23.

Raveling, D. G. and M. E. Heitmeyer. 1989. Relationships of population size and recruitment of pintails to habitat conditions and harvest. *J. Wildl. Manag.* **53**:1088–1103.

Skinner, J. D. 1985. Wildlife management in practice: Conservation of ungulates through protection or utilization. Pp. 25–46 *in* J. P. Hearn and J. K. Hodges (Eds.), *Advances in animal conservation.* Clarendon Press, Oxford, England.

Sparrowe, R. D. and J. H. Patterson. 1987. Conclusions and recommendations from studies under stabilized duck hunting regulations: Management implications and future directions. *Trans. N. Am. Wildl. Nat. Res. Conf.* **52**:320–326.

Storch, I. 1989. Condition in chamois populations under different harvest levels in Bavaria. *J. Wildl. Manag.* **53**:925–928.

CHAPTER 22

Conservation Economics

As human exploitation of nature intensifies, conflicts over the use of resources grow exponentially. Questions arise about the value of bits of nature that have traditionally had no market value, but that some people feel are of infinite value. How does the value of a population of endangered least Bell's vireos in their riparian habitat in southern California stack up against the value of the same land as the bottom of a water-storage reservoir? Unless the vireos and their associated riparian ecosystem can be shown to have real and substantial value, the scales are tipped in favor of the ultimate construction of a reservoir. If such values do not exist, or cannot be shown, laws that now protect vireos are likely to be changed, or perhaps manipulated in devious ways, to permit alternative values to be realized. Thus, a full and accurate economic valuation of ecological resources—natural ecosystems and their member species—becomes more and more necessary.

Determining the full value of ecological resources is difficult, however, since they range so widely in character. Some resources, such as timber trees, are fixed in location and are easily measured; others, such as the great whales, range over vast areas of international oceans and are difficult even to count. Still other environmental features, such as pure air or a scenic natural landscape, are difficult to value in monetary terms. All of these resources, in addition, possess values that are real, but lie largely outside the market systems that define the values of goods and services in modern societies. If the full economic value of such resources is not determined, however, these resources may easily fall victim to unregulated exploitation or destruction.

The objective of the field of conservation economics is to develop ways of determining the full economic value of ecological resources, and gaining the consideration of these values within the political and economic systems of world nations. In this chapter, we shall examine the different sorts of economic values that are associated with ecological resources, and consider some of the new approaches to their comprehensive valuation.

FIGURE 23.4

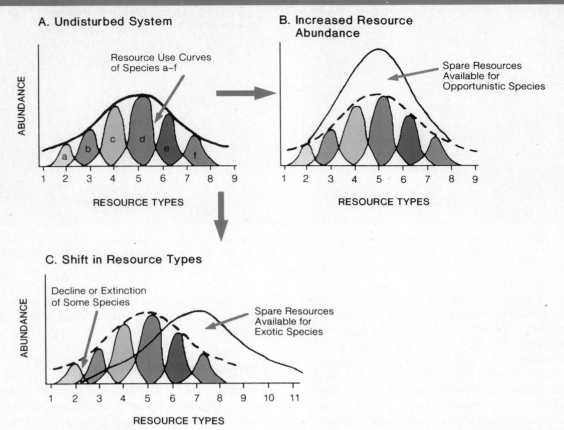

A. Undisturbed System

Resource Use Curves of Species a–f

ABUNDANCE

a b c d e f

RESOURCE TYPES

1 2 3 4 5 6 7 8 9

B. Increased Resource Abundance

Spare Resources Available for Opportunistic Species

RESOURCE TYPES

1 2 3 4 5 6 7 8 9

C. Shift in Resource Types

Decline or Extinction of Some Species

ABUNDANCE

Spare Resources Available for Exotic Species

RESOURCE TYPES

1 2 3 4 5 6 7 8 9 10 11

■ Disturbance of ecosystems can create unutilized resources by increasing total resource availability or shifting the types of resources available in ways that prevent native species from fully utilizing them, thus creating opportunities for establishment of exotics (Modified from Fox and Fox 1986).

Escape from Competitors

The competitive regime that a species encounters in a new area also affects its probability of establishment. In undisturbed, continental ecosystems, native species normally exploit resources almost completely, leaving few "empty niches," or unexploited types of resources, that an exotic species can preempt with ease. In isolated regions, such as oceanic islands, however, species capable of exploiting certain resources may neither have colonized the area nor evolved from earlier colonists. Disturbance of an indigenous ecosystem, in addition, often changes patterns of resource availability (Fig. 23.4), either by altering the overall abundance of resources, or by changing the relative abundance of various types of resources (Fox and Fox 1986). Such changes often create underutilized resources that can be exploited by exotic species.

Analysis of successes and failures of attempted introductions shows that success is greatest when the invaded system has few species (Moulton and Pimm 1986, Brown 1989). In large measure, this probably reflects competitive pressure: the fewer the competitors, the more likely it is that an exotic will succeed. This principle contributes to the high success of exotics on islands and in regions such as Florida,

which has a much less diverse native biota than other, less isolated subtropical areas.

The success of so many urban and agricultural exotics in the New World is certainly due in part to the inability of native North American species to exploit the resources of these habitats fully. Cities have existed in the Old World for thousands of years, and several birds and mammals evolved to exploit this habitat and its resources. In most of North America, cities did not exist in Precolumbian times, and urban species thus did not evolve. European settlement created cities with unoccupied habitats and unexploited resources, assuring the success of urban exotics from the Old World. The same is true of cultivated croplands in North America. On the other hand, repeated attempts to introduce the Eurasian *Coturnix* quail to the New World have probably failed in part because native quail exist in all areas otherwise suitable for *Coturnix*.

MECHANISMS OF INTRODUCTION

Detrimental introductions have been both deliberate and inadvertent (Brown 1989, Heywood 1989). Domestic animals, both livestock and pets, have been carried to all parts of the world. Often they establish **feral populations,** or

FIGURE 23.5

FIGURE 23.6

■ The nutria, brought to the United States from South America for fur farming, has escaped and colonized marshes along the Atlantic and Gulf coasts, often severely disturbing marsh vegetation by its grazing.

■ The red fox, introduced to Australia to control the introduced European rabbit, has also become a serious predator on native small mammals.

populations that live and reproduce in the wild. In many places, feral animals have wreaked havoc on the native vegetation, as ungulates have in the Galapagos Islands, or on the fauna, as cats have in parts of Australia. Undomesticated species, such as the South American nutria (*Myocastor coypus*) that was imported to the United States for fur farming, have also escaped and established wild populations (Fig. 23.5). Predators have often been imported to control previously introduced exotics, such as the red fox to control rabbits in Australia (Fig. 23.6) and mongooses to control rats on various oceanic islands. Often, these predators have become even more serious pests. Settlers to new continents and islands have carried crop and horticultural plants with them, as well. Some of these species, have invaded the native vegetation aggressively, such as the guava and quinine tree have in the Galapagos Islands (Macdonald et al. 1988).

Inadvertent introduction of weeds to new continents or islands has occurred commonly by seed mixed with that of crop plants or carried in the food, bedding material, or pelage of livestock. Some weeds also may have arrived as seed in soil used as ballast by early sailing ships. Animals have inadvertently been carried along with their plant hosts, or, in the case of some of the most destructive rodents, as unwelcome guests on ships and other forms of transport. The brown tree snake probably reached Guam (See chapter 2), for example, in shipments of fruit from areas of Southeast Asia.

DETERMINANTS OF IMPACT

Urban and agricultural ecosystems are particularly prone to invasion of exotics, and these invaders often directly damage the interests of humans: their lawns, farms or gardens, pets or livestock, and homes. The impacts of exotics in natural ecosystems range from direct and obvious to indirect and subtle. The factors that determine their impact include their

ecological distinctiveness and their potentials for competitive displacement, disease or disease transmission, and genetic swamping.

Ecological Distinctiveness

Species that are highly successful in invading natural communities are usually different in structure, physiology, or behavior from native forms. As a result, counteradaptation to them by native species is often absent. Consequently, exotics with distinctive features can become **keystone exotics**—invaders that can cause almost complete biotic reorganization of the ecosystem. Keystone exotics, being so different ecologically from native forms, are often able to invade undisturbed natural systems, effectively pursuing patterns of resource exploitation that native species have not evolved. Keystone exotics can exert their effects at the population, community, or ecosystem level (Macdonald et al. 1989, Vitousek 1990). Some disrupt major patterns of coevolved mutualism among native species. Others exert trophic effects that restructure the food web. Still others modify basic processes of nutrient cycling or hydrology, or encourage unusual patterns of disturbance by fire or other factors.

Among plants, keystone exotics are exemplified by invasive species that become the dominants of the vegetation. In northern Australia, for example, a prickly leguminous shrub, *Mimosa pigra,* has invaded wet ecosystems ranging from open sedge marshes to monsoon woodlands (Braithwaite et al. 1989). Open sedge marshes tend to be invaded first, and the mimosa soon forms a dense, monospecific stand of shrubs up to five meters high (Fig. 23.7). Invasion in some areas has been promoted by soil disturbance by feral water buffalos, another exotic species. Later, woodlands are invaded, the mimosas forming a dense understory that prevents the reproduction of native trees. The almost complete restruc-

FIGURE 23.7

■ *Mimosa pigra*, a spiny legume shrub, has invaded vast areas of Australia's Northern Territory, crowding out native vegetation and changing habitat conditions for many animals.

FIGURE 23.8

■ The pike killifish, a voracious predator, has almost completely eliminated native fish in many areas in Florida.

turing of wetlands vegetation by this species threatens to cause a massive change in animal communities of the region.

In the southeastern United States, vines introduced from the Far East, such as Japanese honeysuckle (*Lonicera japonica*) and kudzu (*Pueraria lobata*), have smothered native forest areas in a manner similar to the banana poka (*Passiflora mollissima*) in Hawaii (See chapter 9). In southern Florida, an Australian paperbark tree (*Melaleuca quinquenervia*), an aggressive invader of the borders of cypress swamps, is replacing the native pond cypress (*Taxodium ascendens*) and other species over wide areas (Ewel 1986). The paperbark is not only tolerant of prolonged flooding, but is highly adapted to recovering from fire. It thus poses a major threat to the vegetation of the Everglades area. The Brazilian peppertree (*Schinus terebinthefolius*) is also an aggressive invader of disturbed upland sites in southern Florida. With bird-dispersed seeds and the ability to resprout after fire, it is a potential invader of almost all upland habitats.

Among animals, introduced livestock have played a keystone role in the transformation of extensive areas of perennial grasslands into annual grasslands in North America, South America, and Australia (Mack 1989). On the Columbia Plateau, annual Eurasian grasses such as cheat (*Bromus tectorum*), which is adapted to both heavy grazing and soil disturbance, have become the dominants of once-perennial grasslands. In this region, where bison were originally absent, the original bunch grasses were adapted to a light grazing regime. These grasses, which do not possess rhizomes, unlike many of the grasses of the Great Plains, are unusually sensitive to the effects of grazing and trampling by large ungulates (Mack and Thompson 1982). A similar transformation followed livestock introduction to California (See chapter 4). Dominance of the grasslands by annuals has lead in turn to significant changes in animal life. In Manitoba, for example, the numbers of several grassland birds were increased in areas dominated by introduced Eurasian plants, whereas the abundance of others was reduced (Wilson and Belcher 1989).

In the Andean foothills of northern Patagonia, introduced red, fallow, and axis deer have severely disturbed evergreen beech and cedar forests of two of Argentina's most important national parks (Veblen et al. 1989). Browsing by deer has nearly eliminated one understory tree species and reduced the abundance of many shrubs and herbaceous plants. Regeneration of the over-story beeches and cedars in tree-fall gaps has also been inhibited. These changes pose a serious threat to efforts to encourage the recovery of populations of two species of small native deer, the pudu (*Pudu pudu*) and the huemul (*Hippocamelus bisuclus*). In New Zealand, damage by red deer to high-elevation tussock grasslands has contributed to the decline of the takahe (*Notornis mantelli*), a flightless rail that feeds on the bulblike bases of these grasses (Mills et al. 1989). Once thought to be extinct, this rail was rediscovered in the mountains of southern South Island, an area now included in Fiordland National Park.

In aquatic ecosystems, introduced plants such as water hyacinth (*Eichornia crassipes*) and hydrilla (*Hydrilla verticillata*) have transformed open waters to weed-choked swamps in many locations in the southern United States. Introduced fish such as the sea lamprey in the St. Lawrence Great Lakes, the peacock bass in Gatun Lake, Panama, and the Nile perch in Lake Victoria, East Africa also illustrate the effects of keystone exotics (See chapter 14). In the United States, more than 40 species of fish have become established by intentional or accidental releases (Taylor et al. 1984). Of these, 37 are of tropical or subtropical origin, their establishment being primarily in the southern and southwestern states. In Florida, some 20 species have become established, including the pike killifish (*Belonesox belizanus*), a voracious predator that has virtually eliminated native fish from many waters (Fig. 23.8).

Potential for Competitive Exclusion

Invaders are sometimes ecologically similar to native species, and able to utilize the same habitats and resources with greater efficiency, particularly if the natural system has been disturbed. Introduced riparian shrubs and trees, such as salt cedars (*Tamarix* spp.) have proved to be formidable competitors to native willows, cottonwoods, and other species, replacing them along streams in much of the southwestern United States. Sika deer (*Cervus nippon*) have been introduced to several locations in North America from Japan. On Assateague Island, Maryland, the Sika deer (Fig. 23.9) has increased in numbers at the apparent expense of the native white-tailed deer (*Odocoileus virginianus*) (Keiper 1985). The impact of feral burros on desert bighorn sheep in Death Valley is also an example of the detrimental competitive influence of an exotic species (See chapter 5). Even the ring-necked pheasant, usually regarded as a beneficial game species, may have a detrimental competitive effect on native prairie chickens. Vance and Westemeier (1979) found that pheasants tended to dominate prairie chickens in behavioral encounters and to displace them from booming grounds. Pheasants often laid eggs in prairie chicken nests, reducing the reproductive success of the latter.

Potential for Disease or Disease Transmission

Introduced species may be reservoirs or vectors of diseases that affect native forms. Exotic birds introduced to Hawaii, for example, carried avian malaria, and introduced mosquitos provided a system of vectors to transmit the disease to native land birds (See chapter 2). Fungal diseases introduced to North America have substantially altered the composition of the eastern deciduous forests (von Broembsen 1989). The chestnut blight (*Endothia parasitica*), accidentally introduced to North America in 1890s on nursery stock of Asian chestnuts, has eliminated the American chestnut as a component of mature forests. The Dutch elm fungus, brought to North America from Europe in the early 1900s on logs for veneer production, has caused extensive mortality of native elm species.

Potential for Genetic Swamping

Some exotics are also genetically similar to native species, and thus able to interbreed with them. The result, genetic swamping, can be loss of the genetic identity of the native species, a form of extinction (See chapter 2). In New Zealand, for example, the introduced mallard duck (*Anas platyrhynchos*) has interbred so extensively with the native grey duck (*Anas superciliosus*) that less than 5 percent of the combined population of these two forms is representative of the pure grey duck (Gillespie 1985). It appears unlikely that the grey duck will survive in the wild, given the extent of interbreeding.

Genetic swamping is a major threat to many fish species. Miller et al. (1989) concluded that hybridization with other native or introduced fish was a contributing factor in

FIGURE 23.9

■ The sika deer has become established in several locations in North America, and is sometimes a strong competitor of the native white-tailed deer.

the extinction of 15 species or subspecies in North America. Several forms of the cutthroat trout (*Oncorhynchus clarki*), for example, have disappeared because of interbreeding with introduced rainbow trout (*O. mykiss*). Eight other forms of the cutthroat are endangered, threatened, or reduced to the point of concern, largely as a result of hybridization (Williams et al. 1989).

DEALING WITH DETRIMENTAL EXOTICS

Several strategies can be used to avoid and reduce the impacts of exotics on natural ecosystems. Nowhere is the need for such measures greater than on oceanic islands, where exotics rank as the most serious threats to survival of endemic species.

1. *Prevention of Entry.*
   This strategy includes prohibition of entry, quarantine of living plants or animals before or at entry, inspection of biological materials to detect unwelcome associates, and fumigation of imported materials to kill hitchhiking invaders. Human inspection at ports of entry is being augmented by use of more effective canine or electronic "sniffers" to detect biological materials that might contain unwelcome exotics (Stone and Loope 1987). Strict rules against the importation of new species of plants and animals have been adopted in some locations. In the Galapagos Islands, most of which constitute an Ecuadorian national park, the importation of exotics is completely prohibited, for example.

2. *Control of Spread.*
   Once an exotic has gained a foothold, its spread can sometimes be slowed by direct control or by control of vectors that assist its dispersal. This approach is

being used against the Argentine ant in several locations. In Hawaii, the Argentine ant has been introduced to most islands, and has invaded Haleakala National Park on Maui and Hawaii Volcanos National Park on Hawaii. The ant preys on native insects and other small invertebrates, many of them endemic. Some of these forms have disappeared from areas invaded by the ant. In South Africa, Argentine ants have invaded fynbos shrublands, which contain many species of plants with ant-dispersed seeds (Bond and Slingsby 1984). Native ants harvest the seeds of these species and store them underground, effectively "planting" them. Argentine ants displace this native ant fauna, but do not bury these seeds, leaving them vulnerable to small mammals and other seed predators. In both regions, efforts are being made to control this ant, which has flightless queens and spreads by the division of colonies, with the use of toxic baits.

3. *Protection of Pristine Areas.*

Location and protection of areas that have not yet been invaded by exotics is also an effective strategy. Small islets lying offshore from inhabited islands are sometimes free of exotic species. In parts of Hawaii, islands of native vegetation, or **kipukas,** exist where extensive lava flows have isolated them from the main vegetated landscape. Fencing, monitoring, and weeding such sites may enable small areas of pristine nature to be maintained with relatively small effort.

4. *Local Eradication.*

Exotics can often be removed and then excluded from small areas, such as small islets or fenced exclosures. Rabbits and goats were eliminated from Round Island, an islet of 151 ha near Mauritius in the Indian Ocean (Atkinson 1988), providing an excellent opportunity for preservation of many endemic plants and animals of the Mauritius Islands. Eradication of goats, dogs, cats, black and Norway rats, and five exotic plants from Nonesuch Island, a six-ha islet off Bermuda, has provided a site for the attempted restoration of native Bermudian ecosystems, as well as for nesting of the endangered cahow petrel (Wingate 1985). In Hawaii, exclosures for goats and pigs have been created in some 51 sites to protect specific examples of native ecosystems or to promote recovery of native vegetation (Stone and Loope 1987). In some of these exclosures plants thought to have been locally extinct have reappeared, apparently from seeds lying dormant in the soil.

5. *Protection of Individuals.*

Exclosures or chemical repellents may be necessary to protect individuals of critically reduced species. Some exclosures in Hawaii function to protect the last individual plants from destruction by goats or pigs (Stone and Loope 1987). Chemical repellents have also been used to deter rats from climbing into canopies of certain native trees to feed on flowers and fruits, in an effort to increase reproduction of critically threatened species.

6. *General Population Reduction.*

Reduction of keystone exotics over large areas has been attempted for only a few species. Eradication is usually difficult, if not impossible, with highly successful invaders of habitats that possess wilderness characteristics. Direct control, by killing or capturing of animals and weeding of undesirable plants, has been successful in a few cases. Efforts have been made to reduce feral goat and pig populations in the Hawaiian Islands by hunting, both by citizen hunters and by hired hunters, for example (Stone and Loope 1987).

Biological Control

Biological control is the establishment of a strong, negative biological interaction against a designated species. Although biological control is sometimes portrayed as "reestablishing the balance of nature," it really involves the creation of a strong "imbalance of nature." The goal is to reduce an undesirable species to such a low abundance that its impact is minor. Biological control is often more desirable than eradication, since as long as the detrimental species survives at a low abundance, its control agent also survives. If a pest is eradicated locally, its biological control agent may also disappear, favoring reinvasion of the pest from a source area.

Biological control employs several different strategies. Protection and promotion of native predators, parasites, and disease species is one strategy. Under this strategy, pesticides are used in ways that do not kill such species, and in ways sought to improve the microhabitats and resources on which they depend. A second strategy is the release of native biological control agents at critical times, either to reestablish populations in habitats where they do not persist well, or to overwhelm the target pest at a critical stage of its life cycle. The third strategy, sometimes termed "classical" biological control, is introduction of exotic biological control species, either from the native area of the pest species or from some other location.

Reduction of the impact of the introduced perennial Klamath weed (*Hypericum perforatum*) is one of the major successes of biological control in North America (Dahlsten 1986). Klamath weed was introduced to western North America from Europe about 1900, and by the 1940s had invaded hundreds of thousands of hectares of rangeland, crowding out desirable range species and creating health problems in livestock that ate the plant. Major infestations of the weed have also occurred in Australia, South America, and South Africa. Eight species of herbivorous insects that fed on Klamath weed were identified in its European homeland, the most effective being a beetle of the Family Chrysomelidae, *Chrysolina quadrigemina*. Introduction of this beetle, and often one or more other species, has led to varying degrees of biological control of Klamath weed. In California, biological control has reduced this weed to less than 1 percent of its former abundance.

CRITERIA FOR DELIBERATE INTRODUCTIONS

Introduction of exotic species should obviously be done only under special circumstances, and then only after careful study. A clear need or benefit should be evident, such as the benefit of achieving biological control of a pest species. Suitable habitat and resources for the proposed introduction must exist. Strong evidence that the introduced form will not have a detrimental impact on other species, or on basic ecosystem properties, must be provided. Finally, it must be shown that the introduced form can be confined to the region for which its suitability has been evaluated.

KEY STRATEGIES FOR MANAGEMENT OF EXOTIC SPECIES

1. Prevent the introduction of species that may become keystone exotics or threaten native species in other ways.

2. Devise techniques to reduce keystone exotics or eradicate them from key conservation areas.

LITERATURE CITED

Atkinson, I. A. E. 1988. Opportunities for ecological restoration. *New Zealand Journal of Ecology* **11**:1–12.

Baker, H. G. 1974. The evolution of weeds. *Ann. Rev. Ecol. Syst.* **5**:1–24.

Bock, C. E. and L. W. Lepthien. 1976. Population growth of the cattle egret. *Auk* **93**:164–166.

Bond, W. and P. Slingsby. 1984. Collapse of an ant-plant mutualism: The Argentine ant (*Iridomyrmex humilis*) and myrmecochorous Proteaceae. *Ecology* **65**:1031–1037.

Braithwaite, R. W., W. M. Lonsdale, and J. A. Estbergs. 1989. Alien vegetation and native biota in tropical Australia: The impact of *Mimosa pigra*. *Biol. Cons.* **48**:189–210.

Brown, J. H. 1989. Patterns, modes and extents of invasions by vertebrates. Pp. 85–109 *in* J. A. Drake, H. A. Mooney, F. di Castri, R. H. Groves, F. J. Kruger, M. Rejmanek, and M. Williamson (Eds.), *Biological invasions: A global perspective.* John Wiley & Sons, Chichester, England.

Coblentz, B. E. 1978. The effects of feral goats (*Capra hircus*) on island ecosystems. *Biol. Conserv.* **13**:279–286.

Dahlsten, D. L. 1986. Control of invaders. Pp. 276–302 *in* H. A. Mooney and J. A. Drake (Eds.), *Ecology of biological invasions of North America and Hawaii.* Ecological Studies, Vol. 58. Springer-Verlag, New York.

Ewel, J. J. 1986. Invasibility: Lessons from South Florida. Pp. 214–230 *in* H. A. Mooney and J. A. Drake (Eds.), *Ecology of biological invasions of North America and Hawaii.* Springer-Verlag, New York.

Fox, M. D. and B. J. Fox. 1986. The susceptibility of natural communities to invasion. Pp. 57–66 *in* R. H. Groves and J. J. Burdon (Eds.), *Ecology of biological invasions.* Cambridge Univ. Press, Cambridge, England.

Gillespie, G. D. 1985. Hybridization, introgression, and morphological differentiation between mallard (*Anas platyrhynchos*) and grey duck (*Anas superciliosus*) in Otago, New Zealand. *Auk* **102**:459–469.

Heywood, V. H. 1989. Patterns, extents and modes of invasions by terrestrial plant. Pp. 31–60 *in* J. A. Drake, H. A. Mooney, F. di Castri, R. H. Groves, F. J. Kruger, M. Rejmanek, and M. Williamson (Eds.), *Biological invasions: A global perspective.* John Wiley & Sons, Chichester, England.

Keiper, R. R. 1985. Are sika deer responsible for the decline of white-tailed deer on Assateague Island, Maryland? *Wildl. Soc. Bull.* **13**:144–146.

Macdonald, I. A. W. 1985. The Australian contribution to southern Africa's alien flora: An ecological analysis. *Proc. Ecol. Soc. Aust.* **14**:225–236.

Macdonald, I. A. W., L. Ortiz, J. E. Lawesson, and J. B. Nowal. 1988. The invasion of highlands in Galapagos by the red quinine-tree *Cinchona succirubra*. *Env. Conserv.* **15**:215–220.

Macdonald, I. A. W., L. L. Loope, M. B. Usher, and O. Hamman. 1989. Wildlife conservation and the invasion of nature reserves by introduced species: A global perspective. Pp. 215–255 *in* J. A. Drake, H. A. Mooney, F. di Castri, R. H. Groves, F. J. Kruger, M. Rejmanek, and M. Williamson (Eds.), *Biological invasions: A global perspective.* John Wiley & Sons, Chichester, England.

Mack, R. N. 1989. Temperate grasslands vulnerable to plant invasions: Characteristics and consequences. Pp. 155–179 *in* J. A. Drake, H. A. Mooney, F. di Castri, R. H. Groves, F. J. Kruger, M. Rejmanek, and M. Williamson (Eds.), *Biological invasions: A global perspective.* John Wiley & Sons, Chichester, England.

Mack, R. N. and J. N. Thompson. 1982. Evolution in steppe with large, hooved mammals. *Amer. Nat.* **119**:757–773.

Miller, R. R., J. D. Williams, and J. E. Williams. 1989. Extinctions of North American fishes during the past century. *Fisheries* **14(6)**:22–38.

Mills, J. A., W. G. Lee, and R. B. Lavers. 1989. Experimental investigations of the effects of takahe and deer grazing on *Chionochloa pallens* grassland, Fiordland, New Zealand. *J. Appl. Ecol.* **26**:397–417.

Moulton, M. P. and S. L. Pimm. 1986. Species introductions to Hawaii. Pp. 231–249 *in* H. A. Mooney and J. A. Drake (Eds.), *Ecology of biological invasions of North America and Hawaii.* Springer-Verlag, New York.

Myers, K. 1986. Introduced vertebrates in Australia, with emphasis on the mammals. Pp. 120–136 *in* R. H. Groves and J. J. Burdon (Eds.), *Ecology of biological invasions.* Cambridge Univ. Press, Cambridge, England.

Norton, D. A. 1991. *Trilepidea adamsii*: An obituary for a species. *Conserv. Biol.* **5**:52–57.

Rees, M. D. 1989. Red wolf recovery effort intensifies. *End. Sp. Tech. Bull.* **14(1–2)**:3.

Shaughnessy, G. 1986. A case study of some woody plant introductions to the Cape Town area. Pp. 37–43 *in* I. A. W. Macdonald, F. J. Kruger, and A. A. Ferrar (Eds.). 1986. *The ecology and management of biological invasions in southern Africa.* Oxford Univ. Press, Cape Town, South Africa.

Stone, C. P. and L. L. Loope. 1987. Reducing negative effects of introduced animals on native biotas in Hawaii: What is being done, what needs doing, and the role of national parks. *Env. Conserv.* **14**:245–258.

Taylor, J. N., W. R. Courtney, Jr., and J. A. McCann. 1984. Known impacts of exotic fishes in the continental United States. Pp. 322–373 *in* W. R. Courtney, Jr. and J. R. Stauffer, Jr. (Eds.), *Distribution, biology, and management of exotic fishes.* Johns Hopkins Univ. Press, Baltimore, MD.

Vance, D. R. and R. L. Westemeier. 1979. Interactions of pheasants and prairie chickens in Illinois. *Wildl. Soc. Bull.* **7**:221–225.

Veblen, T. T., M. Mermoz, C. Martin, and E. Ramilo. 1989. Effects of exotic deer on forest regeneration and composition in northern Patagonia. *J. Appl. Ecol.* **26**:711–724.

Vitousek, P. M. 1990. Biological invasions and ecosystem processes: Towards an integration of population biology and ecosystem studies. *Oikos* **57**:7–13.

von Broembsen, S. L. 1989. Invasions of natural ecosystems by plant pathogens. Pp. 77–83 *in* J. A. Drake, H. A. Mooney, F. di Castri, R. H. Groves, F. J. Kruger, M. Rejmanek, and M. Williamson (Eds.), *Biological invasions: A global perspective.* John Wiley & Sons, Chichester, England.

Williams, J. E., J. E. Johnson, D. A. Hendrickson, S. Contreras-Balderas, J. D. Williams, M. Navarro-Mendoza, D. E. McAllister, and J. E. Deacon. 1989. Fishes of North America endangered, threatened, or of special concern: 1989. *Fisheries* **14(6)**:2–20.

Wilson, S. D. and J. W. Belcher. 1989. Plant and bird communities of native prairie and introduced Eurasian vegetation in Manitoba, Canada. *Cons. Biol.* **3**:39–44.

Wingate, D. B. 1985. The restoration of Nonesuch Island as a living museum of Bermuda's precolonial terrestrial biome. Pp. 225–238 *in* P. J. Moors (Ed.), *Conservation of island birds.* International Council for Bird Preservation, Tech. Pub. No. 3.

CHAPTER 24

Conserving Genetic Diversity

The genetic structure of a population of any species is, in a real sense, its evolutionary life insurance policy. The variability that exists in the gene pool of the population is the raw material for natural selection, and is essential to permitting the species to continue adapting to its changing environment (Selander 1983).

When populations decline to a few individuals, however, much of this genetic variability can be lost. This means that the evolutionary adaptability of the species is also being reduced. In addition, in remnant populations of a few individuals, breeding with close relatives becomes more frequent. Such inbreeding often leads to detrimental genetic effects. **Conservation genetics** is concerned with evaluating the dangers of genetic impoverishment and inbreeding, and with devising ways to maintain as high a degree of genetic variability as possible in populations of species that have been pushed to the edge of extinction. The principles of conservation genetics apply both to populations in nature and to those in captivity.

GENES IN POPULATIONS

Genes are units of the genetic molecule, DNA (deoxyribose nucleic acid), that contain the coded instructions for assembly of particular proteins, such as enzymes, that in turn determine the structure and function of organisms. An organism has many genes, which are located on chromosomes. Humans, for example, have 23 different chromosomes, which contain about 100,000 genes.

The gene for a particular protein may exist in slightly different forms, known as **alleles,** that consequently tend to produce different versions of the protein, or sometimes a protein that is so different that it is non-functional. One human gene, for example, produces an enzyme that mediates the conversion of tyrosine to melanin, a dark pigment present in epidermal cells of the body. The form of the gene that produces the functional enzyme is one allele, but another allele exists that produces a non-functional enzyme. Humans who have only this allele are unable to produce melanin, and are albinos. Alleles are produced by **mutation,** which is nothing more than an accident in the replication of the gene during the process of cell division. Mutations may be spontaneous, or produced through the action of agents, such as ionizing radiation or highly reactive chemicals, that can interfere with molecular replication.

One component of genetic variability is thus whether few or many of the genes in a population are **polymorphic,** having two or more alleles, rather than being **monomorphic,** or having only one allele. Polymorphic genes also vary in the number of alleles they possess. In principle, no limit exists, so that in a large population, many different alleles might exist for a particular gene.

Polymorphism, the percent of genes that have more than one allele, is one common measure of the genetic makeup of a population, or what is termed the **gene pool** of the population. Normally, about 30 to 50 percent of the genes of an organism are polymorphic, although this figure varies somewhat among different groups of plants and animals. An allele that arises by mutation in one location within the range of a species, provided it is not so deleterious that selection eliminates it quickly, can spread to other individuals through dispersal and interbreeding of individuals. The spread of alleles in this way is termed **gene flow,** and is one process that maintains a characteristic level of polymorphism in the overall species population. The movement of only a few individuals from one subpopulation to another each generation, carrying their alleles, can be quite effective in maintaining a high polymorphism level.

The body tissues of higher plants and animals normally contain chromosomes in pairs, one member of which came from each parent. Humans, for example, possess 23 pairs of chromosomes that are distinct in size, shape, and the genes they carry. Each member of a pair holds one version of each of the many genes located on that chromosome (Fig. 24.1). These might be the same allele, in which case the individual is **homozygous** for that allele, or they might be different

alleles, in which case the individual is **heterozygous.** At most, an individual can have two alleles, but in a population, of course, many different alleles can occur.

Mean heterozygosity, the average percent of the genes that are heterozygous in individuals, is thus a second component of genetic variability. Normally, five to seven percent of genes are heterozygous, but again, this varies somewhat from group to group. Although changes in mean heterozygosity in natural populations are not easy to relate to specific selective advantages, many studies show increases in mean heterozygosity with increasing age or following periods of severe weather. This suggests that individuals with high mean heterozygosity survive better than those with low mean heterozygosity (Allendorf and Leary 1986). Watt (1983), for example, found that heterozygosity of the gene for a particular enzyme important in energy mobilization enabled the sulfur butterfly, *Colias philodice,* to be active over a wider range of weather conditions than could homozygous individuals. This resulted in increased survival of heterozygous individuals.

Thus, the components of genetic variability of a population include 1) the fraction of genes that are polymorphic, 2) the average number of alleles for polymorphic genes, and 3) the average percentage of genes that are heterozygous in individuals.

GENETIC CHANGES IN SMALL POPULATIONS

As the overall abundance of a species declines, its population typically changes from one that is relatively continuous and has only small gaps separating subpopulations to one consisting of widely separated, small groups of individuals. With increased isolation of such subpopulations, the rate of gene flow decreases, due to reduced interchange of individuals, and mutation becomes the only process working to increase genetic variability within individual subpopulations.

When the number of individuals in an isolated population declines to less than 100 or so, some of the alleles of polymorphic genes tend to be lost from the population by a process termed **genetic drift,** the random fluctuation in the frequency of an allele due to accidents that affect the survival and reproduction of individuals (Fig. 24.2). At first, genetic drift affects the very rare alleles that may be carried only by a small percentage of the individuals in the population. Death or failure of the few individuals to transmit the rare allele, simply by chance, may cause these alleles to disappear. But as the population becomes reduced to a handful of individuals, such accidents may cause an erratic fluctuation from generation to generation in the frequencies of even the more common alleles. In time, such fluctuations may cause the frequency of one allele to reach 100 percent, for much the same reason that flipping a coin a few times will occasionally yield a run of all heads.

An example of the relation of genetic variability to population size is provided by the coniferous tree, *Halocarpus bidwillii,* in New Zealand (Billington 1991). Isolated populations

FIGURE 24.1

A. GENES AND ALLELES

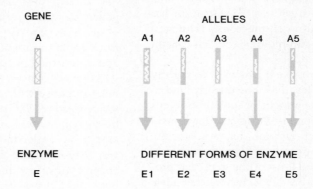

GENE

A

ALLELES

A1   A2   A3   A4   A5

ENZYME

E

DIFFERENT FORMS OF ENZYME

E1   E2   E3   E4   E5

B. ALLELES ON PAIRED CHROMOSOMES

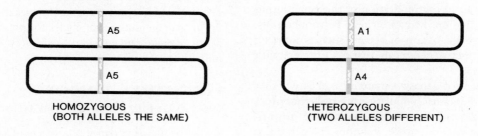

A5

A5

HOMOZYGOUS
(BOTH ALLELES THE SAME)

A1

A4

HETEROZYGOUS
(TWO ALLELES DIFFERENT)

C. GENE POOL OF POPULATION

POLYMORPHISM

GENES

A	5 alleles	
B	2 alleles	
C	1 allele	Polymorphism
D	1 allele	= 40%
E	1 allele	

HETEROZYGOSITY

INDIVIDUALS

A1/A1	
A2/A3	
A1/A3	Heterozygosity
A2/A4	= 60%
A2/A2	

■ Alleles are different forms of the gene for a particular enzyme or other protein. Each individual has two copies of each gene, which might be the same (homozygous condition) or different (heterozygous condition). The gene pool of the population is characterized by the number and relative abundance of the alleles present, and by the mean fraction of heterozygous individuals.

of this tree in the mountains of South Island range from about 400,000 individuals to only 20 to 25. In populations of 10,000 or more individuals, polymorphism is 25 to 35 percent, whereas in the smallest populations it is only zero to five percent (Fig. 24.3).

When the influence of drift is considered in relation to time measured in years or decades, rather than generations, one can see that the rate of loss of variability by genetic drift is faster in species with short generation times than in a population of an equal number of individuals of a long-lived, slow-maturing species (Franklin 1980). Over a century, for ex-

ample, a population of a small, short-lived desert pupfish may be replaced perhaps 50 times, each replacement carrying a certain risk of loss of alleles by genetic drift. Over the same period, a population of elephants would replace itself perhaps four times, running a much lower overall risk of loss of genetic variability.

Over the entire gene pool, the loss of genetic variability by genetic drift is quite predictable. The rate of loss of heterozygosity, H, for example, is predicted by a simple exponential relationship (Lande and Barrowclough 1987). The mean loss per generation depends only on the size of the

FIGURE 24.2

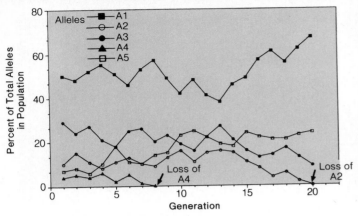

■ Genetic drift is the random fluctuation of relative abundances of alleles due to accidents of reproduction and survival. Drift can often lead to the loss of alleles in small populations.

FIGURE 24.3

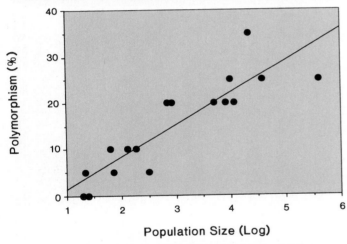

■ Percentage of polymorphic genes in isolated populations of the tree, *Halocarpus bidwillii,* in mountains of South Island, New Zealand, in relation to population size. Source: Data from H. L. Billington, "Effects of Population Size on a Genetic Variation in a Dioecious Conifer" in *Cons. Biol.,* 5:115–119, 1991.

breeding population, *N.* Heterozygosity at a time *t* generations from t_0 is given by the equation:

$$H_t = H_0 \, e^{-t/2N}$$

In this equation, generation time, *t,* corresponds to the mean age of all individuals that are engaged in reproduction.

Using a modeling approach, Lacy (1987) examined the roles of genetic drift, immigration, and mutation, together with that of natural selection, in determining genetic variability in small populations. For populations less than about 100 individuals, drift proved to be a very strong force, much more important, for example, than common intensities of selection. Mutation, because of its low incidence, was unable to

FIGURE 24.4

■ The northern elephant seal is an example of a species that experienced a recent population bottleneck that has apparently caused the loss of most of the genetic variability in its population.

maintain polymorphism by overriding drift, and was essentially unimportant. Immigration, from a large source population, on the other hand, was a strong force for maintaining genetic variability in the face of genetic drift, even when only a few immigrants arrived per generation.

Occasionally, populations of organisms drop to a low level, and then recover. At their low point, however, they may have experienced severe effects of genetic drift, passing, in a sense, through a **"genetic bottleneck."** How much variability is lost depends on the size of the population and the number of generations that the bottleneck condition lasts (Allendorf 1986). During a bottleneck of short duration, say a single generation, some alleles may be lost, but the general level of heterozygosity may not be reduced much. Populations that are reduced to only a few individuals for several generations may lose all variability, however. The northern elephant seal (Fig. 24.4), a species which once occurred along much of the Pacific coastline of North America, was reduced by Russian seal hunters to fewer than 100 individuals—perhaps as few as 20 individuals—in the 1890s (Bonnell and Selander 1974) on Guadalupe Island, Mexico. Among these individuals, reproduction involved perhaps only a few harem males, so that the effective breeding population was certainly very small. Although the population has now recovered to 120,000 or more animals, and the breeding range of the species has spread north to central California waters, the genetic variability within the population is very small. Of the 21 genes that can be analyzed, none shows polymorphism. Southern hemisphere populations of elephant seals were not reduced to such low levels, and, in contrast, show about 30 percent polymorphism and three percent heterozygosity, values that are essentially normal for mammals.

The African cheetah (Fig. 24.5), particularly the population in southern Africa, shows a similar absence of genetic variability (O'Brien et al. 1985, 1987). One evidence of the

FIGURE 24.5

■ The cheetah may have experienced two genetic bottlenecks, one prior to the separation of eastern and southern African populations, and a second in the southern African population in very recent time.

FIGURE 24.6

■ The bison in Badlands National Park, South Dakota, were derived from only about nine founder animals, and show a very low level of genetic polymorphism.

FIGURE 24.7

■ Washington fan palms, limited to small populations in desert oases in the American Southwest, lack any genetic polymorphism.

genetic uniformity of these animals is the fact that skin grafts can be made successfully between unrelated individuals. This indicates that there are essentially no differences in the skin proteins, which might cause antibodies to be produced against the grafts. The low genetic variability suggests that cheetahs in southern Africa passed through one, or possibly two, genetic bottlenecks. The most recent bottleneck might have occurred during the early European settlement period, prior to the establishment of wildlife preserves, when wildlife of all sorts was decimated by settlers.

Many other examples of bottlenecked populations have been reported in both plants and animals. The 75 to 125 lions of Ngorongoro Crater, Tanzania, are derived from an estimated 15 founder individuals; members of this population show lower heterozygosity and higher frequencies of sperm abnormality than animals in the nearby Serengeti Plains (Packer et al. 1991). The Arabian oryx and Pere David's deer, species eliminated in the wild and reduced to small captive populations, show little genetic variation (Woodruff and Ryder 1986). The 700 or so bison in Badlands National Park, South Dakota (Fig. 24.6), which were derived from about nine founder animals, show polymorphism for only one of 24 genes (McClenaghan et al. 1990). Populations of the Brazilian lion tamarin, nearly extinct in the wild, exhibit reduced genetic variability (Forman et al. 1986). Local populations of desert pupfishes in the Death Valley area of California and Nevada (Turner 1983), along with topminnows in small desert springs of Arizona and northern Mexico (Vrijenhoek et al. 1985), also show loss of genetic variability. Among plants, the Torrey pine, a tree restricted to a small range on the southern California coast and on one of the California Channel Islands, shows no genetic variability (Ledig 1986), and the California fan palm (Fig. 24.7), limited to small populations in the deserts of California and nearby areas, has

greatly reduced polymorphism (McClenaghan and Beauchamp 1986).

Loss of variability by genetic drift in small populations is often coupled with an increased degree of **inbreeding,** or mating of closely related individuals. Closely related individuals are much more likely to carry certain rare, recessive alleles for deleterious expressions of a gene, which are usually masked by the presence of a dominant, normal allele. Inbreeding greatly increases the chances of pairing of such recessive alleles.

The effects of paired deleterious alleles are varied, but many lead to poor survival or reproduction, a syndrome known as **inbreeding depression.** Ralls et al. (1988), for example, compared the survival of the offspring of inbred and non-inbred parents for 44 species of animals at the National Zoological Garden in Washington, D.C. They found that survival was higher in non-inbred offspring in 41 of the 44 species, and that survival of offspring of inbred crosses averaged 33 percent lower. The low genetic variability in the cheetah is coupled with some possible inbreeding depression. Cheetahs show very high percentages of abnormal sperm, as well as a relatively high infant mortality rate among captive animals. In addition, captive cheetahs show low resistance to disease (O'Brien et al. 1985, 1987). In the wild, inbreeding problems may occur in the small wolf population on Isle Royale, Michigan (See chapter 12), which now consists of the offspring of a single female (Wayne et al. 1991), and in the relict Florida panther population, in which males show very high levels of abnormal sperm (U.S. Fish and Wildlife Service 1987).

GENETIC PROBLEMS IN CAPTIVE POPULATIONS

When a captive population is established, still another source of reduced genetic variation operates: the **founder effect.** A small group of individuals used to found a captive breeding population will possess only part of the total variability of the wild population. If the individuals are closely related or come from a small part of the total species range, they may lack much of the allelic variation in the population. Thus, if possible, a captive population should be founded with unrelated individuals drawn from various parts of the wild population.

The okapi, an African forest antelope, illustrates how difficult it is to obtain a good representation of the variability in a wild population (Fig. 24.8). Okapis in captivity around the world are derived from only 75 individuals captured in the wild (Lacy 1989). Of these, only 30 have bred, and only 23 have living offspring. Those that have bred vary considerably in the number of offspring they have left. The result is that the present genetic structure of the captive population is equivalent to that which would be produced by only about 12 wild individuals contributing equally in their breeding efforts.

In small captive populations, these problems are also coupled with a peculiar selective environment. Since individuals are artificially protected from predators and are provided food and shelter, selection to maintain natural survival behaviors is non-existent (Lacy 1987). On the other hand, selection may be strong for characteristics relating to tolerance of confined quarters and absence of normal social contacts.

Still another problem that may arise in artificially managed populations is **outbreeding depression,** or reduced fitness of offspring of distantly related individuals (Templeton 1986). The clearest cases of outbreeding depression are hybrid offspring of individuals from populations adapted to different environments. In such situations, local populations may have become adjusted genetically to local conditions. Hybrid offspring of individuals from two such situations may be adapted

FIGURE 24.8

■ Although some 75 okapi, an African forest antelope, have been brought into captivity in zoos around the world, the gene pool of the present zoo population shows variability equivalent to that of only 12 equally productive parent individuals.

to neither. Introduction of individuals of two foreign subspecies to the ibex population in the Tatra Mountains, Czechoslovakia, for example, reportedly resulted in the birth of hybrid young in midwinter (Greig 1979). The result was high mortality, and eventual extinction of the ibex population.

In zoo populations of many animals, the interbreeding of individuals belonging to different subspecies, or occasionally species, has been frequent. Inaccurate understanding of the taxonomy of the species, poor record keeping, and shortage of possible breeding individuals have all contributed to this sort of hybridization. The result, however, is that some zoo populations are unsuitable for use in captive breeding programs aimed at survival of species and their distinct subspecies. Analysis of the genetics of animals in captive populations, to enable breeding programs to avoid such hybridization but maintain genetic diversity within natural taxa, is a major priority.

CONSERVING GENETIC VARIABILITY

The first strategy in conservation of genetic variability, of course, is to prevent populations from reaching "bottleneck" levels. Efforts have been made to determine what constitutes a minimum population size to prevent loss of evolutionary adaptability (Shaffer 1981, Gilpin and Soulé 1986). Obviously, this minimum population size must apply to the breeding individuals. Often, however, the genetically effective population size is much smaller than the actual population (Vrijenhoek 1989). As a formal genetic concept, the **genetically effective population size** is the ideal breeding population (a 1:1 sex ratio, random mating pattern, and random variation in number of offspring per mating) to which an observed population is equivalent (Lande and Barrowclough 1987). Consider a hypothetical population of nine male and nine female elephant seals, in which only one male mates with the nine females each generation. In addition to the fact that eight males are excluded from breeding, the genetically effective breeding population may be reduced still more because of the close relation of individuals. The various females, in this example, would all be full or half siblings. In this example, the chance of losing alleles by genetic drift would be about equal to that in a population of only two males and two females (Frankel and Soulé 1981).

A rough estimate of the genetically effective population size, N_e, of a population in which generations do not overlap (such as insects with a single summer breeding effort) can be obtained from the numbers of breeding males, N_m, and breeding females, N_f, by the equation:

$$N_e = (4N_mN_f)/(N_m + N_f)$$

More precise estimates require information on the variability of reproduction among individuals, and can be calculated for populations with non-overlapping or overlapping generations by more complicated equations (Lande and Barrowclough 1987).

Estimates of the minimum genetically effective population size necessary to maintain the evolutionary vitality of a species vary considerably. Franklin (1980) suggested that a minimum effective breeding population of 50 would be necessary to prevent deleterious short-term effects due to inbreeding—the so-called "rule of 50." Applied to the preservation of genetic variability, however, one must specify the percentage of variability and the length of time in order to determine the minimum population required. For example, one might specify that 95 percent of the existing heterozygosity must be retained for 100 years. In such a case, the minimum genetically effective population is related to the generation length of the organism, and would be much greater for small, short-lived organisms than for large species with long generation times. For an endangered population of a species of kangaroo rat, a century might be equivalent to 50 generations, whereas for the African elephant it might equate to only four generations. Based on the equation for the loss

of heterozygosity through genetic drift, discussed earlier, a population of 487 kangaroo rats would be necessary to retain 95 percent of the original level of genetic variability for 100 years, but a population of only 39 elephants would suffice. Frankel and Soulé (1981) suggested that a minimum of 500 breeding individuals—the total population, including non-breeding individuals, would be larger—is a rough first estimate of a genetically healthy population. Considering both genetic and ecological factors, however, a single minimum population required for long-term survival in nature cannot be specified, and must be determined for each species (Soulé 1987). We shall examine this problem in detail in chapter 25.

Many cases exist, however, of populations that have fallen well below the 50-individual level below which serious inbreeding problems may occur. In North America, populations of the whooping crane, California condor, Florida panther, and black-footed ferret have all dropped below 50 individuals. Some of these populations have recovered to higher levels, but many still remain below the size at which losses of variability by genetic drift are likely to occur.

Captive breeding programs for endangered animal species run the risk of all of the genetic problems noted above (Ralls and Ballou 1986). In establishing such populations, efforts should be made to maximize genetic diversity by selecting unrelated individuals. Careful management of the breeding population can also maximize the genetically effective population size. Breeding exchanges between institutions having separate colonies of species can reduce the chance of inbreeding effects and expand the effective breeding population. Pairings of the most genetically distinct individuals, and regulation of the number of offspring entering the next generation from different pairings can also increase the effective population size. Improved technologies of sperm and embryo banking will eventually allow an effective breeding population to include individuals that cannot be exchanged, or that are no longer alive. Obviously, a well-integrated, international system of breeding records is essential for efforts to create minimum genetically effective populations of species in captivity.

MANAGING THE GENETICS OF SMALL WILD POPULATIONS

Growing human populations and shrinking wildlife habitat mean that, even in nature, increasing numbers of species will need active human management not only to maintain minimum numbers but also to maintain a healthy genetic structure. Fragmentation of the range of species in nature, and the accompanying reduction of gene flow, can be alleviated by translocation of individuals (Griffith et al. 1989). Even though small, relict subpopulations tend individually to lose variability due to genetic drift, altogether they often retain much of the normal variability of the original, large population (Lacy 1987). Regular transplants of a few individuals from one population to another can offset much of the loss of variability

FIGURE 24.9

a.

b.

■ Tiger populations are fragmented throughout much of the species' range, and the small populations in many locations may lead to inbreeding and consequent genetic problems leading to their extinction. Less than 1,000 animals occur in the highly fragmented range on

Sumatra (a), for example. Formal protection exists only in five parks and game preserves (b), themselves small and isolated from each other (Modified from Santiapillai and Ramono 1987).

by genetic drift. The tiger (Fig. 24.9), for example, now survives in small, widely scattered populations in eastern and southeastern Asia. These populations are subject to inbreeding, and it is unlikely that they will survive indefinitely unless deliberate efforts are made to translocate breeding animals among them (Wemmer et al. 1988).

Translocations or breeding exchanges should only be carried out, however, when evidence exists that the isolated populations were once part of a continuous, freely interbreeding ancestral population. Determining the actual degree of genetic divergence among isolated populations and different subspecies is thus essential to conservation management (Lacy 1988). Long-isolated populations that are classified as the same species may actually have evolved many genetic differences that adapt them to their respective areas. Transplants between such populations may thus do more damage, through outbreeding depression, than benefit.

Just how much genetic diversity must be maintained to assure the evolutionary health of populations is uncertain. Soulé et al. (1986) have recommended the goal of preserving heterozygosity of managed populations at 90 percent of normal over a course of 200 years. A goal of this nature is a general guideline, however, and no species should be abandoned because it has fallen below the criterion. The successful recovery of bottlenecked populations such as the northern elephant seal, in fact, suggests that genetic diversity *per se* may be less important than elimination of deleterious recessive alleles that may appear in homozygous condition through inbreeding. Nevertheless, until the genetics of the species in question are thoroughly inventoried, the strategy of preserving as much diversity as possible is the only reasonable alternative.

ZOOS, BOTANIC GARDENS, AND GENE BANKS

Because of the rapid disappearance of natural ecosystems and their species, zoos, aquaria, botanic gardens, arboretums, and other facilities will bear a heavier responsibility for preserving genetic resources and diversity. Zoos, which focus their efforts on terrestrial vertebrates, will probably be faced by the need to maintain about 2,000 species of larger vertebrates by some time within the next 200 years (Soulé et al. 1986). At best, existing zoo facilities have a capacity of housing about 1,000 such species with minimum populations of 250 individuals (Office of Technology Assessment 1988). Considering the space requirements and costs of zoo maintenance, this is a major challenge.

New technologies may assist the conservation of animal germ plasm. **Cryogenic storage,** long-term preservation of semen, ova, or embryos at subfreezing temperatures, is an established technique for domestic ungulates, and has been proved applicable for many other mammals. The temperatures at which these materials are stored range from −160 to −196 degrees C. How long animal materials will remain viable under such storage is uncertain, but domestic animal semen and embryos have been stored successfully for 10 to 30 years. Little has been done to extend this technique to other vertebrates and invertebrates, however.

Botanic gardens and arboreta face a similar challenge. In the United States alone, for example, about 5,000 rare to endangered taxa of plants exist in the wild (Falk 1990). Typically, botanic gardens focus attention on particular families or genera of plants for which their environment is best suited. The plant groups best represented are often those of horticultural interest, such as orchids and palms. Agricultural research institutions often maintain collections of plants of agronomic importance. In such collections, space is a severe constraint on the number of living individuals that can be preserved, particularly for shrubs and trees. In the United States, the **Center for Plant Conservation,** located in Jamaica Plain, Massachusetts, coordinates the activities of 20 botanical gardens and arboreta in the propagation of endangered plants.

Seed storage facilities have also been developed for higher plants, both those of recognized economic value and other species. Conventional facilities, such as the **National Seed Storage Laboratory (NSSL)** in Fort Collins, Colorado, store seeds at about 4.4 degrees C and low relative humidity. The stored samples must be tested for viability at about five-year intervals, and propagated in the field to produce new seed when viability begins to decline. Seeds of most crop plants must be propagated about every 15 years. Conventional facilities like the NSSL exist in many countries, primarily to protect genetic lines of plants of agronomic value (Cox and Atkins 1979).

Cryogenic storage of many seeds is also possible. For this, seeds must be dried to reduce moisture to about 5 percent and stored at temperatures below −160 degrees C. The viability of seeds stored at such temperatures needs to be checked only at about 50-year intervals, and the seeds probably need to be propagated anew only at intervals of hundreds of years. Unfortunately, the seeds of many tropical and aquatic plants cannot tolerate the drying necessary for cryogenic storage. The **Royal Botanic Gardens** at Kew, England has pioneered cryogenic storage of the seeds of wild species (Koopowitz and Kaye 1983). Several regional centers for cryogenic storage of seed of wild species have now been established, such as the **Iberian Gene Bank** in Madrid, Spain, which has become the center for storage of seed of endangered plants for the entire Mediterranean region. In 1979, the **Botanic Gardens Conservation Coordinating Body** was created to help integrate the efforts of such centers on a worldwide basis.

KEY STRATEGIES FOR CONSERVATION OF GENETIC DIVERSITY

1. Prevent the decline of populations to levels at which genetic bottlenecks and inbreeding depression become significant.

2. Combat the effects of isolation due to range fragmentation by transplantation of individuals between subpopulations.

3. Maximize genetic heterogeneity by choosing unrelated individuals for establishment of captive populations and promoting outbreeding among captive individuals.

LITERATURE CITED

Allendorf, F. W. 1986. Genetic drift and the loss of alleles versus heterozygosity. *Zoo Biol.* **5:**181–190.

Allendorf, F. W. and R. F. Leary. 1986. Heterozygosity and fitness in natural populations of animals. Pp. 57–76 *in* M. E. Soulé (Ed.), *Conservation biology: The science of scarcity and diversity.* Sinauer, Sunderland, MA.

Billington, H. L. 1991. Effects of population size on genetic variation in a dioecious conifer. *Cons. Biol.* **5:**115–119.

Bonnell, M. L. and R. K. Selander. 1974. Elephant seals: Genetic variation and near extinctions. *Science* **184:**908–909.

Cox, G. W. and M. D. Atkins. 1979. *Agricultural ecology.* W.H. Freeman Co., San Francisco, CA.

Falk, D. A. 1990. Integrated strategies for conserving plant genetic diversity. *Ann. Missouri Bot. Gard.* **77:**38–47.

Forman, L., D. G. Kleiman, R. M. Bush, J. M. Dietz, J. D. Ballou, L. G. Phillips, A. F. Coimbra Filho, and S. J. O'Brien. 1986. Genetic variation within and among lion tamarins (*Leontopithecus rosalia rosalia*). *Amer. J. Phys. Anthropol.* **71:**1–12.

Frankel, O. H. and M. E. Soulé. 1981. *Conservation and evolution.* Cambridge Univ. Press, Cambridge, England.

Franklin, I. R. 1980. Evolutionary change in small populations. Pp. 135–150 *in* M. E. Soulé and B. A. Wilcox (Eds.), *Conservation biology: An evolutionary-ecological perspective.* Sinauer, Sunderland, MA.

Gilpin, M. E. and M. E. Soulé. 1986. Minimum viable populations: Processes of species extinction. Pp. 19–34 *in* M. E. Soulé (Ed.), *Conservation biology: The science of scarcity and diversity.* Sinauer Associates, Sunderland, MA.

Greig, J. C. 1979. Principles of genetic conservation in relation to wildlife management in southern Africa. *S. Afr. J. Wildl. Res.* **9:**57–78.

Griffith, B., J. M. Scott, J. W. Carpenter, and C. Reed. 1989. Translocation as a species conservation tool: Status and strategy. *Science* **245:**477–480.

Koopowitz, H. and H. Kaye. 1983. *Plant extinction: A global crisis.* Winchester Press, Piscataway, NJ.

Lacy, R. C. 1987. Loss of genetic diversity from managed populations: Interacting effects of drift, mutation, immigration, selection, and population subdivision. *Cons. Biol.* **1:**143–158.

Lacy, R. C. 1988. A report on population genetics in conservation. *Cons. Biol.* **2:**245–248.

Lacy, R. C. 1989. Analysis of founder representation in pedigrees: Founder equivalents and founder genome equivalents. *Zoo Biol.* **8:**111–123.

Lande, R. and G. F. Barrowclough. 1987. Effective population size, genetic variation, and their use in population management. Pp. 87–123 *in* M. E. Soulé (Ed.), *Viable populations for conservation.* Cambridge Univ. Press, Cambridge, England.

Ledig, F. T. 1986. Heterozygosity, heterosis and fitness in outbreeding plants. Pp. 77–104 *in* M. E. Soulé (Ed.), *Conservation biology: The science of scarcity and diversity.* Sinauer, Sunderland, MA.

McClenaghan, L. R., Jr. and A. C. Beauchamp. 1986. Low genic differentiation among isolated populations of the California fan palm (*Washingtonia filifera*). *Evolution* **40**:315–322.

McClenaghan, L. R., Jr., J. Berger, and H. D. Truesdale. 1990. Founding lineages and genic variability in plains bison (*Bison bison*) from Badlands National Park, South Dakota. *Cons. Biol.* **4**:285–289.

O'Brien, S. J., M. E. Roelke, L. Marker, A. Newman, C. A. Winkler, D. Meltzer, L. Colly, J. F. Evermann, M. Bush, and D. E. Wildt. 1985. Genetic basis for species vulnerability in the cheetah. *Science* **227**:1428–1434.

O'Brien, S. J., D. E. Wildt, M. Bush, T. Caro, C. Fitzgibbon, I. Aggundey, and R. E. Leakey. 1987. East African cheetahs: Evidence for two population bottlenecks? *Proc. Nat. Acad. Sci.* **84**:508–511.

Office of Technology Assessment. 1988. *Technologies to maintain biological diversity.* J.B. Lippincott Co., Philadelphia, PA.

Packer, C., A. E. Pusey, H. Rowley, D. A. Gilbert, J. Martenson, and S. J. O'Brien. 1991. Case study of a population bottleneck: Lions of the Ngorongoro Crater. *Cons. Biol.* **5**:219–230.

Ralls, K. and J. Ballou. 1986. Preface to the proceedings of the workshop on genetic management of captive populations. *Zoo Biol.* **5(2)**:81–86.

Ralls, K., J. D. Ballou, and A. Templeton. 1988. Estimates of lethal equivalents and the cost of inbreeding in mammals. *Cons. Biol.* **2**:185–192.

Santiapillai, C. and W. S. Ramono. 1987. Tiger numbers and habitat evaluation in Indonesia. Pp. 85–91 *in* R. L. Tilson and U. S. Seal (Eds.), *Tigers of the world: The biology, biopolitics, management, and conservation of an endangered species.* Noyes Publications, Park Ridge, NJ.

Selander, R. K. 1983. Evolutionary consequences of inbreeding. Pp. 201–215 *in* C. M. Schoenewald-Cox, S. M. Chambers, B. MacBryde, and L. Thomas (Eds.), *Genetics and conservation.* Benjamin/Cummings, Menlo Park, CA.

Shaffer, M. L. 1981. Minimum population sizes for species conservation. *BioScience* **31**:131–134.

Soulé, M. 1987. Where do we go from here? Pp. 175–183 *in* M. E. Soulé (Ed.), *Viable populations for conservation.* Cambridge Univ. Press, Cambridge, England.

Soulé, M., M. Gilpin, W. Conway, and T. Foose. 1986. The millennium ark: How long a voyage, how many staterooms, how many passengers? *Zoo Biol.* **5**:101–113.

Templeton, A. R. 1986. Coadaptation and outbreeding depression. Pp. 105–116 *in* M. E. Soulé (Ed.), *Conservation biology: The science of scarcity and diversity.* Sinauer Associates, Sunderland, MA.

Turner, B. J. 1983. Genic variation and differentiation of remnant natural populations of the desert pupfish, *C. macularius. Evolution* **37**:690–700.

U.S. Fish and Wildlife Service. 1987. *Florida panther (*Felis concolor coryi*) recovery plan.* Florida Panther Interagency Commission, Atlanta, GA

Vrijenhoek, R. C. 1989. Population genetics and conservation. Pp. 89–98 *in* D. Western and M. Pearl (Eds.), *Conservation for the twenty-first century.* Oxford Univ. Press, New York.

Vrijenhoek, R. C., M. E. Douglas, and G. K. Meffe. 1985. Conservation genetics of endangered fish populations in Arizona. *Science* **229**:400–402.

Watt, W. B. 1983. Adaptation at specific loci. II. Demographic and biochemical elements in the maintenance of the *Colias PGI* polymorphism. *Genetics* **103**:691–724.

Wayne, R. K., N. Lehman, D. Girman, P. J. P. Gogan, D. A. Gilbert, K. Hansen, R. O. Peterson, U. S. Seal, A. Eisenhawer, L. D. Mech, and R. J. Krumenaker. 1991. Conservation genetics of the endangered Isle Royale gray wolf. *Cons. Biol.* **5**:41–51.

Wemmer, C., J. L. D. Smith, and H. R. Mishra. 1988. Tigers in the wild: The biopolitical challenges. Pp. 396–404 *in* R. L. Tilson and U. S. Seal (Eds.), *Tigers of the world: The biology, biopolitics, management, and conservation of an endangered species.* Noyes Publications, Park Ridge, NJ.

Woodruff, D. S. and O. A. Ryder. 1986. Genetic characterization and conservation of endangered species: The Arabian oryx and Pere David's deer. *Isozyme Bull.* **19**:35.

25

Endangered Species Preservation

Human activities have pushed many species to the brink of extinction. The recognition that special protection of such species is essential, and also that recovery of some of them is possible, has led to the establishment of endangered species programs by many national and regional governments. The objective of endangered species programs is thus to identify species that are threatened with extinction, provide them with immediate protection, and develop plans for their recovery.

Endangered species management has become one of the most interdisciplinary areas of conservation biology. The fact that socioeconomic and political forces are largely responsible for the endangerment of species has increasingly forced conservation biologists into political and legal arenas (Kellert 1985). In addition, efforts to achieve the recovery of endangered species have forced them to draw on the expertise of agronomy, horticulture, and zoo biology.

In this chapter we shall first examine the history and current status of federal and state endangered species programs in the United States. Next, we shall consider some of the recovery programs that are under way, emphasizing some of the special techniques that are being employed. Finally, we shall examine other private and international programs relating to endangered species.

UNITED STATES GOVERNMENTAL PROGRAMS

Although conservationists have long been concerned about endangered species, the first comprehensive federal policy of protection of such species in the United States was not established until 1966, with the passage of the **Endangered Species Preservation Act** (Bean 1987). In 1969, this act was replaced by the **Endangered Species Conservation Act,** which strengthened several aspects of the program. In 1973, Congress passed a much more comprehensive Endangered Species Act; this act, with various amendments, has been reauthorized on several occasions, most recently in 1988, and currently governs United States federal endangered species policy. Federal endangered species law has become the focus of some of the most bitter political controversy relating to conservation (Tobin 1990).

Under the 1973 act, responsibility for management of endangered species is assigned to the Fish and Wildlife Service (Department of Interior), in the case of terrestrial and freshwater species, and the National Marine Fisheries Service (Department of Commerce), for marine species. The major responsibilities of endangered species offices within these agencies are 1) designation of endangered species, 2) designation of critical habitat for endangered species, and 3) development and implementation of recovery plans for these species.

The way in which species are designated and protected makes the endangered species act a very powerful conservation instrument. Species are designated as **endangered,** if they are determined to be in danger of extinction throughout all or a significant portion of their range, or **threatened,** if they are likely to become endangered within the foreseeable future. In its legal usage, the term *species* is not restricted to the entire biological species, but can refer to any subspecies of animal or plant, or to any distinct population of a vertebrate animal. This broad definition of the term is one of the features that makes the 1973 act so powerful: a portion of the biological species population can be designated as endangered even though other portions are still abundant.

The 1973 Endangered Species Act covers not only vertebrate animals, but also invertebrates (exclusive of pest insects) and plants (Fig. 25.1). Both species occurring within the United States and foreign species are covered (Fig. 25.2). The listing of foreign species as endangered or threatened strengthens the ability of the United States to participate in international agreements regulating commerce in such species. The federal list of endangered and threatened species for 1991 contains about 1,127 species (Table 25.1).

The procedure for placing species on the list of endangered and threatened species can be initiated by the offices of endangered species themselves, or by petitions from private groups. Petitions for placing species on the list are reviewed for merit by scientific panels. If addition of a species to the list is recommended, an announcement of the proposed action is published in the *Federal Register* and other appropriate places

FIGURE 25.1

■ Lang's metalmark is one of several species of California butterflies that are federally listed as threatened or endangered.

FIGURE 25.2

■ The United States listing of endangered and threatened species includes many foreign species, such as the jaguar, which is classified as endangered.

to solicit additional information and public comment. Following a year-long review period a final decision on listing is made by the agency panel.

Other important activities of the offices of endangered species include designation of **critical habitat** and the development of recovery plans. Critical habitat consists of areas, either within or outside the range of the species, that are essential to its conservation. Essential areas within the range consist of occupied habitat; those outside the range might, for example, consist of the watershed on which a small area of aquatic habitat occupied by an endangered species depends. As of 1984, critical habitats had been designated for 64 species. **Recovery plans,** which we shall examine in detail

TABLE 25.1

Numbers of Species Listed as Endangered and Threatened by the United States Office of Endangered Species (1991) and by the California Department of Fish and Game (1990).

	USA				California	
	ENDANGERED		THREATENED		ENDANGERED	THREATENED OR RARE
	USA	Foreign	USA	Foreign		
Mammals	54	249	8	22	6	10
Birds	72	153	12		16	5
Reptiles	16	58	18	14	4	4
Amphibians	6	8	5		2	6
Fish	53	11	33		13	2
Snails	4+	1	6			1
Clams	37	2	2			
Crustaceans	8		2		2	
Insects	11	1	9			
Arachnids	3					
Plants	186	1	60	2	124	85
TOTAL	450	484	155	38	167	113

shortly, are designed to increase the abundance and distribution of species so that they can be removed from the list. As of 1990, such plans had been developed for 321 species. Information on these activities, as well as on the listing activities, is published monthly or bimonthly in the *Endangered Species Technical Bulletin.*

The federal endangered species program of the United States has become a model for many other governmental programs. In the United States, almost all state and territorial governments have developed some type of endangered species program, and most of these have established cooperative agreements with the federal endangered species offices. California, for example, has a system for designating species as *endangered, threatened,* and in the case of plants, *rare* (Table 25.1).

Private conservation organizations also contribute to efforts to identify and protect endangered species in a variety of ways. *American Birds,* a magazine published by the National Audubon Society and aimed primarily at bird-watching enthusiasts, periodically publishes the **Blue List,** a listing of bird species that seem to be experiencing non-cyclic population declines or range contractions. None of the species on this list are on the federal list of endangered or threatened species. The Blue List, in effect, constitutes an "early warning notice," the purpose of which is to focus attention of the birding community on these species, in the hope that such attention will reveal the extent and cause of declining species before they reach a crisis stage. The 1986 Blue List for North America (Tate 1986) contained 22 species, including seven water birds, six hawks and owls, and nine songbirds (Fig. 25.3).

FIGURE 25.3

■ The Blue List for 1986, published in *American Birds,* identified the upland sandpiper as a species that may be declining in abundance in North America because of the disappearance of old-field habitats through biotic succession.

TABLE 25.2

Estimated Initial Populations of Species Belonging to Various Mammal Groups in National Parks in Western North America, Summarized Separately for Populations that have Subsequently Become Extinct and those that have Survived.

	EXTINCT POPULATIONS		SURVIVING POPULATIONS	
GROUP	MEDIAN	95% CONFIDENCE LIMIT	MEDIAN	95% CONFIDENCE LIMIT
Lagomorphs	3,276	702–56,952	70,889	34,720–173,150
Artiodactyls	241	3–1,273	792	429–1,504
Small Carnivores	256	122–880	1,203	908–1,704
Large Carnivores	24	14–68	108	70–146

Source: Data from W. D. Newmark, "Mammalian Richness, Colonization, and Extinction in Western North American National Parks" Ph.D. Dissertation, University of Michigan, Ann Arbor, MI.

INTERNATIONAL ENDANGERED SPECIES LISTINGS

The **Convention on International Trade in Endangered Species of Wild Fauna and Flora (CITES)** is an international agreement that regulates or prohibits commercial trade in globally endangered species or their products. This convention was developed in 1973, and by 1989 some 95 countries were party to its provisions. Appendix I of CITES lists about 700 species for which trade is prohibited, and Appendix II lists about 2,350 species of animals and about 24,000 species of plants for which trade is regulated. In the United States, responsibilities for general management, scientific decisions, and enforcement activities relating to CITES are assigned to separate divisions within the Fish and Wildlife Service.

The **World Conservation Union,** located in Gland, Switzerland, is a private organization that coordinates many global efforts in wildlife conservation. The World Conservation Union sponsors the publication of the **Red Data Books,** which are worldwide listings of species that indicate their degree of endangerment, together with summaries of information on distribution, populations, habitat ecology, cause of endangerment, and conservation measures taken and proposed. Red Data Books are available for birds (King 1981), mammals of the New World and Australasia (Thornback and Jenkins 1982), invertebrates (Wells et al. 1983), plants (Lucas and Synge 1988), and several more specific regional or taxonomic groups of animals.

POPULATION VULNERABILITY ANALYSIS

Central to a long-term management of endangered species is the determination of what constitutes a "safe" population size. This effort, known as **population vulnerability analysis (PVA),** is designed to determine the **minimum viable population (MVP),** or the smallest number of breeding individuals that has a specified probability of surviving for a certain time, without losing its evolutionary adaptability (Gilpin

and Soulé 1986). Specific probabilities and time periods are an essential feature of PVA. Indeed, PVA is an effort to force endangered species management to become objective in its procedures. Following the PVA approach, for example, one must define the population of Kirtland's warblers that must be maintained to, say, ensure the survival of the species with 95 percent probability for 100 years.

PVA assumes that the population in question is protected against factors that might drive it progressively toward extinction (See chapter 2). Thus, given favorable habitat conditions, an MVP is the population size necessary to prevent extinction from various accidental or random processes. These include accidental variation in natality and mortality, the impacts of random variation in environmental conditions, and chance events affecting genetic makeup of the population (See chapter 24).

Several approaches have been taken to PVA. One approach has been the empirical determination of the size of populations that have persisted for a known period following isolation. Newmark (1986), for example, estimated the population sizes of various mammals in national parks in western North America at the time of their establishment (Table 25.2). He then compared populations of those that have become extinct with those that have survived, the length of intervening time being roughly 75 years. The median sizes of surviving populations varied considerably with the size and trophic level of the mammal. Populations of hundreds, thousands, or tens of thousands were not always enough to guarantee survival of lagomorphs (hares and rabbits), which tend to show high levels of natural fluctuation (including major cycles of abundance and scarcity). On the other hand, the median size of surviving populations of large carnivores was only 108 individuals. Perhaps systems of territoriality of these animals stabilize their populations well within the limits of prey availability, making them less prone to extinction. More detailed analysis of data sets such as this may allow more precise estimation of probability of survival of populations of different size.

A second approach to determining MVP's has been through modeling population demography (Goodman 1987,

FIGURE 25.4

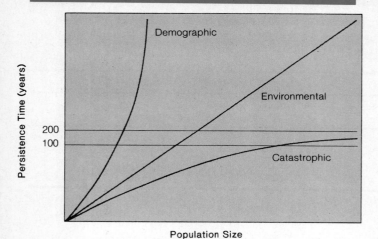

■ The relationship of survival probability to population size for populations subject to demographic stochasticity, moderate environmental stochasticity, and catastrophic environmental effects (Modified from Shaffer 1987).

Belovsky 1987). The basic strategy of this approach has been to work with r, the per capita population growth rate, and an estimate of the variability of r due to demographic or environmental influences. Results of such modeling (Fig. 25.4) indicate that different types of variability affect the MVP estimate to very different degrees. The risk of extinction from random accidents affecting births and deaths can be minimized at a relatively small population size. Moderate levels of environmental variability, in contrast, lead to a more or less linear reduction in extinction risk with increase in population size. Catastrophic patterns of environmental variation, however, create an extinction risk that declines very little with increase in population size above a certain level (Shaffer 1987).

Belovsky (1987) attempted to estimate rough MVP's for mammals of various body sizes, using allometric relationships of body mass with r and its measure of variability (Fig. 25.5). These estimates show that the MVP's of small animals can sometimes be very large. For a 25–g mammal, for instance, the MVP would range from about 8,000 individuals, if the variability in r were low, to about 150,000, if variation in r were high. For an animal with a body mass of 100 kg,

FIGURE 25.5

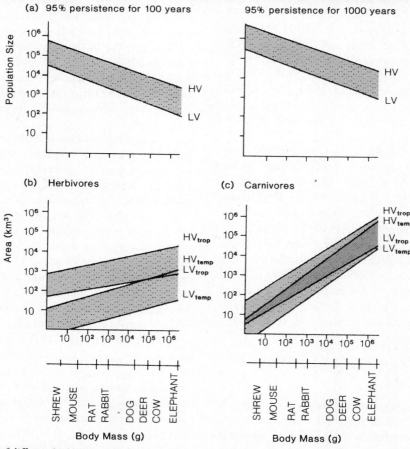

■ a. Populations of mammals of different body size required to provide 95 percent probability of survival for 100 and 1,000 years. A range of values is given for species with low (LV) to high (HV) population variability. b and c. Habitat areas needed to sustain herbivorous and carnivorous mammals for 100 years with 95 percent probability. Different ranges are shown for temperate (temp) and tropical (trop) species. Modified from Belovsky (1987).

the comparable range would be 300 individuals (low variability in r) to 8,000 individuals (high variability in r).

A third approach to estimating MVP's has been to determine the minimum population required to maintain most of the genetic variability of the species. By this approach, an MVP corresponds to the genetically effective population size (See chapter 24) necessary to retain, say, 95 percent of the original genetic variability (e.g., heterozygosity) for 100 years. This approach is not really an alternative to the first two methods, but rather a second basic criterion: an MVP must satisfy criteria both of demographic adequacy and of genetic and evolutionary health.

Although it is obvious from the above that MVP's must be determined for individual species, the very rough rule of thumb that a genetically effective population of at least 50 breeding individuals must exist to avoid serious inbreeding problems and at least 500 to prevent serious loss of variability by genetic drift has become incorporated in many species recovery plans. The recovery plan for the red-cockaded woodpecker, for example, incorporates such a requirement (Reed et al. 1988). Analysis of the populations of woodpeckers on major areas of protected habitat, however, showed, first, that an adult population of about 1,323 total birds is needed to give a genetically effective population of 500, because of the breeding system and demography of the species. Further, the analysis revealed that genetically effective populations in the nine major protected habitats of the species ranged only from 95 to 378 individuals.

RECOVERY OF ENDANGERED SPECIES

The long-range objective of endangered species programs is to develop programs for the recovery of endangered species, so that they can eventually be removed from the list. Recovery programs for species that have declined to the brink of extinction are often forced to utilize unusual or extreme techniques in efforts to encourage the reestablishment of viable populations. These techniques include captive propagation, reintroduction to the wild, and translocation of individuals to new habitat areas. These activities, in turn, employ several special techniques to increase reproduction and to encourage establishment of individuals in the wild.

Ex Situ Care of Endangered Species

Ex situ care refers to the maintenance of populations outside their native habitat, as in botanical gardens, zoological parks, and aquariums. Increasingly, these institutions are being called on to maintain populations of species that are being driven to extinction in nature. Relative to the growing need, however,

the space in these facilities is extremely limited, and coordination of the efforts of institutions is needed to increase their effectiveness.

An **International Species Inventory System** has been established to maintain records of animal species and subspecies that exist in captivity. Beginning with this inventory system, the American Association of Zoological Parks and Aquariums has developed the **Species Survival Plan,** an outline for the coordination of the captive management of endangered animals. This plan provides for identification of priority species for captive management, and for the integration of captive breeding activities by participating institutions to maximize the retention of genetic variability in the total captive population. By animal exchanges, breeding loans, sperm and embryo banks, and the maintenance of comprehensive stud books, genetically effective populations that can reduce the loss of genetic variability to a very low level can be achieved with many fewer individuals than in wild populations.

Captive Propagation and Reintroduction to the Wild

Captive propagation is the breeding of species in confinement for their preservation, display in zoos, and reintroduction to the wild. Several species of mammals and birds have been reduced to populations that survived only in captivity. In North America these include the black-footed ferret and California condor. The black-footed ferret, native to central and western North America, was reduced to low population levels by control of prairie dog populations (See chapter 4). Ultimately, only a single population of ferrets remained, and when a plague epidemic devastated the prairie dog population, an emergency captive breeding program was begun. Unfortunately, many of the ferrets captured died of distemper, reducing the total population of the species to 18 individuals in 1987, all in captivity. By 1990, however, the captive population had grown to 180 in breeding colonies in six locations, and plans were being developed for reintroduction of the species to the wild. This plan calls for release in 1991 of about 50 individuals in a 160 square kilometers white-tailed prairie dog town complex in Shirley Basin, south of Casper, Wyoming (Anon. 1991).

In several instances, captive populations of birds have been established by removal of the first clutches of eggs from nests in the wild, stimulating **recycling** by the wild parents (Simons et al. 1988). Recycling is simply initiation of a new breeding cycle, leading to the production of a replacement clutch. This procedure is being used extensively with the bald eagle, and was employed with California condors before the last wild birds were brought into captivity.

THE WHOOPING CRANE

The recovery program for the whooping crane (Fig. 25.6) illustrates the use of these techniques, as well as the overall nature of recovery plans for endangered species. The whooping crane is a large bird with a low reproductive potential. Adults typically lay two eggs annually, but normally only one chick can be raised successfully. This species was originally widespread in North America (Fig. 25.7), breeding from Illinois northwestward through the Great Plains to the Northwest Territory of Canada, and wintering in several locations along the Atlantic and Gulf coasts and in north-central Mexico (McMillen 1988). The decline of whooping crane populations was due to hunting in wintering areas and on

FIGURE 25.6

■ The whooping crane declined to 14 to 15 individuals between 1938 and 1941, before beginning a gradual recovery in numbers.

migration, conversion of southern wintering habitats to rice fields and the drainage of marshlands breeding areas in the Great Plains, and egg collecting. The decline was rapid from 1890 through 1910, and by 1918 the total estimated population was only 47 individuals. This population migrated between a breeding range in the southern Mackenzie District, Northwest Territory, Canada (now included in Wood Buffalo National Park) and a wintering range in coastal Texas (now centered on Aransas National Wildlife Refuge).

By 1938–1941, the population of whooping cranes had dropped to only 14 to 15 individuals. At this point, protection

FIGURE 25.7

■ Original and present ranges of the whooping crane in North America. Source: Data from J. L. Macmillan, "Conservation of North American Birds" in *American Birds,* 42: 1212–1221, 1988 National Audoban Society, New York, NY. Based on data from U.S. Fish and Wildlife Service (1986).

and publicity gained the upper hand, and a slow recovery began. A captive nesting colony was also established at the United States Fish and Wildlife Service research center at Patuxent, Maryland. With the creation of the federal endangered species program, a recovery plan was developed for the species. This plan has two specific objectives. First, the Wood Buffalo/Aransas Refuge population is to be increased to at least 40 breeding pairs. Second, two additional breeding populations with at least 20 breeding pairs are to be established.

To increase the population breeding in Wood Buffalo National Park and wintering at Aransas Refuge, efforts have concentrated on reducing mortality from predators in nesting areas, and expansion and improvement of wintering habitat on the Texas coast (Fig. 25.8). In wintering areas, pairs of cranes defend large feeding territories, limiting the number of birds that can be accommodated on a given refuge area. The use of controlled burning to improve winter habitat, as well as the acquisition of additional coastal marshlands to be managed as crane winter habitat are two efforts aimed at increasing the carrying capacity of the wintering area. By 1989–1990, this population had increased to about 145 birds.

An effort was also made to establish a second wild breeding population, using the cross-fostering technique. Whooping crane eggs were taken from nests at Wood Buffalo National Park (one egg from each set of two) and from the captive breeding colony at Patuxent, Maryland, and placed in nests of sandhill cranes at a breeding colony at Gray's Lake National Wildlife Refuge, Idaho. The young whooping cranes were reared by sandhill cranes, and followed them in migration south to a wintering area centered on Bosque del Apache National Wildlife Refuge, New Mexico. From 1975 through 1989, 288 whooping crane eggs were taken to the Gray's Lake site. Of these, 210 hatched and 85 fledged. This population built up to 33 individuals in 1984–1985, but the young whooping cranes suffered heavy mortality from accidents, disease, and predation. No breeding efforts were attempted and this program was terminated in 1989. Other sites, such as Kissimmee Prairie, Florida, have been proposed for establishment of a second wild population.

In addition, following the mortality of four whooping cranes in the Patuxent colony from disease in 1987, a plan to divide the captive breeding population was developed. In 1989, 22 of the 54 birds in captivity were moved to facilities of the International Crane Foundation in Wisconsin, where breeding populations of several other species of cranes exist.

FIGURE 25.8

■ On the winter range of whooping cranes at Aransas National Wildlife Refuge, Texas, prescribed burning is used to expand the area of feeding habitat. The cranes are strongly attracted to burned upland areas, where they feed heavily on acorns.

The California condor recovery plan illustrates the technical difficulties and political problems that confront a recovery effort for a species that has reached the brink of extinction.

The California condor (Fig. 25.9) is the largest raptor in North America, reaching a weight of over nine kg, and a wingspan of 2.75 m. California condors lay a single egg at a time, and care for the young for over a year, so that successful nestings can occur only at two-year intervals. The young birds require six to eight years to mature, but adults are believed to live for 20 to 40 years.

Like the whooping crane, the California condor has been at a critically low level for decades. The condor appears to be a relic of the Pleistocene megafauna that survived in North America until about 11,000 years ago. Fossil remains show that the species occurred in Utah and Arizona, and the species was probably widespread in mountainous areas of western North America. In this century, however, the condor population declined rapidly, largely because of poisoning, shooting, collision of birds with power lines, and lead poisoning (Snyder and Snyder 1989). Annual census efforts beginning in 1965, however, showed that the population was restricted to a limited U-shaped range in the southern coast ranges, Tehachapi Mountains, and southern Sierra Nevadas in California, and that total numbers were about 50 to 60. Two condor sanctuaries were established in areas of the Los Padres National Forest where the birds were known to nest.

A condor recovery plan was prepared in 1975. This plan had the objective of creating at least two wild populations, one in the area of the existing population and one somewhere else within the historical range. The plan also stated an objective of a total population of at least 100 birds in the wild. These objectives were to be achieved by reducing mortality, protecting nesting habitat, and improving feeding habitat. Captive propagation and the release of birds to the wild was one of the techniques proposed.

In the late 1970s, it became apparent that the wild population was continuing to decline, and that shell thinning from DDT,

FIGURE 25.9

■ The California condor, reduced to a small population in the Los Padres National Forest in California, is the object of a major captive breeding and reintroduction program.

FIGURE 25.10

■ A captive breeding program for Andean condors helped provide information on the techniques of captive propagation of California condors.

lead poisoning from shot in carcasses eaten by condors, and illegal shooting were major causes of mortality and poor reproductive performance. The recovery plan was revised in 1980 to intensify efforts at captive propagation. Propagation centers were established at the Los Angeles and San Diego Zoos. Initially, young birds and eggs from nests in the wild were brought into these centers, but eventually the decision was made to bring all remaining wild birds into captivity. This decision was bitterly contested by some conservation groups, in part because it removed all condors from areas considered important for management as condor range.

Reason for optimism existed in regard to the captive breeding program. In the early 1900s, two female condors at the National Zoo in Washington, D.C. had regularly laid eggs (infertile), indicating that conditions of captivity did not inhibit reproductive behavior of females. Begin-

FIGURE 25.11

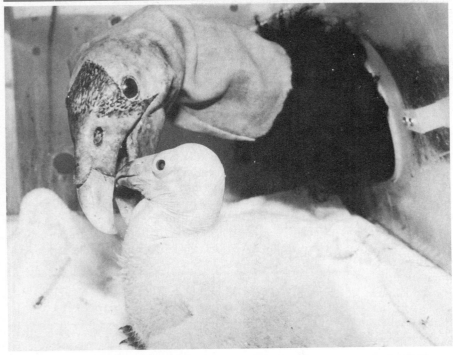

■ At the San Diego Wild Animal Park, young California condors are fed by hand puppets designed to simulate the appearance of an adult condor. This chick, named Molloko, was the first offspring of birds paired in captivity.

FIGURE 25.12

■ Andean condors were released to the wild at a hacking site in the Los Padres National Forest as a prelude to the reintroduction of California condors to the wild in 1991.

ning in 1965, the United States Fish and Wildlife Service had conducted a captive breeding program for Andean condors, a different but related species, with the objective of learning techniques of propagation and release of birds of this type (Fig. 25.10). This program has successfully reared many birds, and has successfully returned several birds to the wild in South America (Wallace and Temple 1987).

Captive breeding efforts have had good initial success. In 1991, 40 birds existed in captivity, 19 at the Los Angeles Zoo and 21 at the San Diego Wild Animal Park. Twelve pairs of breeding age were represented, and since 1988 young birds have been hatched from eggs laid in captivity (Fig. 25.11). About 10 to 20 young were expected to be hatched in 1991.

Successful reintroduction of California condors to the wild, scheduled to begin in the fall of 1991, will be a difficult undertaking. Major uncertainties exist about whether the recent California range is a suitable reintroduction site, inasmuch as the population was declining rapidly in this area. How to reintroduce the species to an environment lacking any wild birds is a technical challenge, since the effort must enable birds to forage for food, find appropriate roosting and nesting sites, and establish a social group that permits them to breed successfully. One effort to study these problems has been the release of captive-reared Andean condors in the Sespe Condor Sanctuary in Ventura County, California. Beginning in 1988, groups of juvenile female Andean condors have been gradually introduced to the wild, using the hacking technique (Fig. 25.12), in Sespe Condor Sanctuary. These birds are intended to provide information about the behavior and survival of condors released into an area where no adults exist, and to form a surrogate population into which California condors can be released. Eventually, the Andean condors will be recaptured and released in Colombia as part of a program to reintroduce this species to South American areas from which it has disappeared.

A few species have been propagated in captivity for a number of years, and then reintroduced to their original ranges. The red wolf, extinct in the wild as a purebred species since the 1970s, has been bred in captivity since 1975 at the Point Defiance Zoo captive breeding facility in Tacoma, Washington. Animals from this population were experimentally reintroduced to Alligator River National Wildlife Refuge, located on a peninsula on the North Carolina coast (Rees 1989). Several desert ungulates have been reintroduced to their original ranges in Africa and the Middle East (See chapter 5). The Socorro Island dove, a species closely related to the mourning dove of continental North America, was reintroduced to Socorro Island, off the west coast of Mexico, in 1988.

Captive propagation has also contributed to the recovery of other species, such as the peregrine falcon in North America. Captive breeding facilities for this species are located in New York, Idaho, and California, and since the mid-1970s over a thousand birds have been released to the wild (Peakall 1990). These releases are credited with establishing 30 to 40 breeding pairs in regions from which the species had disappeared due to DDT contamination (See chapter 13).

A research and propagation center for raptors, known as *The World Center for Birds of Prey* has been established in Boise, Idaho. Captive propagation of several species of hawks and owls is being carried out at this location.

Reintroduction of individuals produced in captivity to the wild is usually done gradually. Individuals are first placed in large pens or flight cages in a seminatural setting to enable them to acclimate to outdoor conditions. After this, they are released, but food is provided regularly at the acclimation site, a procedure termed **hacking** (a falconry term) for birds.

A second technique, **cross-fostering,** is the rearing of young of one species by adults of another. For birds, cross-fostering involves the exchange of eggs between species, so that foster parents incubate the eggs and rear the young. For mammals, cross-fostering may involve gestation by foster parents, accomplished by means of embryo implants.

Population Translocation

Translocation is the release of individuals into the wild to create or enlarge a wild population. Translocation sites are usually part of the original range of the species. Most animal translocations have involved game species, and most have involved individuals recently captured from the wild (Griffith et al. 1989). When the individuals are fully adapted to life in the wild, translocations are much easier to accomplish than when captive-bred individuals are used. In several instances, large ungulates, such as bison, elk, and mountain sheep, have been reintroduced to parts of their former ranges in western North America. Muskoxen have been translocated successfully to areas in Siberia, Alaska, eastern Canada, and western Greenland (Klein 1988). The sea otter was successfully reintroduced to areas of its original range along the coast of Oregon, and recent efforts have been made to reestablish a population on one of the California Channel Islands by translocation. An endangered, nearly flightless bird, the saddleback, was probably saved from extinction in New Zealand by translocation of populations to small islands that were free of predators (Merton 1975).

Translocation has been used successfully with endangered plant species, as well. In California, where temporary pond habitats support a rich, highly endemic flora, artificial pond basins have been created and successfully inoculated with members of this flora, including a federally endangered annual plant, the San Diego mesa mint (P. Zedler, personal communication).

Careful evaluation of possible consequences of translocation must be conducted, especially if the area lies outside the range of the species (Conant 1988). A proposal was made to translocate the Nihoa millerbird, an insectivorous bird endemic to Nihoa Island in the western Hawaiian Islands, to nearby Necker Island. Necker Island, however, possesses a fauna of highly endemic arthropods which has never been exposed to an avian insectivore. This proposal, fortunately rejected, might have led to the loss of some of these arthropods.

KEY MANAGEMENT STRATEGIES FOR ENDANGERED SPECIES

1. Legal protection of populations of endangered and threatened species and their essential habitat.

2. Development of recovery plans for creation of populations that have a minimal risk of extinction due to random demographic, environmental, and genetic processes.

3. Use of ex situ care, captive propagation, and translocation to supplement protection of the species and their habitats in nature.

LITERATURE CITED

Anonymous. 1991. Black-footed ferret recovery effort progresses toward reintroduction. *End. Sp. Tech. Bull.* **16(1):**1, 3–5.

Bean, M. J. 1987. The federal endangered species program. Pp. 147–160 *in* R. L. Di Silvestro, W. J. Chandler, K. Barton, and L. Labate (Eds.), *Audubon wildlife report 1987.* Academic Press Orlando, Florida.

Belovsky, G. E. 1987. Extinction models and mammalian persistence. Pp. 34–57 *in* M. E. Soulé (Ed.), *Viable populations for conservation.* Cambridge Univ. Press, Cambridge, England.

Conant, S. 1988. Saving endangered species by translocation. *Bio-Science* **38:**254–257.

Gilpin, M. E. and M. E. Soulé. 1986. Minimum viable populations: Processes of species extinction. Pp. 19–34 *in* M. E. Soulé (Ed.), *Conservation biology: The science of scarcity and diversity.* Sinauer Associates, Sunderland, MA.

Goodman, D. 1987. The demography of chance extinction. Pp. 11–34 *in* M. E. Soulé (Ed.), *Viable populations for conservation.* Cambridge Univ. Press, Cambridge, England.

Griffith, B., J. M. Scott, J. W. Carpenter, and C. Reed. 1989. Translocation as a species conservation tool: Status and strategy. *Science* **245**:477–480.

Kellert, S. R. 1985. Social and perceptual factors in endangered species management. *J. Wildl. Manag.* **49**:528–536.

King, W. B. 1981. *Endangered birds of the world. The ICBP bird Red Data Book.* Smithsonian Institution Press, Washington, D.C.

Klein, D. R. 1988. The establishment of muskox populations by translocation. Pp. 298–318 *in* L. Nielsen and R. D. Brown (Eds.), *Translocation of wild animals.* Wisconsin Humane Society, Madison, and Caesar Kleberg Wildlife Research Institute, Kingsville, TX.

Lucas, G. and H. Synge. 1988. *Plant Red Data Book.* International Union for the Conservation of Nature and Natural Resources, Gland, Switzerland.

McMillen, J. L. 1988. Conservation of North American cranes. *Amer. Birds* **42**:1212–1221.

Merton, D. V. 1975. The saddleback: Its status and conservation. Pp. 61–64 *in* R. D. Martin (Ed.), *Breeding endangered wildlife in captivity.* Academic Press, London.

Newmark, W. D. 1986. *Mammalian richness, colonization, and extinction in western North American national parks.* Ph. D. Dissertation, University of Michigan, Ann Arbor, MI.

Peakall, D. B. 1990. Prospects for the peregrine falcon, *Falco peregrinus,* in the nineties. *Can. Field Nat.* **104**:168–173.

Reed, J. M., P. D. Doerr, and J. R. Walters. 1988. Minimum viable population size of the red-cockaded woodpecker. *J. Wildl. Manag.* **52**:385–391.

Rees, M. D. 1989. Red wolf recovery effort intensifies. *End. Sp. Tech. Bull.* **14(1–2)**:3.

Shaffer, M. 1987. Minimum viable populations: Coping with uncertainty. Pp. 69–86 *in* M. E. Soulé (Ed.), *Viable populations for conservation.* Cambridge Univ. Press, Cambridge, England.

Simons, T., S. K. Sherrod, M. W. Collopy, and M. A. Jenkins. 1988. Restoring the bald eagle. *Amer. Sci.* **76**:253–260.

Snyder, N. F. R. and H. A. Snyder. 1989. Biology and conservation of the California condor. *Current Ornithology* **6**:175–267.

Tate, J. R., Jr. 1986. The Blue List for 1986. *American Birds* **40**:227–236.

Thornback, J. and M. Jenkins. 1982. *The IUCN mammal Red Data Book.* IUCN, Gland, Switzerland.

Tobin, R. J. 1990. *The expendable future: U.S. politics and the protection of biological diversity.* Duke Univ. Press, Durham, NC.

Wallace, M. P. and S. A. Temple. 1987. Releasing captive-reared Andean condors to the wild. *J. Wildl. Manag.* **51**:541–550.

Wells, S. M., R. M. Pyle, and N. M. Collins. 1983. *IUCN invertebrate Red Data Book.* IUCN, Gland, Switzerland.

CHAPTER 26

The Design of Natural Preserves

As human populations grow and the exploitation of natural resources intensifies, natural ecosystems are becoming relegated to a small fraction of the biosphere. Once a continuous matrix that surrounded islands of landscape where humans were active, natural ecosystems have mostly been reduced to island-like preserves and remnants in a sea of disturbance and development. Yet we expect these small, isolated areas to bear the responsibility for preserving the bulk of natural diversity. In addition, most preserves were established in an *ad hoc* manner, with little consideration to how they contributed to conservation on a regional basis. We now realize that systems of preserves must be designed carefully, in order to prevent large numbers of species from slipping into endangerment, where their survival depends on extraordinary and expensive efforts. Many features of preserves, including size, shape, structure, and location, influence their effectiveness in protecting biodiversity.

The design of systems of natural preserves falls in the emerging field of **landscape ecology,** which focusses on the structure and function of portions of the biosphere containing a mosaic of different ecosystems. Much of the theory about the design of natural preserves comes from the field of **island biogeography,** which is concerned with the processes of colonization, evolution, and extinction of species in insular environments. We shall first examine some of the principles of island biogeography, and then consider how these principles are being applied to the design of adequate systems of natural preserves. Finally, we shall consider how modern systems of data acquisition, storage, and analysis can contribute to designing effective systems of preserves.

ISLAND BIOGEOGRAPHIC THEORY

In small island areas where speciation is unimportant in the addition of new species to the biota, the number of species is determined by the processes of colonization and local extinction, a fact first emphasized by MacArthur and Wilson (1967). This relationship can be illustrated by considering how the rates of colonization of an island and of extinction of species on the same island are related to the number of species present (Fig. 26.1). If an island is stripped of its plant and animal life by, say, a volcanic explosion, but nearby islands or continental areas are unaffected, we expect that recovery will soon begin, as dispersal brings potential colonists to the island. If conditions are relatively favorable, many of these will probably become established. Species that are well adapted for overwater dispersal and for occupation of severely disturbed habitats will probably be the first colonists. Later, species with weaker dispersal mechanisms and specialized habitat requirements will arrive. Eventually, as the number of species on the island approaches that on the nearby continent, the rate of colonization necessarily declines. Thus, the rate of colonization should tend to be high in early stages of recovery, and should theoretically decline to zero when all species from the source areas have colonized the island.

Once species have colonized, however, some of them may then suffer extinction. When only a few hardy, abundant species are present, the rate of extinction will obviously be low. As the number of species increases, the abundance of each tends to become smaller, each becomes more specialized in its habitat and niche, and each encounters stronger biotic challenges such as competition, predation, and disease. As a result, the rate of extinction increases. On our hypothetical island, colonization rate—high in the early stages of recovery—declines through time, and extinction rate—low in the early stages—increases. Eventually, these two rates become equal, so that the number of species on the island reaches a steady state value. In actuality, at this steady state, the biota fluctuates randomly about an average number of species defined by average rates of colonization and extinction. In addition, even though the number of species remains relatively constant, the composition of the biota is constantly changing, as certain species disappear by extinction, and others arrive and colonize the island.

These **colonization and extinction curves,** and the equilibrium biotas they define, vary with factors such as island size and distance from the sources of potential colonists (Fig. 26.2). Islands that are very small, thus presenting a small "target" for dispersal propagules, or very far from a dispersal source will have colonization curves beginning at much lower rates than those for islands that are large or near source areas. Extinction curves for very small islands, where the individuals of any species are few in number and confined in space, will rise at a much faster rate that those for large islands. A small, far island, therefore, will have a low colonization rate, a high extinction rate, and a low equilibrium number of species. A large, near island will have just the opposite.

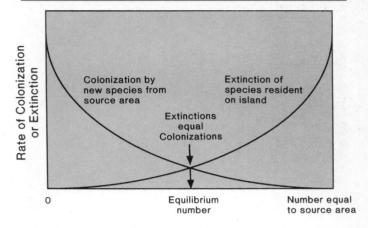

FIGURE 26.1

■ The rate of colonization by new species from a source area, the extinction rate of resident species, and the equilibrium number of species on an island as predicted by basic island biogeographic theory.

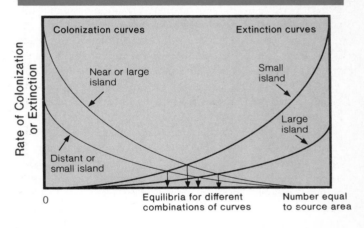

FIGURE 26.2

■ The effect of island size and isolation from a source area of colonists on the equilibrium number of species on an island.

A major corollary of this relationship is the principle of **faunal relaxation.** If an island is suddenly created from an area that was once part of a much larger region of similar habitat, it probably contains more species than can be maintained by the dynamics of colonization and extinction. As a result, extinction will run ahead of colonization, and the number of species will decline toward the steady state number. Case (1975) showed that relaxation of the number of species of lizards had apparently occurred on islands created in the Gulf of California by rising post-Pleistocene sea levels. In Panama, Karr (1982) estimated that 50 to 60 species of forest birds have disappeared from Barro Colorado Island, a 15-square kilometers area of upland tropical forest made into an

FIGURE 26.3

■ Barro Colorado Island, Panama, 15 square kilometers in area, has lost perhaps 50 to 60 species of breeding birds since the flooding of Gatun Lake isolated it during the construction of the Panama Canal in 1915.

FIGURE 26.4

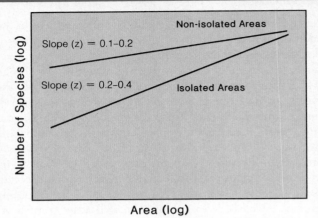

■ Species-area curves for biotas in which the areas examined for numbers of species are isolated have a steeper slope than those for areas of non-isolated, continuous habitat of similar nature.

island by the impoundment of Gatun Lake during construction of the Panama Canal in 1915 (Fig. 26.3). This represents about 19 to 22 percent of the 268 or so species that were likely to have been resident on the area before it became an island. As we shall see below, faunal relaxation has major implications for the design of biotic reserves.

The combined effects of area and isolation on steady state species number can be seen by examining **species-area curves** (Fig. 26.4). These are simply curves of the number of species found against size of the area examined. When both values are plotted as logarithms, the result is usually a straight line, the slope, z, of which is a measure of how fast the number of species changes with a unit change in area. Species-area curves for island situations, where each point represents an island of a certain area, tend to be steeper than those for continental situations, where each point represents an arbitrarily bounded area of certain size. For insular species-area curves z tends to range from 0.20 to 0.40, whereas z-values for continental areas lie in the range from 0.10 to 0.20. The steeper slope of the line for island areas means simply that as one goes to smaller and smaller islands the number of species drops off faster than expected for a continental situation. The reason for this difference is that when an extinction occurs on an area of certain size on a continent, it is easily offset by recolonization, since no habitat barrier has to be crossed. On an island, the presence of a water barrier increases the difficulty of recolonization, and an extinction is not offset as quickly. This means that on small islands, where extinctions are frequent, fewer species are present, on average, than on a mainland area the same size.

ISLAND BIOGEOGRAPHY AND HABITAT ISLANDS

The fragmentation of once continuous ecosystems into patches surrounded by ecosystems of a very different type creates "habitat islands" that are similar in many ways to oceanic islands. Habitat islands, for example, experience a **turnover of species** due to the processes of local extinction and recolonization. This turnover can be seen in long-term census data for bird populations in isolated woodlots in eastern North America. In Trelease Woods, a 22–ha area of forest near Champaign, Illinois, censuses of breeding birds have been conducted nearly every year since 1927 (Kendeigh 1982). Over this period, some 62 species of breeding birds have been recorded, but only nine species have been present every year. On the average, the species composition shows about a 13.6 percent change from one year to the next. Each year, some species disappear, perhaps for only a year or two, and new species appear.

Faunal relaxation can also be seen in habitat islands. Mount Rainier National Park in Washington, for example, falls within the original ranges of 68 species of native mammals. With settlement and development of land surrounding this park, many species have disappeared. By 1920, only 50 species remained, 73.5 percent of the original forms. By 1976, only 37 species, 54.4 percent of the original mammals, remained. Some of these species, such as the wolf, may have disappeared because of direct persecution by humans, but many have become extinct because the area has now effectively become a small habitat island with a high extinction rate, and areas from which recolonization might occur are now quite distant.

Island effects are quite evident in species-area curves for habitat islands of different size. In eastern Illinois, Blake and Karr (1984) censused breeding bird communities in woodlots varying in size from 1.6 ha to 600 ha. For forest interior bird species, they found a species-area slope (z) of 0.57, and for long-distance migrants a slope of 0.30, indication that these groups declined in species number at a rate typical of oceanic islands. Observations like these have led to the recognition of **area-sensitive species,** or species that are likely to disappear when an insular habitat reaches some critical minimum size, even though it may still be much larger than the activity area of individuals of the species itself (Robbins et al. 1989). Analysis of the occurrence of bird species in forest areas ranging from 0.1 to over 3,000 ha in Maryland and neighboring states, for example, revealed that 26 species showed significant decreases in relative abundance with decrease in forest area (Robbins et al. 1989). These included permanent residents such as the pileated woodpecker, short-distance migrants such as the white-breasted nuthatch, and long-distance migrants such as the scarlet tanager.

EVALUATING THE ADEQUACY OF PRESERVES

Examination of species-area relationships for preserves can reveal how strong the island effect is for various groups of species. Kitchener and his colleagues (1980a, 1980b, 1982) examined these relationships for mammals, birds, and lizards in 23 preserves in the wheatbelt region of Western Australia. The preserves, varying in size from 34 to 5,119 ha, exist in an intensively farmed landscape. For birds, the species-area slope (z) was 0.18, indicating that the faunas of the various preserves were not strongly affected by their isolation. Many of these birds are adapted to relatively open vegetation types, and are perhaps able to use shrubby growth and trees along roads, fence lines, and streams, or surrounding farmsteads, enabling them to recolonize preserves where extinctions have occurred. For lizards the z value was 0.25, indicating that preserves were showing a slight island effect. For non-flying mammals, however, the z value was 0.39, indicating that these preserves were suffering very severely from isolation in a fashion similar to oceanic islands (Fig. 26.5). This high z value means that many species had, in fact, become extinct on the smaller preserves since they were isolated.

Other workers have used species-area curves to examine how transforming reserves into habitat islands affects the net rate of species loss due to the excess of extinction over colonization. Miller (1978), for example, estimated that the Mkomazi Game Reserve in northern Tanzania, which now contains 39 species of large mammals, would suffer severe faunal relaxation if it became isolated from neighboring reserves. Although it is 3,276 square kilometers in area, Miller estimated that it would likely lose 17 species over the next 300 years.

The species-area relation can also be used to estimate the size of the area required to support a specified fraction of the regional biota. Care must be used in formulating such estimates, however, since the slopes of species-area relations may be influenced greatly by the position of a few data points (Boecklin and Gotelli 1984). For mammals of the Australian wheatbelt, however, Kitchener and his colleagues (1980b) estimated that a preserve 43,000 ha in area would be necessary to support all 25 species, or a preserve 32,700 ha in area to support 90 percent of them. This suggests that the existing preserves, the largest being 5,119 ha, are inadequate to the

FIGURE 26.5

■ The western gray kangaroo, *Macropus fuliginosus,* is absent from Australian wheat belt preserves smaller than about 200 hectares.

task of preserving the mammal fauna of this region. In the United States, Robbins et al. (1989) similarly found that forest areas 3,000 ha in area would be needed to retain all of the forest interior birds of the Middle Atlantic States. Similar analyses of savanna preserves in Africa suggest that preserves must be at least 10,000 square kilometers in area to maintain the large mammal faunas of this region (East 1981). Fortunately, a number of the existing parks and game preserves are larger than this minimum value. Maintaining these large park areas in the face of human population growth will be very difficult, however, and it is possible that lands will be removed from some of the large parks, reducing their effectiveness as preserves for the large mammals.

Terborgh (1975) used data on the apparent loss of bird species on forested tropical islands isolated from the adjacent continent by rising sea level to estimate extinction rates for tropical islands or preserves of different size. The islands in this analysis ranged from Trinidad, off the coast of Venezuela, to small continental islands in the Gulf of Panama. These islands were all connected to the mainland of South or Central America about 10,000 years ago during low sea level periods of the Pleistocene. Terborgh assumed that the Pleistocene faunas of these islands were equal to the present faunas of areas of equal size on the mainland. From these very rough data, Terborgh calculated that an area of about 2,500 square kilometers was needed to reduce extinction rates to less than 1 percent per century. Based on this estimate for birds, other workers suggested that a doubling of this area would achieve a similarly low extinction rate for large or wide-ranging tropical forest animals in general. For lack of any better estimate, this value, 5,000 square kilometers, has become a standard minimum size for major tropical forest preserves in areas such as the Amazon Basin.

Experimental studies are now in progress in the central Amazon Basin to obtain a more exact picture of the effects of creating habitat islands. These studies, originally known as

FIGURE 26.6

■ In the Biological Dynamics of Forest Fragments Project, tropical forest areas, such as these woodlots one and ten ha in area, were created and are being censused regularly to determine the changes induced by isolation.

FIGURE 26.7

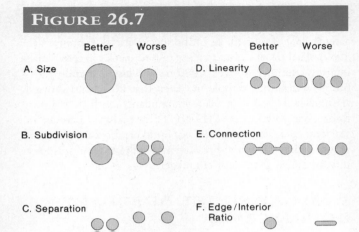

■ Some "better-worse" comparisons of design of preserve systems as derived from island biogeographic theory (Modified from Diamond 1975).

the **Minimum Critical Size of Ecosystems Project** (Lovejoy et al. 1986) and now as the **Biological Dynamics of Forest Fragments Project,** rely on the Brazilian law that 50 percent of land under private ownership in Amazonia must be left in forest. With the cooperation of the operators of new ranches in a region north of Manaus, Brazil, 20 forest patches of one, ten, 100, 1,000, and 10,000 ha have been established (Fig. 26.6). The biotas of these areas were censused prior to the isolation of certain patches by land clearing. Censuses will be continued over at least a 20–year period to reveal how isolation affects the numbers and kinds of species that survive in them. The data already available (Lovejoy et al. 1986) indicate that newly isolated forest fragments experience severe climatic and biotic change at their edges.

SIZE, SHAPE, AND SPACING OF PRESERVES

Island biogeographic theory was first used (e.g., Diamond 1975) to try to predict desirable features in systems of natural preserves (Fig. 26.7). First, as we have seen, large preserves were predicted to contain more species and to show lower extinction rates than small preserves, and thus be more desirable. Second, assuming a homogeneous distribution of species throughout a region, a single large preserve was predicted to protect more species than several small preserves with the same total area (however, see below). Sets of preserves were predicted to be more effective if they were near each other and spaced in an equidistant manner than if they were widely spaced or strung out in a linear fashion. Close, equidistant spacing was presumed to facilitate the recolonization of species, should extinction occur in one of the preserves. Similarly, habitat corridors, or stepping stone arrangements of small preserves between larger preserves, were predicted to facili-

tate biotic interchanges and recolonizations within a system of preserves. Finally, a circular shape that minimizes edge influences was predicted to be better than an elongate shape for survival of obligate interior species of individual preserves. These simplistic predictions, many of them assuming biotic homogeneity of the region in which the preserves are located, stimulated several controversies.

THE SLOSS CONTROVERSY

The suggestion that large preserves are most desirable for conservation purposes spawned a major controversy: the **SLOSS argument,** or whether a **S**ingle **L**arge **O**r **S**everal **S**mall preserves guarantees the survival of most species (e. g., Simberloff and Abele 1976, Diamond et al. 1976). Simple island biogeographic theory, as outlined above, assumes that the biota is uniform throughout the region in question. Under this assumption, a single large preserve does, in fact, contain more species than several small preserves. When a region contains major habitat gradients or several centers of biotic diversity resulting from historical influences, however, a single large preserve obviously will not hold more species that several smaller preserves located in different parts of the region (Simberloff and Abele 1976, 1982). A greater total number of species in small islands or preserves than in a single area of equal total size is usually the case, in fact. Blake and Karr (1984), in their study of breeding bird communities of eastern Illinois woodlots, found that two woodlots of a certain size usually supported more species than a single woodlot of equal total size. Quinn and Harrison (1988) found that this was true both for several groups of organisms of the biotas of oceanic islands and for the faunas of United States National Parks.

Spreading the population of a species among several preserves, rather than concentrating it in a single large preserve, may also reduce the risk of complete extinction of the species. Several workers have studied this question by mathematical modeling. Kobayashi (1985) concluded that smaller reserves protected more species if extinction rates were low and if different groups of species tended to occur in the different small preserves. On the other hand, if rates of extinction were high, a single large preserve protected more species unless the restriction of different groups of species to different preserves was very strong. Quinn and Hastings (1987) found that when only random fluctuation in population sizes due to accidents of reproduction and mortality was considered, species survived best in a single large preserve. When the populations were affected by chronic or catastrophic environmental conditions, however, a certain degree of subdivision of preserves provided the best guarantee that species would survive somewhere in the set of preserves.

In addition to the number of species, the kinds of species that are protected are an important consideration. Using total species diversity as an index of preserve value is certainly not always appropriate (Diamond 1976, Jarvinen 1982). Preserves are not needed for many species adapted to disturbed or developed landscapes, yet in diversity analyses these species are often counted equal to species with special requirements for undisturbed natural habitats. In other words, preserves are not needed for weedy species such as the house mouse or English sparrow. Native species that are endemic to a region, particularly those with restricted distributions, specialized habitat requirements, or low population densities, are the real focus of conservation concern. In addition, large mammals, such as elk and mountain lions, cannot be maintained on preserves a few square kilometers in area—viable populations of such species require larger areas. Thus, considerations of the size and spacing of preserves must really focus on those native species that are adversely affected by disturbance and development, and on their particular space requirements.

SPACING AND INTERCONNECTION OF PRESERVE UNITS

Increasingly, study is being directed at the potential value of designs that favor interchange of organisms between preserve units. In Maryland, for example, MacClintock et al. (1977) found that many forest birds were able to occupy small woodlots, but that this depended largely on dispersal from larger areas of forest located nearby. Corridors of trees linking small forest fragments with larger areas of forest also allowed forest interior species to occupy small woodlots.

On a larger scale, linking natural preserves of various specific types into **conservation networks** may allow viable populations of large or scarce animals to be maintained more effectively (Salwasser et al. 1987). In the United States, ten potential networks involving national parks, monuments, forests, recreation areas, and wildlife refuges can be outlined (Fig. 26.8). These units would range in size from 5,650 to

FIGURE 26.8

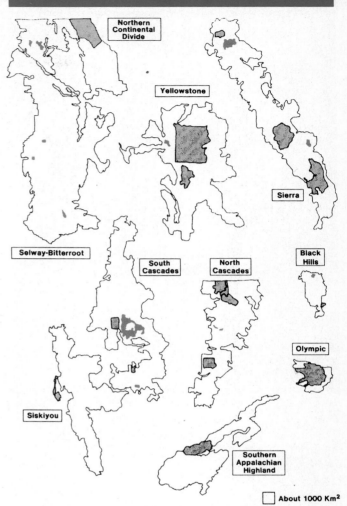

About 1000 Km²

■ Proposed conservation networks of natural areas now under federal management in the United States. Shaded areas are national parks, unshaded areas are other federal lands, and black areas are private holdings (Modified from Salwasser et al. 1987).

75,350 square kilometers. Networking state and private natural areas into such systems could allow further enlargement of these networks and the development of still others. The larger effective units created by such interconnection might support populations of species that cannot now survive in the separate units. Harris (1988) has outlined how the acquisition of critical lands or easements could create such networks in the southeastern United States. Contiguous tracts of at least 2,000 square kilometers are required in this region, for example, to maintain populations of black bears, and a still larger minimum area to maintain the Florida panther (Fig. 26.9). By linking several federal and state areas, four areas of 2,000 square kilometers could be created where only one now exists.

Noss and Harris (1986) have suggested a more formal strategy for structuring networks of preserves. This proposal

FIGURE 26.9

■ Networking various areas of natural habitat in southern Florida may permit the survival of a viable population of the Florida panther.

envisions a network of **multiple-use modules,** or conservation units with a fully protected core area surrounded by zones of natural areas utilized in progressively more intense fashion for recreation, timber production, and other uses. These modules would represent "nodes" in a network tied together by habitat corridors adequate to allow movement of animals between modules. On the Georgia-Florida border, for example, 15 or so existing natural areas could be linked by habitat corridors, some of them centered on major rivers, to create a network extending from Okefenokee National Wildlife Refuge in Georgia to the Florida Gulf Coast. Such a network should be adequate for the Florida panther, for example, and could allow reintroduction of the species to this region.

EDGE EFFECTS AND PRESERVE SURROUNDINGS

The nature of the edges of preserves—the borders of natural habitat islands and their surroundings—has become a major conservation concern. For decades, the **ecotone,** or border zone between distinct ecosystems, has been regarded as a habitat generally beneficial to wildlife (Yahner 1988). In fact, such edge zones usually show higher species diversity and productivity than areas away from them. In some cases, particular species are adapted specifically to edge habitats. Game management practices, geared to early or middle successional species (See chapter 3), have often encouraged the increase of edge habitats by the cutting of clearings or corridors in continuous stands of forest or brushland.

From a conservation standpoint, however, extensive edge habitat may not always be desirable. Edge habitats often support high populations of predators, and the effects of predation may extend well into the areas away from the edge proper. For forest birds, nest predation and parasitism are high near the edges of woodlots, and in small woodlots the entire area may be affected (See chapter 3). Increasing edge habitats can also increase the numbers of herbivores, such as white-tailed deer, that may then permeate and overbrowse forest interior areas (Alverson et al. 1988). Where enhancing edge effects leads to a high edge-to-interior ratio for forest fragments, these effects can cause regional extinction of forest interior species (Temple and Cary 1988).

The nature of the areas surrounding preserves is also an important consideration for preserve design (Jensen 1986). Agricultural activities that foster species such as nest-parasitic cowbirds and omnivorous birds and mammals that can invade nearby forest areas may be incompatible with efforts to preserve forest species. Land uses that allow polluted runoff and eroded sediment to be carried into wetland preserves can effectively destroy the preserve from outside. Thus, establishment of preserves may need to be coupled with zoning or other control of the surrounding area to assure effectiveness of the preserve.

SHORTCOMINGS OF ISLAND BIOGEOGRAPHIC THEORY

Dealing, as it does, with numbers of species and rates of colonizations and extinctions, island biogeographic theory tends to lose sight of the specific ecological requirements of individual species (Boecklen and Simberloff 1986). The seemingly simple colonization and extinction curves of MacArthur and Wilson (1967) have also spawned a mass of theory that can yield widely divergent predictions when applied to a particular case. Boecklen and Gotelli (1984), for example, show that species-area curves often do not account for a high percentage of variability in data, and are strongly affected by unusual data values. These authors question the use of such weak mathematical relations in predicting the features that preserves should possess. Nevertheless, island biogeographic theory does address questions that are of basic importance to long-term preservation of biotic diversity. Like all mathematical models, as well, those of survival of species in insular habitats are subject to improvement, and are making a significant contribution to conservation strategy.

ECOLOGICAL STRATEGIES FOR DESIGN OF BIOTIC PRESERVES

New techniques are being sought for designing preserves in areas where none exist, and for strengthening preserve systems in other areas. Ideally, systems of preserves should include substantial populations of most of the species in a region, and thus should minimize the decline of species to endangered status (Scott et al. 1987a). This means that preserves should be located where major populations of endemic species are concentrated. The identification of centers of endemicity in the Amazon Basin (See chapter 6) is an example of this approach. Another example is provided by the island of New Guinea, divided between the independent nation of Papua, New Guinea and Irian Jaya, part of Indonesia. In this still largely undeveloped region, centers of endemism and biotic diversity can be identified and used as a major criterion for location of preserves (Diamond 1986).

Geographic information systems (GIS) technology promises to assist in the identification of new preserve sites (Scott et al. 1987a). GIS technology is simply the computerized recording of data for a region, using geographic coordinates as the primary indexing system. Ideally, such a system permits data on a particular feature to be stored for all the geographic units included in the indexing system. The kinds of data that can be stored for each coordinate unit include presence or absence of a species, abundance of that species where it is present, ecosystem type, soil type, geology and physiography, land protection status, and many other variables. From such a data base, many specific questions about where species that need protection are concentrated can be answered (Fig. 26.10), an effort now known as **gap analysis** (Scott et al. 1988). With modern GIS systems, it should be possible to develop comprehensive data bases for vegetation types, together with their vertebrate animals, higher plants, and certain groups of invertebrates, such as butterflies, of the scale of geographical units 200 ha in area (Scott et al. 1987a; J. M. Scott, pers. comm.).

GIS systems are well adapted to using data from remote sensing sources. Detailed data on the actual vegetation of a geographical area are difficult to obtain from traditional vegetation maps, which often show the potential climax vegetation thought to characterize a region, rather than the vegetation actually present. With improved techniques of satellite imagery and analysis, detailed data on the vegetation that actually exists can soon be determined on a grid scale such as that indicated above.

This GIS approach has been utilized in an elementary fashion in evaluating conservation reserves for endemic forest birds in Hawaii (Scott et al. 1987b). Over a period of eight years, surveys of the occurrence of forest birds were made at almost 10,000 sites throughout the main islands—a project known as the Hawaii Forest Bird Survey (Scott et al. 1986, 1989). These data enabled isolines of occurrence of different numbers of native species to be drawn. Locations where the greatest numbers of species occurred together were then identified and used as a guideline for the acquisition of new preserve lands. More than 17 square kilometers of new forest preserves, or easements to protect existing forests, resulted from this organized effort. The application of GIS to gap analysis of biodiversity preserves in California and other regions is now being initiated (Davis et al. 1990).

FIGURE 26.10

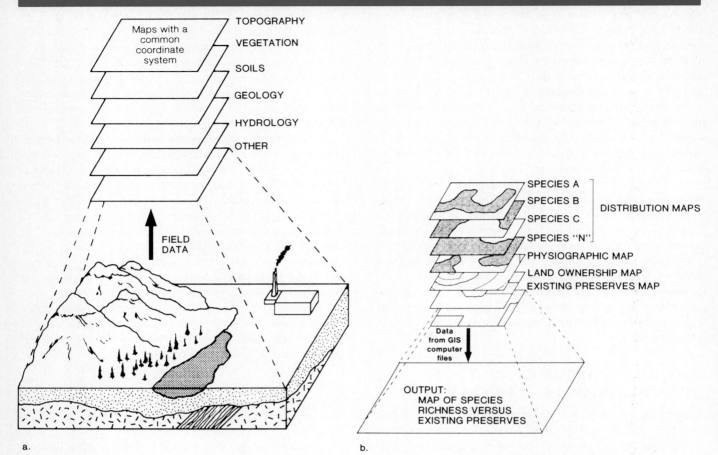

a.

- Outline of a geographic information systems (GIS) database for evaluation of adequacy of existing preserves and the best sites for new preserves. a. Data on various physical and biotic features of a region are stored in computer data files. b. Information from the database is used to create new maps that evaluate preserve adequacy by revealing gaps in the protection of species or ecosystems by existing preserves (Modified from Scott et al. 1987a).

KEY ECOLOGICAL PRINCIPLES FOR PRESERVE DESIGN

1. Determine the ecosystems and their member species that must be protected by designated preserves in a geographical region.

2. Locate major preserve units to maximize the diversity of native species initially included.

3. Structure and manage the overall system of preserves to minimize the extinction rate of endemic species.

LITERATURE CITED

Alverson, W. S., D. M. Waller, and S. L. Solheim. 1988. Forests too deer: Edge effects in northern Wisconsin. *Cons. Biol.* **2**:348–358.

Blake, J. G. and J. R. Karr. 1984. Species composition of bird communities and the conservation benefit of large versus small reserves. *Biol. Cons.* **30**:173–187.

Boecklen, W. J. and N. J. Gotelli. 1984. Island biogeographic theory and conservation practice: Species-area or specious-area relationships? *Biol. Cons.* **29**:63–80.

Boecklen, W. J. and D. Simberloff. 1986. Area-based extinction models in conservation. Pp. 247–276 *in* D. K. Elliott (Ed.), *Dynamics of extinction.* John Wiley and Sons, New York.

Case, T. J. 1975. Species numbers, density compensation, and colonizing ability of lizards in the Gulf of California. *Ecology* **56**:3–18.

Davis, F. W., D. M. Stoms, J. E. Estes, J. Scepan, and J. M. Scott. 1990. An information systems approach to the preservation of biological diversity. *Int. J. Geogr. Inform. Syst.* **4**:55–78.

Diamond, J. M. 1975. The island dilemma: Lessons of modern biogeography studies for the design of nature preserves. *Biol. Cons.* **7**:129–146.

Diamond, J. M. 1976. Island biogeography and conservation: Strategy and limitations. *Science* **193**:1027–1029.

Diamond, J. M. 1986. The design of a nature reserve system for Indonesian New Guinea. Pp. 485–503 in M. E. Soulé (Ed.), *Conservation Biology: The science of scarcity and diversity.* Sinauer Associates, Sunderland, MA.

Diamond, J. M., J. Terborgh, R. F. Whitcomb, J. F. Lynch, P. A. Opler, and C. S. Robbins. 1976. Island biogeography and conservation: Strategy and limitations. *Science* **193**:1027–1033.

East, R. 1981. Species-area curves and populations of large mammals in African savanna reserves. *Biol. Cons.* **21**:111–126.

Harris, L. D. 1988. Landscape linkages: The dispersal corridor approach to wildlife conservation. *Trans. N. A. Wildl. Nat. Res. Conf.* **53**:595–607.

Jarvinen, O. 1982. Conservation of endangered plant populations: Single large or several small reserves? *Oikos* **38**:301–307.

Jensen, D. B. 1986. Concepts of preserve design: What we have learned. Pp. 595–603 in T. S. Elias (Ed.), *Conservation and management of rare and endangered plants.* California Native Plant Society, Sacramento.

Karr, J. R. 1982. Avian extinction on Barro Colorado Island, Panama: A reassessment. *Amer. Nat.* **119**:220–239.

Kendeigh, S. C. 1982. Bird populations in east-central Illinois: Fluctuations, variations, and development over a half-century. *Ill. Biol. Monogr.* **52**:1–136.

Kitchener, D. J., A. Chapman, and B. J. Muir. 1980a. Lizard assemblage and reserve size and structure in the western Australian wheatbelt—some implications for conservation. *Biol. Cons.* **17**:25–62.

Kitchener, D. J., A. Chapman, and B. J. Muir. 1980b. The conservation value for mammals of reserves in the western Australian wheatbelt. *Biol. Cons.* **18**:179–207.

Kitchener, D. J., J. Dell, and B. G. Muir. 1982. Birds in western Australian wheatbelt reserves—implications for conservation. *Biol. Cons.* **22**:127–163.

Kobayashi, S. 1985. Species diversity preserved in different numbers of nature reserves of the same total area. *Res. Pop. Ecol.* **27**:137–143.

Lovejoy, T. E., R. O. Bierregaard, Jr., A. B. Rylands, J. R. Malcolm, C. E. Quintela, L. H. Harper, K. S. Brown, Jr., A. H. Powell, G. V. N. Powell, H. O. R. Shubart, and M. B. Hays. 1986. Edge and other effects of isolation on Amazon forest fragments. Pp. 257–285 in M. E. Soulé (Ed.), *Conservation biology: The science of scarcity and diversity.* Sinauer, Sunderland, MA.

MacArthur, J. W. and E. O. Wilson. 1967. *The theory of island biogeography.* Princeton Univ. Press, Princeton, NJ.

MacClintock, L., R. F. Whitcomb, and B. L. Whitcomb. 1977. Island biogeography and "habitat islands" of eastern forest. II. Evidence for the value of corridors and minimization of isolation in preservation of biotic diversity. *Amer. Birds* **31**:6–12.

Miller, R. I. 1978. Applying island biogeographic theory to an East African reserve. *Env. Cons.* **5**:191–196.

Noss, R. F. and L. D. Harris. 1986. Nodes, networks, and MUMs: Preserving diversity at all scales. *Environ. Manag.* **19**:299–309.

Quinn, J. F. and S. P. Harrison. 1988. Effects of habitat fragmentation and isolation on species richness: Evidence from biogeographic patterns. *Oecologia* **75**:132–140.

Quinn, J. F. and A. Hastings. 1987. Extinction in subdivided habitats. *Cons. Biol.* **1**:198–208.

Robbins, C. S., D. K. Dawson, and B. A. Dowell. 1989. Habitat area requirements of breeding forest birds of the Middle Atlantic States. *Wildl. Monogr.* **103**:1–34.

Salwasser, H., C. Schonewald-Cox, and R. Baker. 1987. The role of interagency cooperation in managing for viable populations. Pp. 159–173 in M. E. Soulé (Ed.), *Viable populations for conservation.* Cambridge Univ. Press, Cambridge, England.

Scott, J. M., S. Mountainspring, F. L. Ramsey, and C. B. Kepler. 1986. *Forest bird communities of the Hawaiian Islands: Their dynamics, ecology, and conservation.* Studies in Avian Biology, No. 9.

Scott, J. M., B. Csuti, J. D. Jacobi, and J. E. Estes. 1987a. Species richness. *BioScience* **37**:782–788.

Scott, J. M., C. B. Kepler, P. Stine, H. Little, and K. Taketa. 1987b. Protecting endangered forest birds in Hawaii: The development of a conservation strategy. *Trans. N. A. Wildl. Nat. Res. Conf.* **52**:348–363.

Scott, J. M., B. Csuti, K. Smith, J. E. Estes, and S. Caicco. 1988. Beyond endangered species: An integrated conservation strategy for the preservation of biological diversity. *End. Sp. UPDATE* **5**:43–48.

Scott, J. M., C. B. Kepler, C. Van Riper III, and S. I. Fefer. 1989. Conservation of Hawaii's vanishing avifauna. *BioScience* **38**:238–253.

Simberloff, D. and L. G. Abele. 1976. Island biogeography theory and conservation practice. *Science* **191**:285–286.

Simberloff, D. and L. G. Abele. 1982. Refuge design and island biogeographic theory: Effects of fragmentation. *Amer. Nat.* **120**:41–50.

Temple, S. A. and J. R. Cary. 1988. Modeling dynamics of habitat-interior bird populations in fragmented landscapes. *Cons. Biol.* **2**:340–347.

Terborgh, J. W. 1975. Faunal equilibria and the design of wildlife preserves. Pp. 369–380 in F. B. Golley and E. Medina (Eds.), *Tropical ecological systems: Trends in terrestrial and aquatic research.* Springer-Verlag, New York.

Yahner, R. H. 1988. Changes in wildlife communities near edges. *Cons. Biol.* **2**:333–339.

Conservation – Past and Present

Both our conservation attitudes and our institutions are products of history, and this history deserves our examination. The settlement and economic development of the New World have transformed ecosystems only lightly influenced by human activity into systems dominated by human influences at a pace unknown in the Old World. The speed at which change took place has given North Americans a deep insight into the interaction of humans with the rest of nature, an insight basic to the leading role that North American conservationists are playing in global efforts to protect biodiversity. The present rests on the past, but leads into the future, and we must strive to improve the attitudes and institutions that now exist. We shall finish our survey of conservation ecology by examining some of the global programs that are now in place, and describing the major strategies that are being explored to strengthen these programs and nurture new ones.

CHAPTER 27

History of Conservation

Conservation of natural ecosystems and their plant and animal inhabitants has a long and complicated history. Modern conservation attitudes and practices have evolved largely with the context of western society, and have been molded definitively by the major political, economic, and intellectual revolutions that western society has experienced. These forces continue to shape the practice of conservation worldwide.

In North America, societal concern for conservation of wildlife and natural environments has been cyclical (Strong 1988, Nash 1989*a*). Periods of conservation activism have occurred at times when crises due to intensified human impact on nature have coincided with breakthroughs in our understanding of the natural world and its significance to human well-being. In this chapter, we shall examine some of the major influences that have shaped conservation activity in western society, and consider the major phases of conservation activism that we have experienced. Finally, we shall look ahead, and speculate on how major trends in environmental conditions may shape the next conservation movement.

Background of Western Conservation Attitudes

Western viewpoints about conservation are conditioned by basic western philosophy, rooted in the Judeo-Christian view of man and nature and conditioned by political, economic, and intellectual attitudes that grew out of the democratic revolutions of the 1700s and 1800s (White 1967).

Judeo-Christian religious philosophy combines two ideas: the right of exploitation, and the responsibility of stewardship. A fundamental Judeo-Christian belief, outlined in the biblical book of Genesis, is that nature was created to serve the human race, so that the exploitation of nature is thus a legitimate and natural pursuit. Judeo-Christianity does not endow the environment and its inhabitants with protective spirits that prohibit exploitation, or that must be appeased before the environment is exploited. Later, in the 1600s, this view found strong support in the philosophy of René Descartes, who argued that humans were "masters and possessors of nature" (Nash 1989b). On the other hand, Judeo-Christian dogma also asserts that human beings are only temporary occupants of the earth, and must be responsible stewards of the natural heritage on which future generations depend. Although many members of western societies may not espouse the "master of nature" view in a biblically fundamentalist way, Judeo-Christian philosophy creates a permissive attitude toward exploitation of nature that is different from that in many eastern societies.

Until the 1700s, however, the rights and rewards of exploitation of the natural world lay largely in the hands of an elite aristocracy. The democratic revolutions of the late 1700s, including the American Revolution of 1775–1783 and the French Revolution of 1789–1799, triggered a restructuring of the framework of society throughout most western societies. With this change came increased access of individuals to productive resources, and increased ability to use them for improving economic and social status. The legitimate right of exploiting nature was now extended to individuals at large in society.

At nearly the same time, the mid-1700s, the Industrial and Scientific Revolution began. On the industrial side, machines and sources of energy were harnessed to enable resources to be processed on new scales and with increased efficiency. With the steam engine came railroads and steamships that opened new regions to exploitation. On the scientific side, discoveries lead to a revolutionary change in the basic concept of nature: a world in which all was created and overseen by God was replaced by a world that functioned according to the operation of basic laws of physics and chemistry. The Industrial and Scientific Revolution gave individuals an enormously expanded ability to exploit resources and create material wealth.

Finally, the 1800s were the culmination of a period of worldwide spread of western culture through colonialism and establishment of world trade. The western system of environmental exploitation was thus spread widely, so that it became the operational system, even in areas where the basic philosophical view of humans and nature was quite different (White 1967).

Conservation: Cycles of Crisis and Activity

In North America, four major eras of conservation activism have occurred in the past century and a half. These were periods of social and political activism, each about 10 to 15 years in duration, when concern about the natural environment spread through much of society, enabling new laws and institutions to be created. These revolutionary periods share four important features. Each, first of all, was triggered by patterns of national growth and development that reached some limit of structure and function of regional ecosystems. Second, each represented a time of emergence of basic new ideas in ecology and evolution. Third, in each period, one or more key naturalist-writers emerged as influential spokespersons that dramatized the issues and approaches to their resolution (Strong 1988). Finally, each period resulted in the establishment of new conservation institutions.

Disappearance of the Eastern Wilderness, 1850–1865

By the middle decades of the 19th century, European settlement was well into its third century. The landscape of eastern North America had experienced its greatest degree of agricultural clearing and ecological transformation. Coupled with clearing of the land was the disappearance of the original forest, referred to as the "wilderness," together with many of the larger forms of wildlife. What remained was a cleared, tamed, urban and agricultural landscape. Small farms covered the landscape in New England and the middle Atlantic states, and cotton and tobacco plantations the lands of the southern states. Recognition of this transformation brought with it a nostalgia for the wilderness as it had originally existed. In 1850, for example, James Fenimore Cooper had just finished the last of the "Leatherstocking Tales," a popular series of novels set in eastern frontier days, then over a century earlier.

At the same time, biology was undergoing a revolutionary change in its view of the natural world—the replacement of a static, creationist view of life by an evolving, mechanistic view. This change is best exemplified by the emergence of the **theory of evolution by natural selection,** presented jointly by Charles Darwin and Alfred Wallace to the Linnean Society of London in 1858. Darwin's "Origin of Species," which documented the process exhaustively, appeared in 1859. The concept of natural selection replaced the creationist view of the origin of living species with a mechanistic process of interactions within nature. Moreover, it placed humans within the world of nature rather than separate from nature. The evolutionary view also opened the eyes of many to the fact that change in the environment, including change caused by humans, could bring about the ex-

FIGURE 27.1

■ Henry D. Thoreau, who died two months short of his 45th birthday, found meaning in the natural world of New England, which was rapidly disappearing through human settlement and clearing for farms.

tinction of many kinds of organisms, as the fossil record demonstrated. The intellectual dynamism of this era infused all of the natural sciences. Other noted scientists of the period—not all evolutionists—included the paleontologist Louis Agassiz and the botanist Asa Gray.

The taming of the eastern landscape stimulated an aesthetic appreciation of the natural areas that remained, as well as a nostalgia for the wilderness that had disappeared. The aesthetic view of nature was evident in the work of Ralph Waldo Emerson (b 1803, d 1882), one of the period's most influential speakers and essayists. Emerson espoused the philosophy of **transcendentalism,** the rejection of material goals and a seeking of harmony and beauty through the contemplation of nature. Emerson's protege, Henry David Thoreau (b 1817, d 1862), adopted and lived this philosophy (Fig. 27.1). He lived simply, observed nature and society in detail, and faithfully recorded his observations and interpretations. For 24 years, Thoreau maintained a journal, begun at Emerson's suggestion, in which he recorded notes from material he had read, personal observations, philosophical musings, and poetry. Drawing on this journal, Thoreau published many newspaper and magazine articles, but, during his lifetime, only two books, both with transcendentalist themes. *A Week on*

the Concord and Merrimack Rivers, published in 1849, used a boat trip by Thoreau and his brother as a unifying vehicle for transcendentalist musings. The book sold very poorly, and Thoreau ended up with over 700 of the 1,000 copies printed. *Walden,* published in 1854, was the account of 26 months of living at Walden Pond, near Concord, Massachusetts. This book, a modest but short-lived publishing success at the time, described the virtue of a simple, contemplative life, close to nature. Two other books, *Excursions* and *The Maine Woods,* were published shortly after his death, with the editorial aid of Ralph Waldo Emerson, Emerson's sister Sophia, and William E. Channing, a close friend. His journals, in edited form, were eventually published as a series of 14 volumes (Torrey and Allen 1906). Thoreau also compiled an extensive set of notebooks, most of which remain unpublished, on the lives of Indians and other aboriginal peoples.

Thoreau sought meaning both by philosophical contemplation and scientific study of nature, however. His early journal entries are strongly philosophical. "The fact will one day flower out into a truth," he wrote in his journal in 1837. In 1856, as his appreciation of nature matured, he lamented the disappearance of the New England wilderness and its wildlife, leaving "a tamer, and as it were, emasculated country," and wished "to know an entire heaven and an entire earth." Later, in 1859, he advocated the establishment of parks, suggesting that every township "should have a park, or rather a primitive forest, of five hundred or a thousand acres, . . . for instruction and recreation." By his later years, Thoreau had become, through his own study, a perceptive ecologist (although the term "ecology" had not yet been coined), and his journal records many scientific observations. His observations of the kinds of trees that predominated in various woodlots near Concord, for example, led him in 1860 to recognize forest succession, anticipating a major concept that was not formally stated until more than 35 years later.

Thoreau's idea of park development was, in fact, carried forward by others. Frederick Law Olmsted (b 1822, d 1903) grew up in a rural, farming area near New York City, acquiring a knowledge of farming, horticulture, and landscaping. A visit to Europe stirred an interest in parks, and in the relation of people to the landscapes they inhabited. Ultimately, by winning a competition among landscape architects for the best design, he gained the opportunity to create Central Park in New York City. Carried out between 1857 and 1861, the Central Park project was a deliberate effort to restore an example of eastern wilderness within the city: "a specimen of God's handiwork" to serve the equivalent of "a month or two in the White Mountains, or the Adirondacks. . . ." Olmsted participated in the development of other urban and non-urban parks in the East, and in 1864 helped the effort to gain protection by the State of California for Yosemite Valley, an action that proved to be the first step in western park development.

This period saw the awakening of an appreciation of the value of nature, its dynamic character, and the need for stewardship of the natural world by humanity. In a sense, the period culminated with the appearance in 1864 of the book *Man and Nature* by George Perkins Marsh. This book described in explicit fashion the worldwide ecological impact of humans on the natural environment. Although it did not influence events of the time greatly, it did influence those of later conservation eras profoundly.

Closing of the Western Frontier, 1890–1905

By the end of the 19th century, western settlement in the United States and Canada had reached the Pacific Ocean; a western frontier beyond which unknown and untapped resources might be found no longer existed. With the recognition that further territorial expansion in mainland North America was impossible came the realization that the forest, water, and mineral resources of North America were finite. Dramatic effects of careless exploitation of some of these vital resources also drew the attention of society to the need for conservation. Disastrous fires raged through wastefully logged forests of the Lake States, destroying towns and killing thousands of persons (Biswell 1989). The effects of severe blizzards and droughts also emphasized the degree to which a healthy national economy depended on a benign environment.

In science, the 1890s saw the emergence of ecology as a recognized branch of biology. Ecology gained recognition, as distinct from natural history, with the growth of a body of theories about interactions in nature. The process of **biotic succession** was the most dramatic example of such theory. Henry C. Cowles (b 1869, d 1939), a botanist at the University of Chicago, studied the sequence of plant communities that occupied the successively higher and older beach terraces at the southern end of Lake Michigan. These beachlines represent lake levels that existed at various times during recession of Pleistocene glaciers over the past 10,000 or so years. From this time sequence, Cowles inferred the process of plant succession and its causes (Cowles 1899). Others soon recognized that succession was also shown by animal communities. Succession revealed that natural communities were dynamic and changing on a time scale that allowed them to be influenced for the better or worse by human activities. It also provided a theoretical framework that proved to be applicable in management of resources such as forests, rangeland, and wildlife.

The importance of living resources, and the need for their sound management, was stated most eloquently by the naturalist-writer John Muir (b 1838, d 1914). Muir (Fig. 27.2), born in Scotland, came to the midwestern United States as a boy. He was strongly influenced by the writings of Emerson and Thoreau (Fleck 1985). The forests and mountains of the West, particularly of California and Alaska, ultimately captured his interest. Muir was appalled at the damage being done to the California Sierras by grazing, logging, and burning, and turned to popular writing as a means of raising public concern about these issues. Through books such as *The Mountains of California* (1894) and *Our National Parks* (1901), he

FIGURE 27.2

■ John Muir, known as "John o' the Mountains," was an articulate proponent of the preservation of wilderness, and influenced the conservation efforts of President Theodore Roosevelt. Here, both men are seen at Glacier Point, overlooking Yosemite Valley.

stimulated an interest in the natural history of the western mountains. He actively promoted federal programs for protecting lands and forests. As the final statement in his book on national parks, he argued, " . . . God has cared for these trees, . . . but He cannot save them from fools—only Uncle Sam can do that." Muir's activism and love for the California Sierras led to the establishment of the Sierra Club, of which he was the first president.

This period saw the establishment of several federal agencies devoted to resource management. In 1891, a system of national forests was established under the administration of the United States Department of Interior, and in 1898 a Division of Forestry created within the United States Department of Agriculture. The first director of the Division of Forestry was Gifford Pinchot (b 1865, d 1946), a forester trained in Europe, where forest management had a long history. In 1905, the national forests themselves were transferred to the United States Department of Agriculture, and the present administrative structure of the **United States Forest Service** was established. Between 1889 and 1897, Presidents Benjamin Harrison and Grover Cleveland designated 36 million acres of national forest. Between 1901 and 1909, Theodore Roosevelt added 148 million acres, creating a system of 150 national forests throughout the United States.

FIGURE 27.3

■ Pelican Island, Florida, the first National Wildlife Refuge, was established in 1903 to protect a nesting area of egrets and other water birds that had been killed for their plumes. A proposal has recently been made to enhance public appreciation of this historic refuge by development of a viewing tower and interpretive facilities on the nearby mainland.

Another new resource agency was the **United States Bureau of Reclamation,** created in 1902, in response to the need for improved management of land and water resources. The **United States National Wildlife Refuge** system was also initiated with the establishment of Pelican Island Refuge (Fig. 27.3) in Florida in 1903. This refuge provided for the protection of breeding colonies of egrets and other water birds. These birds had been exploited heavily for their plumes, which were used in the production of ladies' hats.

This period saw enormous expansion of the system of **United States National Parks and Monuments.** Prior to 1890, the only national park was Yellowstone, created in 1872. Sequoia, Yosemite, and Mt. Rainier National Parks were created in the 1890s, and some 20 national monuments, including Mt. Lassen and the Grand Canyon, were designated, initiating a phase of national park development that continued for three decades.

The era of the closing of the western frontier thus saw recognition of the dynamic nature of living resources, the emergence of a branch of science—ecology—that could provide the theory needed to manage them, and the creation of major federal institutions to oversee such management at a national scale.

The Dust Bowl Era, 1930–1940

Following the turn of the century, crop farming spread onto marginal lands in areas such as the Tennessee Valley and the Great Plains, spurred by population growth, poorly regulated land settlement schemes, and, in 1929, the collapse of the urban-industrial economy that signalled the start of the Great Depression. In the 1930s, water erosion and flooding plagued the midwestern and southern states, while droughts and wind erosion afflicted the central plains. The Dust Bowl of the Texas and Oklahoma panhandles, southeastern Colorado, and southwestern Kansas came to symbolize these conditions, which reached their extreme on Black Sunday, April 14, 1935, in one of the largest and most severe dust storms on record (Fig. 27.4). Agricultural development had reached the limit of tolerance of the land to farming practices that did not recognize the intimate interactions between the living and nonliving components of the environment.

The 1930s were also a time of widespread recognition of a new concept in ecology, that of the **ecosystem.** The concept itself had begun to crystalize in the 1920s, with growing recognition by ecologists that the living and non-living components of natural systems influenced each other in a reciprocal fashion. The activities of organisms modified their non-living environment, as well as being influenced by its physical and chemical conditions. By the 1930s, recognition of this concept was widespread. Aldo Leopold (1933a) wrote, for example, that civilization depended on " . . . *mutual and interdependent cooperation* between human animals, other animals, plants, and soil," a clear, humanistic statement of the ecosystem concept. In 1935, a British ecologist, A. G. Tansley, coined the term *ecosystem,* noting that, "Though the organisms may claim our primary interest, . . . we cannot separate them from their special environment, with which they form one physical system." Scientists in both basic and applied areas of ecology thus recognized a deeper level of dynamics in the natural environment, specifically the powerful influences that living organisms, including humans, could have on the physical landscape.

This period saw the emergence of scientific wildlife management, largely through the effort and inspiration of Aldo Leopold (b 1887, d 1962) (Hedgepeth 1989). Trained as a forester, Leopold (Fig. 27.5) spent his early career with the United States Forest Service in Arizona and New Mexico. There, he gained field experience in wilderness areas with large game populations. Gradually, his interests shifted to the management of wildlife in forests and other ecosystems. At first, his approach was traditional, espousing predator control as a primary technique to encourage game species (Flader 1974). Gradually, however, Leopold and many others began to appreciate the intricacy of relationships among the living and non-living components of nature. This appreciation brought with it a new view of the role of predators and predation in natural communities (Dunlap 1985). After leaving the Forest Service Leopold devoted most of his attention to game ecology, bringing to this field an ecosystems philosophy. This approach is perhaps best expressed in Leopold's (1949) essay, *Thinking Like a Mountain,* in which he describes how deer overpopulation could decimate the natural vegetation and writes, "I now suspect that just as a deer lives in mortal fear of its wolves, so does a mountain live in mortal fear of its deer." Leopold became professor of game

FIGURE 27.4

■ The dust storms of Black Sunday, April 14, 1935, symbolized the misuse of land and water resources that precipitated the conservation crisis of the 1930s.

FIGURE 27.5

■ Aldo Leopold, regarded as the father of scientific wildlife management, saw the concept of the ecosystem as central to management of nature.

management at the University of Wisconsin. The textbook, *Game Management,* published in 1933 (Leopold 1933b), is still a basic reference for students of wildlife ecology. Leopold also supported the establishment of wilderness areas, and in 1935 was one of the founders of the Wilderness Society. In the eyes of environmentalists, one of his most important contributions was to begin the effort to define an ethical relationship between humans and the rest of nature (Flader 1987), an effort embodied in his early essay on "The Conservation Ethic" (Leopold 1933a).

Paul B. Sears was perhaps the most influential writer-naturalist of this era, however. Sears (b 1891, d 1989) was born in Ohio and attended Ohio Wesleyan University, where one spring day in 1913 he found his route to classes blocked by the flooding Olentangy River. His awareness of how humans were mistreating soil and water resources, in fact, might have been awakened by this event. Sears is best remembered for the book *Deserts on the March,* published in 1935 and in print continuously since then. In this book he dramatized the crises of land and water mismanagement that much of North America faced in the 1930s, including the "sea of swirling, swishing liquid mud" of his college experience and the "slow, chilling, and pervasive horror" of the dust storms of the Great Plains. Sears later chaired the Yale

Conservation Program, the first graduate program in conservation science in the United States.

Under the presidency of Franklin Roosevelt, with Harold Ickes as Secretary of Interior, the mid-1930s saw the establishment of federal institutions aimed at the crises of water and land management (Strong 1988). The **Tennessee Valley Authority,** organized in 1933, sought to revitalize one of the most impoverished and degraded regions of the United States by impoundments to control flooding and produce hydro-electric power, coupled with soil conservation and reforestation programs to rehabilitate the land. The **Soil Conservation Service,** formed in 1935, was also created to devise and promote techniques of evaluating land capability and controlling erosion.

The dust bowl era revealed patterns of ecological disruptions on a scale greater than ever before, and showed that the physical environment was not an unmanageable control over the living world, but one both affecting and affected by the plants and animals occupying it. It firmly established the need for an ecosystem approach to the resolution of conservation problems.

The Era of Population and Environmental Pollution, 1960–1975

World War II suddenly diverted attention from conservation issues. It also initiated an era of unparalleled economic expansion, and spawned explosive growth of technology and human population. Technology provided "Better Living through Chemistry" in the form of synthetic fibers, plastics, inorganic pesticides, leaded fuels, and many other products that sooner or later caused the introduction of toxic and biologically active materials into the environment. The availability of cheap petroleum encouraged the growth of a society centered on the internal combustion engine. To this explosion of industrial technology was added that of the human population, in the post-war "baby boom." The result was exponential growth in the pollution of air, land, and water by chemicals and chemical wastes. Soon, natural mechanisms of detoxification of pollutants and of homeostatic adjustment to their effects were overwhelmed by these inputs. The limits of ecosystem function were being challenged by the chemical products of human activity.

Postwar technologies had also given ecology new ways of working with the dynamics of ecosystems. Geiger counters for tracing the movement of radioisotopes, spectrometers for analyzing the composition of organic and inorganic matter, radiometers for measuring the wave length and intensity of solar radiation, calorimeters for determining energy content of organic materials, and computers that permitted rapid analysis of masses of data enabled ecologists to follow the flows of energy and cycles of chemical substances through complex pathways in ecosystems. **Ecosystem analysis,** using these tools, led the way in studies of environmental pollutants and their ecological effects. Stimulated largely by the pioneering efforts of Eugene and Howard Odum (1955) in the 1950s,

FIGURE 27.6

■ At the Hubbard Brook Experimental Forest in New Hampshire, studies beginning in the 1960s revealed the dynamics of nutrient cycling in forested watersheds for which inputs could be measured and outputs monitored by weirs on outflowing streams.

many ecologists concentrated on measuring and modeling the flow of energy through whole ecosystems during the 1960s. At the Hubbard Brook Experimental Forest in New Hampshire, other pioneering studies by F. H. Bormann and Gene Likens (1967) stimulated interest in studies of nutrient cycling in whole ecosystems (Fig. 27.6). These efforts culminated in the decade-long International Biological Program, begun in 1966, one goal of which was to develop predictive mathematical models of ecosystem energy and nutrient dynamics. Although this ambitious goal was not fully attained, ecological modeling emerged as an important tool in the study of environmental pollution.

Rachel Carson (b 1907, d 1964), a writer by training and a naturalist by avocation, was editor-in-chief of publications of the United States Fish and Wildlife Service during this era (Fig. 27.7). Originally, her interests had centered on the sea, and she had written several books about marine science, the most popular being *The Sea Around Us,* published in 1951. In the 1950s, however, her work with the Fish and Wildlife Service made her aware of the growing impacts of environmental pollution, particularly by pesticides, on wildlife. She brought her concerns about pesticides to public attention in 1962 in the book *Silent Spring.* In the preface to

FIGURE 27.7

■ Rachel Carson drew public attention to the problem of environmental contamination by pesticides through publication of *Silent Spring* in 1963.

this book she warned of the possible fate of wildlife: "On the mornings that had once throbbed with the dawn chorus of robins, catbirds, doves, jays, wrens and scores of other bird voices there was now no sound; only silence lay over the fields and woods and marsh." This book, which proved to be a best-seller, was at first regarded as heretical by many biologists. The concerns that it raised, however, proved to be real, and an unparalleled effort to investigate the effects of environmental pollution began. Newly crystallized concepts of energy flow and nutrient cycling played a key role in revealing how pesticides and many other pollutants exerted their impacts. The near extinction of the peregrine falcon in North America due to pesticide pollution, and declines of populations of many other species, also raised concern for endangered species in general.

A second influential scientist and writer of the period was Paul Ehrlich, professor of biology at Stanford University. Ehrlich's concerns centered on the growth of human populations as the ultimate driving force behind environmental pollution. A charismatic speaker and prolific writer, Ehrlich expressed his gravest prognosis in the book, *The Population*

Bomb, published in 1968. Ehrlich warned: "Cancer is an uncontrolled multiplication of cells; the population explosion is an uncontrolled multiplication of people."

Television, radio, and newspapers brought environmental issues to public attention with greater impact than in previous environmental eras. Ultimately, the efforts of environmentalists such as Barry Commoner, David Brower, and many others, extended concern to almost every facet of human relation to the natural environment. Even in competition with the enormous issues of the Viet Nam War, environmental quality became a major societal priority. **Environmentalism,** both as a philosophy of living and as a strong political force, emerged from its roots in philosophy, ecology, and the appreciation of nature (Scheffer 1991). Environmentalists led the drive that culminated in the **Wilderness Act** of 1964, which established an effective system of wilderness areas in national forests, the national park system, and national wildlife refuges. As a result of concern over pollution, the **U.S. Environmental Protection Agency** was established in 1970, and charged with maintaining an environment safe for humans and wildlife, specifically by regulating the discharge of toxic materials into air, water, and soil. Other major items of federal legislation included the **National Environmental Policy Act** of 1970, which established the requirement for environmental impact assessment for projects involving federal support or approval, and the **Endangered Species Conservation Act** of 1973, which created the current system of designation, protection, and recovery of endangered species.

The era of pollution and population brought the realization that human intrusion into ecosystem processes could lead to catastrophic indirect effects, many of which were not intuitively predictable. It also raised the question of global carrying capacity, setting the stage for the concerns about the global environment that have come to the fore in the last decade of the 20th century.

The Era of Global Environmental Change

The activities of humanity are now approaching a new set of limits set by the overall regulatory capabilities of the biosphere. Basic atmospheric processes are being modified by massive urban-industrial discharges of greenhouse gases that produce lower atmospheric warming (Fig. 27.8), chlorofluorocarbons that affect the stratospheric ozone layer, and sulfur and nitrogen oxides that lead to acid deposition. Tropical deforestation and overgrazing of arid lands are changing water and energy relationships of vast areas of the land surface. Directly, and through the climatic changes they are likely to induce, these processes are promoting forest declines in temperate and subarctic regions, intensifying desertification in tropical and subtropical areas, and increasing erosion of soil in mountain areas. Several of these processes have positive

FIGURE 27.8

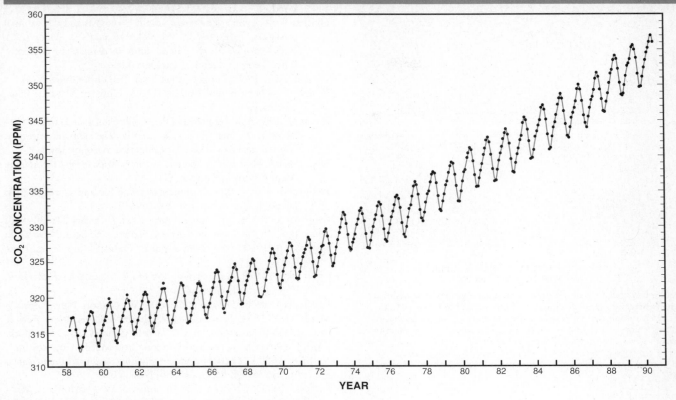

■ Burning of fossil fuels and deforestation of vast areas of the tropics are contributing to a rising concentration of carbon dioxide in the atmosphere, as measured at Mauna Loa Observatory in Hawaii (Figure courtesy of C. D. Keeling).

feedbacks that can increase the severity of disturbance. Global changes such as these compound the problems of conserving biotic diversity in the rapidly declining areas of natural eco-systems that now remain. The combination of these global changes threatens a catastrophic loss of biotic diversity throughout the biosphere.

Will new scientific concepts and technologies emerge to enable environmental scientists to meet these challenges? Will influential individuals who can present these issues dra-matically by modern media appear to mobilize public con-cern? What sorts of national and international institutions will be needed to surmount these global environmental threats?

Might the **Gaia Hypothesis** be a springboard to a new, global concept of ecology that can help address this crisis? The Gaia (GUY-uh) Hypothesis, named for the greek god-dess of Earth, was first proposed by James Lovelock, an in-dependent British scientist and inventor, in 1979. Lovelock proposed that the biosphere is a self-regulating entity that has evolved, in a purposeful manner, to produce and maintain an optimum physical and chemical environment for life. This suggestion stirred immense interest in its suggestion of ho-meostatic regulation at the biosphere level, as well as intense

criticism of its teleological nature—the purposeful evolution of the biosphere toward a goal of optimum conditions. More recently, Lovelock himself has abandoned the teleological as-pects of the hypothesis, and has emphasized the hypothesis that the physical and chemical conditions of the biosphere are regulated within relatively narrow limits by the living com-ponents of the system (Fig. 27.9), the so-called "homeostatic Gaia Hypothesis" (Lovelock 1988). Polunin and Grinevald (1988) have pointed out that this idea of homeostasis is deeply rooted in the ideas of biosphere function suggested by earlier scientists, especially the Russian scientist Vladimir Ver-nadsky, who in 1926 published *Biosfera* (*The Biosphere*), a book viewing the earth not as a planet with life, but as a "living planet." This book was also published in French in 1929, but not translated into English until recently.

The serious scientific questions spawned by this hy-pothesis center on whether global climate is regulated by ho-meostatic biotic controls on the composition of the atmosphere. If such regulation does occur, further questions concern the time scale on which controls operate, the limits within which they can operate, and the consequences for life of those limits being exceeded.

FIGURE 27.9

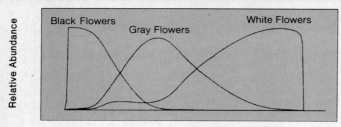

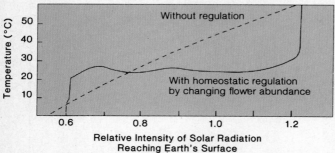

■ James Lovelock's "Daisy World" is an illustration of how the biota of the earth might regulate climate and other conditions favorable to life. In this simplistic model, an earth vegetation consists of flowers of different colors, which change in abundance in response to changes in intensity of solar radiation. Changes in flower abundance in turn increase or decrease reflection of radiation, thus regulating surface temperature over a wide range of conditions (Modified from Lovelock 1988).

KEY CONCEPTS OF THE HISTORY OF CONSERVATION

1. Societal concern about the natural environment has been cyclical, with activism concentrated in periods of environmental crisis and scientific revolution.

2. The crises that have precipitated environmental concern have increased in scale and complexity with each cycle.

3. In the face of growing human populations and economies, environmental laws and institutions have not been able to prevent escalation of environmental crises.

LITERATURE CITED

Biswell, H. H. 1989. *Prescribed burning in California wildlands vegetation management.* Univ. of Calif. Press, Berkeley.

Bormann, F. H. and G. E. Likens. 1967. Nutrient cycling. *Science* **155**:424–429.

Carson, R. 1951. *The sea around us.* Oxford Univ. Press, New York.

Carson, R. 1962. *Silent spring.* Houghton-Mifflin, New York.

Cowles, H. C. 1899. The ecological relationships of the vegetation on sand dunes of Lake Michigan. *Bot. Gaz.* **27**:95–117, 167–202, 281–308, 361–391.

Darwin, C. R. 1859. *The origin of species by means of natural selection or the preservation of favored races in the struggle for life.* Murray, London.

Dunlap, T. R. 1985. "The coyote itself: Ecologists and the value of predators, 1900–1972." Pp. 594–618 *in* K. E. Bailes (Ed.), *Environmental history: Critical issues in comparative perspective.* Univ. Press of America, Lanham, MD.

Ehrlich, P. R. 1968. *The population bomb.* Ballantine, New York.

Flader, S. 1974. *Thinking like a mountain.* Univ. of Missouri Press, Columbia.

Flader, S. 1987. Aldo Leopold and the evolution of a land ethic. Pp. 3–24 *in* T. Tanner (Ed.), *Aldo Leopold: The man and his legacy.* Soil Conservation Society of America, Ankeny, Iowa.

Fleck, R. F. 1985. *Henry Thoreau and John Muir among the Indians.* Archon Books, Hamden, CT.

Hedgepeth, J. W. 1989. Commentary: The life and works of Aldo Leopold. *Quart. Rev. Biol.* **64**:169–173.

Leopold, A. 1933a. The conservation ethic. *J. Forestry* **31**:634–643.

Leopold, A. 1933b. *Game management.* Scribner and Sons, New York.

Leopold, A. 1949. *Sand county almanac.* Oxford Univ. Press, New York.

Lovelock, J. 1979. *Gaia: A new look at life on earth.* Oxford Univ. Press, Oxford, England.

Lovelock, J. 1988. *The ages of Gaia.* W.W. Norton, New York.

Marsh, G. P. 1864. *Man and nature.* Charles Scribner, New York.

Muir, J. 1894. *The mountains of California.* Century Co., New York.

Muir, J. 1901. *Our national parks.* Houghton Mifflin Co., Boston.

Nash, R. 1989a. *American environmentalism: Readings in conservation history.* 3rd ed. McGraw-Hill, New York.

Nash, R. 1989b. *The rights of nature.* Univ. of Wisconsin Press, Madison.

Odum, H. T. and E. P. Odum. 1955. Trophic structure and productivity of a windward coral reef community on Eniwetok Atoll. *Ecol. Monogr.* **25**:291–320.

Polunin, N. and J. Grinevald. 1988. Vernadsky and biospheral ecology. *Env. Cons.* **15**:117–122.

Scheffer, V. B. 1991. *The shaping of environmentalism in America.* Univ. Washington Press, Seattle.

Sears, P. B. 1935. *Deserts on the march.* Univ. of Oklahoma Press, Norman.

Strong, D. H. 1988. *Dreamers & defenders: American conservationists.* Univ. of Nebraska Press, Lincoln.

Tansley, A. G. 1935. The use and abuse of vegetation concepts and terms. *Ecology* **16**:284–307.

Thoreau, H. D. 1849. *A week on the Concord and Merrimack Rivers.* James Munroe, Boston.

Thoreau, H. D. 1854. *Walden.* Ticknor and Fields, Boston.

Torrey, B. and F. H. Allen. 1906. *The journal of Henry D. Thoreau.* Vol. 1–14. Houghton Mifflin Co., Boston.

Vernadsky, V. I. 1929. *La biosphere.* Felix Alcan, Paris. xii + 232 pp. (English edition: 1986. *The biosphere.* Synergetic Press, Oracle, AZ.)

White, L., Jr. 1967. The historical roots of our ecologic crisis. *Science* **155**:1203–1207.

CHAPTER 28

Global Conservation Programs

As human populations continue to grow, and as national economies continue to expand and to develop complex patterns of international trade, the problems of conservation increasingly become an international issue. Many countries, for example, now possess marine fishing fleets capable of exploiting fisheries anywhere in the world oceans. Intensified land use in dozens of countries from the tropics to the arctic affects the populations of migratory birds. Consumer demand for lumber, ivory, skins, and many other products in countries in one part of the world threatens species populations, and in some cases entire ecosystems, in far-away areas of the world. Pollution from urban-industrial regions creates acid deposition that spreads over broad geographic areas, without respect to international borders.

Recognition is growing, as well, that biotic diversity is a resource of the entire human race. African savannas, with their rich diversity of wildlife, are the heritage of all humankind, not just of the countries in which they are located. Vast areas of healthy tropical forests are important to the future of the entire human race, not just the handful of countries in which the bulk of these forests is located. Sustained harvests of marine fisheries contribute to balancing the protein needs of human populations worldwide. These facts demand much more effective international coordination of conservation efforts.

In this chapter we shall describe the major international organizations that are working to preserve biotic diversity. We shall consider some of the major programs that now exist, and outline new efforts that are being planned for the decade of the 1990s.

AREAS UNDER PROTECTION AND SUSTAINED MANAGEMENT

About 4 percent of the world's land area is now in national parks and preserves in which wildlife species, and usually all natural ecosystems, are protected from exploitation (Table 28.1). In North America and Europe, more than 6 percent of land is so protected, while in Asia, Central America, and the Soviet Union much smaller fractions of land are preserved. In Central America, South America, and Africa, individual countries have made major commitments to preservation of natural ecosystems. Costa Rica, for example, has placed 12.0 percent of its land in parks, Chile 16.0 percent, and Tanzania 13.4 percent. The survival of parks in countries like these, however, is threatened by human population growth and the desperate efforts of impoverished peoples to meet their daily needs.

Still other areas are protected by private organizations or local governments, or are exploited under sustained-yield management. In the United States, for example, national forests are supposed to be managed for sustained yields of timber, wildlife, water, and recreation. The extent of natural ecosystems managed in sustainable fashion is unknown, but is probably greater in developed than in developing parts of the world.

Conservation ecologists agree that the areas of strictly protected ecosystems that now exist or are likely to be established in the near future are inadequate to preserve the biotic diversity on which a healthy global ecosystem depends. Thus, international conservation organizations are striving to find ways to exploit ecosystems, particularly those that retain much of their natural structure, in sustainable fashion.

These programs must also develop ways to match costs and benefits of conservation on an international scale. For example, the citizens and government of Costa Rica have supported the development of a comprehensive network of tropical parks and preserves—a system of international benefit and significance. But should this nation be expected to bear the entire cost of maintaining this system, subsidizing the rest of the world in the process?

INTERNATIONAL CONSERVATION ORGANIZATIONS

Three agencies of the United Nations are active in global conservation. The **United Nations Educational, Scientific, and Cultural Organization (UNESCO)** and the **Food and Agricultural Organization (FAO)** were founded in 1945 as original branches of the United Nations, and the **United Nations Environment Program (UNEP)** was organized in 1972.

UNESCO, headquartered in Paris, has been active from its beginning in environmental affairs such as park development and scientific research. UNESCO is the coordinating organization for the **Man and the Biosphere Program (MAB),** initiated in 1971 as a successor to the **International Biological Program (IBP)** of the 1960s and early 1970s. The IBP emphasized research on ecology and evolution at the ecosystem level, one of the stated aims being the development of predictive systems models of the dynamics of major ecosystem types such as grasslands, tundra, and tropical forests. The studies of the IBP were thus highly theoretical in nature. In contrast, MAB has emphasized more applied studies of natural and exploited ecosystems, with the goal of improving their management for human welfare. The activities carried out under both IBP and MAB have been supported from sources within individual countries. During its first decade, MAB tended to carry on some of the theoretical emphasis of the IBP period (Di Castri et al. 1981). In 1981,

TABLE 28.1

Distribution of Parks and Preserves of Major Importance to the Preservation of Natural Diversity, as of 1989.

REGION	TOTAL AREA IN PRESERVES (PERCENT)	NUMBER OF BIOSPHERE RESERVES	NUMBER OF NATURAL WORLD HERITAGE SITES	NUMBER OF WETLANDS OF INTERNATIONAL IMPORTANCE
North America	6.2	50	17	37
Central America	2.0	14	5	1
South America	4.6	23	8	4
Europe	6.6	81	11	277
USSR	0.9	19	0	12
Asia	2.3	34	8	34
Africa	3.4	40	20	31
Oceania	4.7	12	10	31
WORLD	4.0	276	78	432

Source: Data from World Conservation Monitoring Center, Cambridge, England, as appeared in *World Resources 1990–1991,* World Resources Institute.

however, a review of MAB activities concluded that much greater stress should be placed on sustainable development of resources for human welfare. The efforts of MAB have been centered largely on the Biosphere Reserve System, which is examined in detail below. The World Heritage Program, another major UNESCO effort, is also examined in detail below.

The FAO, with head offices in Rome, Italy, is responsible for United Nations activities relating to food and fiber production, including not only crop agriculture but the status and yields of world fisheries, forests, and rangelands. In addition to sponsoring technical studies of fisheries and their management, FAO monitors world fisheries harvests, publishing these statistics annually. The FAO also supports forestry research, and monitors the status of world forests and the rate of deforestation. In addition, the FAO is the front-line agency for United Nations activities relating to parks and protected areas. FAO activities in this last area have largely consisted of aid for planning preserves, training personnel, and developing national agencies to manage such areas.

UNEP, headquartered in Nairobi, Kenya, focused most of its early efforts on environmental factors bearing closely on human health, particularly nutrition, disease, and pollution by toxic chemicals. In recent years, as the recognition of ecosystem disruption and its relation to human welfare has heightened, UNEP has become involved with more basic problems of environmental quality, such as desertification, acid deposition, marine pollution, and global climatic change. In this arena, one of UNEP's major programs is the **Global Environmental Monitoring System (GEMS).** GEMS is designed to feed information into a global resource database, using geographical information systems technology (See chapter 26). These data cover climate, oceanic conditions, pollution levels, human health, and natural resources, including biotic diversity. Using this database, UNEP has issued reports on the state of the world environment, the most recent in 1987 (United Nations Environment Programme 1987). Desertification continues to be a major focus of activity (Fig. 28.1), and UNEP supports a wide variety of projects to alleviate desertification in sub-Saharan Africa.

UNEP also sponsors programs dealing with international aspects of endangered species, migratory species, and marine mammals. UNEP secretariats exist for the administration of the **Convention on International Trade in Endangered Species** (See chapter 25) and the **Convention on Migratory Species.** In 1990, the Ocean and Coastal Areas Programme presented a worldwide plan for conservation of marine mammals (See chapter 18).

Private organizations also play an important role in global conservation efforts. The **World Conservation Union,** formally known as the **International Union for the Conservation of Nature and Natural Resources (IUCN),** is one of the most active. The IUCN was founded in 1948 and has its headquarters in Gland, Switzerland. It is a non-governmental body that coordinates conservation activities on behalf of its member organizations. The membership of the IUCN in 1989 included 61 countries, 121

FIGURE 28.1

■ The United Nations Environment Program supports projects such as tree planting in Senegal in efforts to combat desertification.

governmental agencies of many countries, and 394 private organizations. Altogether, 120 countries participate in IUCN efforts through this membership. Active involvement in IUCN activities by developing countries in Africa, Asia, and Latin America is still relatively weak, however.

The annual IUCN budget of about $12.2 million supports a variety of activities. The IUCN plays a key role in the formation of global conservation policy, largely by coordinating the planning and operation of programs that are supported financially by United Nations agencies and by private groups such as the World Wide Fund for Nature. It furnishes technical aid to countries and other conservation organizations, emphasizing ways to integrate conservation with economic development. Regional offices are maintained in Latin America, Africa, and Asia to facilitate these efforts. The IUCN also maintains an **Environmental Law Center** that offers technical assistance in the development of laws and conventions and serves as a clearinghouse of information on national and international regulations relating to endangered species. Under support from UNEP, the IUCN also operates the **World Conservation Monitoring Center (WCMC),** in Cambridge, England. The WCMC publishes the *U.N. List of National Parks and Other Protected Areas,* a compilation of natural preserves that meet basic standards of protection and financial support. This list is the basis for the percentages of preserve areas in different world regions, given in table 28.1. The WCMC also oversees publication of the various **Red Data Books** on endangered species of animals and plants (See

FIGURE 28.2

■ The World Wide Fund for Nature (WWF) has supported studies of many endangered species. Here, a snow leopard is being fitted with a radio collar to enable investigators to follow its movements.

chapter 25). For the Red Data Books relating to birds, the WCMC collaborates with the **International Council for Bird Preservation.** Through these and various other technical documents, the WCMC provides up-to-date information on the status of conservation activities worldwide. The IUCN also publishes the *IUCN Bulletin* and *Parks* magazine, which give current information on world conservation activities.

In 1975, UNESCO, FAO, UNEP, and the IUCN established a coordinating committee, the **Ecosystem Conservation Group,** to assist their joint conservation efforts. Recently, the **International Board for Plant Genetic Resources,** the **United Nations Development Program,** and the **World Bank** have become members of the committee. This coordinating committee thus has a very broad perspective on preservation of biotic diversity, both in nature and at off-site locations, and on the means to support this effort.

Several private organizations with individual membership are also active in international conservation. The **World Wildlife Fund (WWF),** better known internationally as the **World Wide Fund for Nature,** is the largest such organization. Historically, WWF has primarily been concerned with endangered plants and animals, and has supported research activities designed to promote their survival and recovery (Fig. 28.2). In 1990, WWF merged with the **Conservation**

Foundation, an organization that had focussed on environmental policy relating to pollution, wetlands protection, and public land use.

Two major private organizations pursue international activities of assisting national governments in developing their own strong, ecosystem focused, conservation organizations. **The Nature Conservancy (TNC),** an outgrowth of the conservation arm of the Ecological Society of America, was organized in 1951. Its international program, established in 1980, has concentrated on the encouragement of **Conservation Data Centers (CDCs)** in various Latin American countries. These CDCs are designed to build inventories of natural diversity for individual countries, and serve as centers of action for the growth of a comprehensive network of preserves to protect this diversity. **Conservation International,** established in 1987, has similar objectives. Conservation International pioneered the use of **debt-for-nature swaps** to support conservation efforts in developing countries. This technique involves the purchase of portions of the foreign debt owed by a nation to international banks, and the cancellation of this debt in exchange for the government's action in establishing preserves or supporting conservation in other ways. In 1987, Conservation International arranged for the purchase of $650,000 of Bolivian foreign debt at a cost of $100,000. In exchange for the cancellation of this debt, the Bolivian government agreed to establish over 16,000 square kilometers of preserves, including the 1,336 square kilometer Beni Biosphere Reserve, an area of wet savannas on the Rio Beni, an Amazon tributary (Fig. 28.3). As of 1990, Conservation International, TNC International, and WWF had arranged debt-for-nature swaps in five countries, retiring about $100 million in their foreign debt.

The **World Resources Institute,** in Washington, D.C., works with other international organizations to analyze environmental problem areas and prepare summaries of resource status and trend.

BIOSPHERE RESERVE SYSTEM

The **Biosphere Reserve System** is a global network of sites that combine preservation with research on sustainable management for human welfare. This system is being developed as one aspect of UNESCO's Man and the Biosphere Program (Batisse 1986). Most biosphere reserves are chosen to be ecologically representative of major types of regional ecosystems, and include large protected areas of these ecosystems. Many biosphere reserves, however, include ecosystems that have been modified or exploited by humans, such as rangelands, subsistence farmlands, or areas used for hunting or fishing. A few biosphere preserves include human settlements and environments that have been exploited extensively by traditional patterns of land use.

FIGURE 28.10

■ Grand Canyon National Park is one of the United States World Heritage sites.

FIGURE 28.11

■ Kakadu National Park, in the Northern Territory of Australia, was designated a World Heritage Site partly because of the cultural importance of the area to aborigines, who have created numerous pictographs on the park's cliffs.

FIGURE 28.12

■ Kakadu National Park also contains large areas of woodland, riparian forest, and seasonal wetlands representative of the ecosystems of northern Australia's region of monsoonal climate.

FIGURE 28.13

■ Ngorongoro Crater in Tanzania is a World Heritage Site that has been placed on the World Heritage in Danger List due to management problems relating to land use by pastoralist peoples and high levels of poaching of animals such as the black rhinoceros.

were spent for these efforts. Threatened sites are placed on the **List of the World Heritage in Danger.** Seven sites were listed as in danger in 1989 (Anonymous 1989), including Ngorongoro Conservation Area in Tanzania (Fig. 28.13), Garamba National Park in Zaire, and Djoudj National Park, Senegal. Ngorongoro and Garamba lie in the savanna zone of eastern and central Africa. Ngorongoro Conservation Area has suffered from a shortage of funds for management and research, as well as from poaching and increase in the numbers of pastoral people and their herds. Garamba National Park holds the last wild population of the northern white rhinoceros, but lies in an area difficult to protect against poaching. Djoudj National Park encompasses seasonally flooded wetlands of the delta of the Senegal River, and supports an estimated three million migratory water birds. It is being threatened by construction of an upstream dam on the Senegal River.

A continuing issue within the World Heritage Program has been the balance of cultural versus natural sites that are admitted (Hales 1984). The 3:1 ratio of cultural to natural sites reflects the dominance of nations with primarily cultural natural interests on the committee that evaluates proposed sites. Efforts have been made in recent years to improve the representation of natural areas by adopting new guidelines for structure and organization of the review committee, and for systematic identification of types of areas that are underrepresented on the World Heritage List.

THE INTERNATIONAL GEOSPHERE-BIOSPHERE PROGRAM

The **International Geosphere-Biosphere Program (IGBP)** was initiated in 1986 by the International Council of Scientific Unions, the coordinating committee for national academies and international unions of science (World Resources Institute 1987). Headquarters for the program have been established in Stockholm, Sweden. The IGBP is envisioned as a 10 to 20 year program to determine how conditions of the biosphere influence and are influenced by their interaction with the earth's physical and chemical environment. Foremost among the objectives of the program is evaluating the role of human activities in modifying the global environment. The IGBP is projected to be the most ambitious and expensive international program yet undertaken, and will require funding from many national and international sources to achieve its goals. Major emphasis will be placed on the use of remote sensing and geographic information systems to gain global data on physical and biological conditions. Full-scale efforts in the program are planned to be under way by 1992. By 1995, this program should begin to use the first of a series of large space platforms that will form the **Earth Observing System** (World Resources Institute 1990). These platforms, orbiting at an elevation of 705 km, will be equipped with advanced remote sensing devices that can measure many characteristics of the atmosphere, land, and oceans simultaneously.

WORLD CONSERVATION STRATEGY

Developed by the IUCN, with support from the WWF, FAO, UNESCO and UNEP, the **World Conservation Strategy (WCS)** was launched in 1980 (Allen 1980a). The WCS was a response to the gap that has existed between conservation and economic development (Halle 1985). The strategy had three basic objectives: 1) to maintain the ecological processes on which life depends, 2) to ensure the sustained utilization of ecosystems and their member species, and 3) to conserve genetic diversity (Talbot 1984). It recognized that economic development is necessary for successful conservation, and that failure to protect ecological resources limits the level of economic development that can be achieved. The WCS focussed on the developing nations, where resources are being depleted rapidly, and where institutions to manage their exploitation are weak or absent.

The WCS, first of all, documented the need for protection of basic ecosystem processes, for the preservation of genetic diversity, and for the development of sustained use management of natural ecosystems. It also outlined ways to incorporate conservation planning as a positive component in national development plans, rather than, as often portrayed, a force opposing development. The rationale of the WCS has been outlined in detail for the general public (Allen 1980b). The IUCN has worked with various countries in using these materials to develop national conservation strategies. About 30 national strategies were completed or under development by the late 1980s.

The revised WCS, entitled "Caring for the Earth. A Strategy for Sustainable Living," focuses much more directly on ways to achieve sustainable resource development, and emphasizes the need to integrate economic and ecological efforts fully, both at the local and global levels. A major emphasis is the need to foster a global ethic of sustainable

interaction with the biosphere. A companion plan, the **Global Strategy for Conserving Biological Resources,** will treat the protection of biotic diversity in detail. This companion strategy is being developed by the World Resources Institute, IUCN, and UNEP (McNeely 1990).

FUTURE EMPHASIS IN GLOBAL CONSERVATION EFFORTS

Success in preserving the earth's natural diversity depends on the establishment of institutions that hold a long-term view of the values of natural diversity, and possess the strength to protect these values. These institutions must be stable enough to withstand the challenges of unwise decisions made in response to immediate political pressures at local, regional, and national levels. They must be able, for example, to ensure the survival of designated preserves in the face of decisions that could lead to their destruction or impoverishment. To do this, conservation institutions must be able to bring strong political and economic pressure to bear on the side of preservation.

Conservation efforts at the global level, pursued by various international agencies and private groups, still remain poorly coordinated. National park development still depends primarily on the initiative of individual countries, and is still financed by the countries involved. Under the Biosphere Reserve and World Heritage Programs, some parks have gained a degree of international recognition, but very little financial support comes from these arrangements. Some international organizations focus their efforts on species, others on ecosystems. Much can be accomplished by coordination of the activities of the varied organizations through sharing of data bases, identification of critical issues for joint action, and publicizing conservation issues in coordinated fashion. This is the major goal of the **Sustainable Biosphere Initiative** adopted by the Ecological Society of America (1991). This initiative identifies priority areas of basic ecological research dealing with global environmental change, preservation of biotic diversity, and sustainable management of ecological systems. It also calls for the society to participate actively in education and in policy development at all national and international levels.

To be successful, conservation efforts must make reality out of the contention that biotic diversity is a valuable resource (Western et al. 1989). Viable conservation strategies must be developed at national, regional, and local levels. These strategies must show ways of utilizing natural diversity to improve economic and social conditions, yet protect the natural diversity resource itself. Using and protecting this resource require education, both as part of basic school curricula and through extension programs. The educational aspects of conservation have not yet received adequate attention.

These issues will be the agenda for the **United Nations Conference on Environment and Development,** to be held in Brazil in 1992. This conference will attempt to establish an **"Earth Charter,"** defining basic principles of conduct relating to environment and development, as well as outlining an action agenda reaching into the 21st century.

The tools for global integration of such an effort are rapidly becoming available: satellite systems for remote sensing of environmental change, computer networks for storing and sharing ecological data, communications systems for transmitting materials to people throughout the world, and a growing, international community of ecological scientists and environmental activists.

KEY STRATEGIES FOR GLOBAL CONSERVATION

1. Develop comprehensive networks of protected areas for scientific study and educational use.

2. Formulate strategies for sustained use of ecological resources to promote national economic and social development.

3. Devise effective techniques of conservation education for use in school curricula and extension activities.

LITERATURE CITED

Agardy, T. and J. M. Broadus. 1989. Coastal and marine biosphere reserve nominations in the Acadian Boreal Region: Results of a cooperative effort between the U.S. and Canada. Pp. 98–105 in W. P. Gregg, S. L. Krugman, and J. D. Wood (Eds.), *Proceedings of the symposium on biosphere reserves*. USDI National Park Service, Atlanta, GA.

Allen, R. 1980a. The world conservation strategy: What it is and what it means for parks. *Parks* 5(2):1–5.

Allen, R. 1980b. *How to save the world: Strategy for work conservation*. Kogan Page, London.

Anonymous. 1989. World heritage: From Da Vinci to the white rhino. *UN Chronicle* 26(2):48–53.

Batisse, M. 1986. Developing and focusing the biosphere reserve concept. *Nature and Resources* 22:1–10.

Di Castri, F., M. Hadley, and J. Damlamian. 1981. MAB: The Man and the Biosphere Program as an evolving system. *Ambio* 10:52–57.

Ecological Society of America. 1991. The sustainable biosphere initiative: An ecological research agenda. *Ecology* 72:371–412.

Hales, D. F. 1984. The World Heritage Convention: Status and directions. Pp. 744–750 in J. A. McNeely and K. R. Miller (Eds.), *National parks, conservation, and development: The role of protected areas in sustaining society*. Smithsonian Univ. Press, Washington, D.C.

Halle, M. 1985. The World Conservation Strategy—An historical perspective. Pp. 241–257 in J. P. Hearn and J. K. Hodges (Eds.), *Advances in animal conservation*. Clarendon Press, Oxford, England.

IUCN/UNEP/WWF. 1991. Caring for the earth. A Strategy for Sustainability. Gland, Switzerland.

Kenchington, R. A. and M. T. Agardy. 1990. Achieving marine conservation through biosphere reserve planning and management. *Env. Cons.* 17:39–44.

McNeely, J. A. 1990. How conservation strategies contribute to sustainable development. *Env. Cons.* 17:9–13.

Office of Technology Assessment. 1988. *Technologies to maintain biological diversity*. J.B. Lippincott Co., Philadelphia, PA.

Ray, G. C. and M. G. McCormick-Ray. 1989. Coastal and marine biosphere reserves. Pp. 68–78 in W. P. Gregg, S. L. Krugman, and J. D. Wood (Eds.), *Proceedings of the symposium on biosphere reserves*. USDI National Park Service, Atlanta, GA.

Slatyer, R. O. 1983. The origin and evolution of the World Heritage Convention. *Ambio* 12:138–145

Talbot, L. M. 1984. The role of protected areas in the implementation of the World Conservation Strategy. Pp. 15–19 in J. A. McNeely and K. R. Miller (Eds.), *National parks, conservation, and development: The role of protected areas in sustaining society*. Smithsonian Univ. Press, Washington, D.C.

United Nations Environment Programme. 1987. *The state of the world environment*. UNEP/GC.14/6, Nairobi, Kenya.

Von Droste, B. 1989. The role of biosphere reserves at a time of increasing globalization. Pp. 1–6 in W. P. Gregg, S. L. Krugman, and J. D. Wood (Eds.), *Proceedings of the symposium on biosphere reserves*. USDI National Park Service, Atlanta, GA.

Western, D., M. C. Pearl, S. L. Pimm, B. Walker, I. Atkinson, and D. S. Woodruff. 1989. An agenda for conservation action. Pp. 304–323 in D. Western, and M. Pearl. *Conservation for the twenty-first century*. Oxford Univ. Press, New York.

World Resources Institute. 1987. *World resources 1987*. Basic Books, Inc., New York.

World Resources Institute. 1990. *World resources 1990–91*. Oxford Univ. Press, New York.

GLOSSARY

A

Acid deposition: The increased input of acids to ecosystems through rain, fog, snow, dry particulate fallout, and capture of gaseous compounds.

Acid mine drainage: Pollution of streams with acids and heavy metals due to leaching of spoil material from coal or ore mines.

Acid precipitation: Rain or snow with a pH less than 5.0.

Acre-foot: An amount of water equal to one acre, or about the area of a football field, covered to the depth of one foot.

Adaptive radiation: The evolutionary diversification of species for different ways of life.

Age-structured populations: Populations in which reproductive potential differs for individuals of different age.

Agroforestry: Management of multi-specied, tree-dominated communities of plants that yield timber, fuelwood, basic foods, and industrial products such as rubber and oils.

Algal turfs: Low mats of diverse species of algae that flourish in sheltered shallow waters of coral reefs.

Alleles: Slightly different forms of a gene that consequently tend to produce different versions of the coded protein.

Alpine tundra: Treeless vegetation at high altitudes, dominated by perennial herbaceous plants and dwarf shrubs.

Ambergris: A substance formed in the intestine of the sperm whale and used as a chemical stabilizer in perfumes.

Anadromous: Entering rivers from the ocean to spawn.

Anagenesis: Evolutionary change in a species until it has become so different that taxonomists no longer regard it as the same taxonomic unit.

Antarctic Marine Ecosystem: The ocean region lying between Antarctica and the antarctic convergence.

Aquaculture: The husbandry of aquatic plants or animals in a fashion comparable to that of domesticated agricultural species.

Arctic tundra: Treeless vegetation at high latitudes, dominated by perennial sedges, broad-leafed herbs, and dwarf shrubs.

Area-sensitive species: Species that are likely to disappear when an insular habitat reaches some critical minimum size.

Area strip mining: See *strip mining*.

Arribada: Mass arrival of female sea turtles at nesting beaches, where they concentrate their egg-laying into a period of a few nights.

B

Baleen apparatus: A set of fibrous plates suspended from the upper jaws of toothless whales, which acts as a filter to capture small animal foods.

Barrier beaches: See *barrier islands*.

Barrier islands: Long, narrow islands formed of beach sediments and lying more or less parallel to the seacoast. **Barrier beaches** are similar, differing only in that they are connected to the mainland at one end.

Benefit-cost analysis: Evaluation of the economic values to be gained by an action in comparison to the costs of carrying out the action.

Bioaccumulation: The concentration of a chemical from an organism's physical environment.

Biological control: The establishment of a strong, negative biological interaction against a designated pest species.

Biological Dynamics of Forest Fragments Project: The continuing experimental study in the Amazon Basin to determine how recently created tropical forest fragments of different sizes respond to changed conditions.

Biological oxygen demand: The quantity of oxygen needed to decompose the organic matter in one liter of water.

Biomagnification: The increase in concentration of a chemical in tissues of organisms from one link to the next in a food chain.

Biome: A regional ecosystem type characterized by the distinctive appearance of the dominant plants.

Biosphere: The zone of the earth's surface occupied or influenced by living organisms.

Biosphere Reserve System: A global network of sites, developed under the *Man and the Biosphere Program,* that combine preservation of biodiversity with research in sustainable use of natural ecosystems by humans.

Biotic community: The different species of organisms that occur together and interact with each other in an area of habitat.

Biotic interactions: Actions between organisms of the same or different species, including predation, parasitism, competition, and mutual benefit.

Biotic relaxation: Decline in the number of species on a newly formed island due to an excess of extinctions over colonizations by new species.

Biotic succession: The change in the composition and structure of an ecosystem through time, beginning on a site previously unoccupied, in the case of **primary succession,** or on a site from which a community was removed by some disturbance, in the case of **secondary succession.**

Blowout: A wind erosion basin formed in a sandy substrate following disturbance of the stabilizing vegetation.

Blue List: Compilation of bird species that seem to be experiencing non-cyclic population declines or range contractions.

Botanic Gardens Conservation Coordinating Committee: An international committee that coordinates collections of living plant materials by all types of institutions.

Bromeliads: Members of a family of largely tropical flowering plants, many of which grow as epiphytes.

Brucellosis: A bacterial disease of ungulates that can be very destructive to herds of livestock and can also be transmitted to humans.

Bunch grasses: Species of grasses that grow in dense clumps or large patches, rather than as a continuous turf.

C

Caatinga forest: A forest type of the northern Amazon basin, with soils that are highly leached, quartzitic sands without appreciable quantities of nutrients.

Caliche: A cemented horizon of calcium carbonate and other minerals deposited by leaching at a certain depth in soils of arid regions.

California Desert Conservation Area Plan: The Bureau of Land Management's zoning plan for regulating vehicular and other human activity in its desert land in California.

Calipee: The cartilaginous material lining the inner side of the ventral shell of the green sea turtle, an essential ingredient of green turtle soup.

Canids: Members of the dog family.

Captive propagation: The breeding of species in confinement for their preservation, display in zoos, and reintroduction to the wild.

Carbamates: A chemical group of pesticides that have very high toxicity, but a short environmental life.

335

Carnivores: Organisms that feed exclusively on herbivorous animals, **top carnivores** being carnivores that themselves feed exclusively on other carnivores.

Carrying capacity: The maximum population that can exist in equilibrium with average conditions of food, nest sites, cover, and other resources.

Catadromous: Living in fresh water but returning to the sea to spawn.

Ceilometer: A fixed-beam or rotating light used at airports to show the height of the cloud ceiling.

Center for Plant Conservation: An organization in Massachusetts that coordinates propagation of endangered plant species by botanical gardens and arboreta.

Channelization: Straightening a meandering stream channel to speed flood water outflow.

Chemosynthesis: The process by which some producer organisms use energy from chemical reactions for manufacturing organic molecules.

Chlorinated hydrocarbons: A group of petroleum-derivative chemicals with a structure of carbon rings to which chlorine atoms are attached.

Chocolate mousse: Heavier components of crude oil that have been churned into a viscous emulsion in sea water.

Cladocera: A sub-class of crustaceans, commonly known as water fleas, that are common in freshwater zooplankton.

Clean Air Act: The United States federal law regulating the emission of pollutants into the air from various sources.

Clearcutting: Cutting of all trees and removal of all harvestable timber at one time.

Climax community: A biotic community consisting of species that are able to reproduce and replace themselves under existing conditions.

Cloud forests: High mountain tropical forests that are bathed in mist much of the time.

Coastal Barrier Resources Act: A U.S. federal law making highway and disaster insurance funds unavailable for undeveloped barrier islands, and designating them as a **Coastal Barrier Resource System.**

Coastal Barrier Resource System: See *Coastal Barrier Resource Act.*

Cold deserts: See *desert.*

Colonization and extinction curves: Graphs of the rates of establishment of new species and disappearance of established species on islands relative to the number of species present.

Commensal: A species that lives in close association with another species without benefitting or hurting the latter.

Common property resource: A resource owned by society at large, and typically is not fully valued in market transactions affecting it.

Conservation Data Centers (CDCs): Organizations established in Latin American countries through efforts of *The Nature Conservancy* to collect information on biodiversity and promote its preservation.

Conservation Foundation: Private organization, now merged with the *World Wildlife Fund,* that has concentrated its effort on public policy relating to pollution, wetlands protection, and public land use.

Conservation genetics: The branch of conservation biology concerned with maintaining an essential level of genetic variability in species of conservation concern.

Conservation International: A private organization whose efforts are centered on protection and sustainable use of tropical forests.

Conservation networks: Natural and semi-natural areas managed by different organizations or public agencies, but linked by habitat corridors to maximize effective total habitat area.

Consumer's surplus: See *ecological resources.*

Consumptive value: See *ecological resources.*

Continental islands: Islands that once possessed a continental connection.

Contingent valuation method: A technique of estimating non-market economic values by asking people their willingness to pay to preserve such values.

Contour strip mining: See *strip mining.*

Convention Concerning the Protection of the World Cultural and Natural Heritage: See *World Heritage Program.*

Convention on International Trade in Endangered Species of Wild Flora and Fauna: The international agreement regulating worldwide trade in endangered species or their products.

Convention on Migratory Species: International agreement, administered by UNEP, on protection of animals that migrate across or outside national boundaries.

Coevolution: The mutual evolutionary adaptation of living members of ecosystems to each other.

Compound 1080: Sodium fluoroacetate, a broad-spectrum poison sometimes used to kill mammals, such as coyotes, judged to be pests.

Conservation ecology: The science concerned with the origin and maintenance of biotic diversity.

Continental shelf: Areas of continental crust lying under shallow seas adjacent to exposed land portions of continental crust.

Cool deserts: See *desert.*

Copepods: A sub-class of small freshwater and marine crustaceans that occur commonly in zooplankton.

Coral reefs: Tropical or subtropical, shallow water ecosystems dominated by stony corals and coralline algae.

Counteradaptation: Evolutionary adjustments by the member species of a community to each other.

Critical habitat: Areas within or outside the range of an endangered species that are formally designated as essential to its survival.

Cross-fostering: The rearing of young of one species by adults of another.

Cryogenic storage: Long-term preservation of semen, ova, or embryos at sub-freezing temperatures.

D

Debris flows: Mass slippages of destabilized, saturated soil on hillsides.

Debt-for-nature swap: The purchase and cancellation of portions of the foreign debt owed by a nation to international banks in exchange for a government's establishment of preserves or support of conservation.

Decomposers: Organisms that carry out the breakdown of dead organic matter.

Decreasers: Native rangeland plants that decline in abundance under heavy grazing by livestock.

Demographic failure: The extinction of a population by the death of all its members.

Demographic stochasticity: See *stochastic variability.*

Density dependent factors: Agents that affect a population with an intensity determined by its density relative to carrying capacity.

Density independent factors: Agents that affect a population with an intensity unrelated to its density relative to carrying capacity.

Desert: The dominant ecosystem type of arid regions. **Hot deserts** occupy regions where freezing temperatures rarely occur, **cool deserts** regions where freezing conditions occur in winter, and **cold deserts** polar or alpine regions where freezing temperatures can be prolonged and intense.

Desert grassland: The native prairie vegetation of southern Arizona, New Mexico, and northern Mexico.

Desertification: Decline in the productivity of arid or semi-arid lands due to human mismanagement.

Desert pavement: A surface layer of stones in deserts that protects the underlying soil from wind erosion.

Differential dispersal: A mechanism of biotic succession involving the arrival of seeds, spores, eggs, or other reproductive structures at different times after a new habitat is formed.

Differential survival: A mechanism of biotic succession in which initial colonizing species in a new habitat disappear at different times because they are unable to tolerate the range of habitat conditions.

Direct nutrient cycling: The passage of nutrients directly from dead matter into mycorrhizal fungal filaments and then into plant rootlets without entering the mineral soil.

Direct use values: See *ecological resources.*

Dominants: Members of a biotic community that receive the impact of external conditions and determine the internal environment of an ecosystem.

Drift gill netting: Open-sea fishing with gill nets that are suspended from floats, hang downward in the surface waters, and extend for many kilometers.

Dry dystrophic savannas: See *tropical savannas.*

Dry eutrophic savannas: See *tropical savannas.*

Dynamic pool models: Sets of equations describing numbers or biomass of different age (or size) classes as a function of variables that influence their growth, survival, and reproduction.

Dystrophic: Having low nutrient availability.

E

Early successional species: Species that are adapted to the beginning stages of biotic succession.

"Earth Charter": Basic principles for sustainable development and sound environmental management that are a major objective of the 1992 United Nations Conference on Environment and Development.

Earth Observing System: A proposed set of space platforms that would monitor global environmental conditions by remote sensing.

Ecogenetic variation: Differentiation of populations within a species in their genetic adaptations to local environmental conditions.

Ecological niche: The resources that a species exploits to survive and reproduce.

Ecological release: The expansion of the ecological niche to include more habitats and more kinds of resources when competitor species are fewer.

Ecological resources: Organisms, ecosystems, and landscape features or qualities that have value for human well-being. Values that are **consumptive** require the destruction or depletion of the resource, **non-consumptive use** does not. **Direct use values** are those gained by first-hand utilization, **indirect use values** those received second-hand, and **non-use values** those involving no current use of the resource. **On-site use value** is that gained at the resource location, and includes the **on-site economic impact,** or the money spent to use the resource, and the **consumer's surplus,** or value a user receives in excess of actual expenditures. **Off-site values** are those gained at other locations. Other types of values include: **harvestable value,** the monetary value of materials taken from the environment and used consumptively; **recreational value,** all the consumptive and non-consumptive use values involving pleasure uses of a resource; **ecosystem stabilization value,** the value of species in maintaining beneficial conditions in ecosystems; **habitat reconstruction value,** the utility of species for restoring or recreating valuable ecosystems; **environmental monitoring value,** the usefulness of a species as an indicator of environmental stress; **scientific research value,** the importance of a species or ecological feature for answering scientific questions; **educational value,** the utility of an ecological resource for teaching purposes; **future use value,** the benefit of knowing a resource can be used at a later time; and **undiscovered value,** potential usefulness that is not yet identified.

Ecology: The science concerned with relationships of organisms with each other and with their non-living environment.

Ecosystem: A unit of the environment, consisting of living and non-living components that interact with exchanges of energy and nutrients.

Ecosystem analysis: The branch of ecology that deals with the dynamics of energy flow and nutrient cycling in ecosystems.

Ecosystem Conservation Group: Coordinating committee for United Nations agencies and private international organizations involved in global conservation efforts.

Ecosystem dynamics: The functional processes that occur in ecosystems, such as energy flow and nutrient cycling.

Ecosystem stabilization value: See *ecological resources.*

Ecosystem structure: The physical, chemical, and biotic conditions that exist in an ecosystem at a given instant.

Ecotone: The border zone between distinct ecosystems.

Educational value: See *ecological resources.*

El Niño: The influx of warm, nutrient-poor tropical water into areas of normally cold, upwelling water in tropical or temperate oceans.

Endangered species: Species, subspecies, or populations determined to be in danger of extinction in all or a significant portion of their range.

Endangered Species Conservation Act: The 1969 law that was the forerunner to the present United States federal legislation governing designation and recovery of endangered species.

Endangered Species Preservation Act: The 1966 United States federal law that first gave formal protection to endangered species.

Endemic species: Species that have evolved in, and are restricted to, a particular area.

Energy flow: The capture of solar energy by producers, its transfer along food chains, and its loss as heat energy from living organisms.

Environmental impact assessment: Evaluation of the total effect of a proposed action on aspects of the natural and human environment.

Environmentalism: The philosophy of striving to live in harmony within the global ecosystem.

Environmental Law Center: Branch of the World Conservation Union that provides information and assistance in international law relating to endangered species.

Environmental monitoring value: See *ecological resources.*

Environmental stochasticity: See *stochastic variability.*

Epilimnion: Surface zone of warm, low-density water in a lake.

Epiphylls: A thin layer of algae, lichens, and mosses covering leaves of tropical forest plants.

Epiphytes: Plants that grow on the trunks and branches of larger trees and shrubs, especially in tropical forests.

Estuary: Bodies of water occurring where rivers discharge into the ocean and having salinity regimes intermediate between fresh waters and the ocean.

Eutrophic: Having high nutrient availability and high primary productivity.

Eutrophication: The increase in fertility and productivity of an ecosystem due to an increased rate of nutrient input.

Existence value: The economic value many people attach to the "right-of survival" of natural ecosystems and their members.

Ex situ care: The maintenance of populations outside their native habitat, as in botanical gardens, zoological parks, and aquariums.

Extinction: The termination of an evolutionary lineage due to death or genetic modification of all of its members.

Extractive reserves: Areas of tropical forest that are protected from logging, but are open to harvesting of tree products such as rubber, oils, fruits, nuts, tubers, and other substances.

F

Facilitation: A mechanism of biotic succession in which habitat changes caused by one species favor the establishment of other species.

Faunal relaxation: The decline in number of species on a newly created island due to the fact that the initial number exceeds the equilibrium that can be maintained by colonization-extinction relationships.

Federal Environmental Pesticide Act: The 1972 U.S. law giving federal authority over intrastate use of pesticides.

Federal Insecticide, Fungicide, and Rodenticide Act (FIFRA): The 1947 U.S. law requiring federal approval for use of specific pesticides.

Feral populations: Populations of domestic animals that live and reproduce in the wild.

Fish and Wildlife Conservation Act: The 1980 federal law requiring the U.S. Fish and Wildlife Service to determine the conservation needs of non-game birds.

Fishery Conservation and Management Act: The 1976 federal law establishing the United States' 320-kilometer exclusive economic zone in adjacent oceans.

Fitness: The contribution of an individual to the genetic composition of subsequent generations in a population.

Flyways: Major pathways of migration of birds, particularly the **Atlantic, Mississippi, Central,** and **Pacific** waterfowl flyways.

Food and Agriculture Organization (FAO): United Nations agency, founded in 1945, that has concentrated on human food and fiber production systems.

Food supply—predation hypothesis: The postulate that cycles of small mammals are caused both by their interaction with plant foods and the influence of predators.

Foredune ridge: Elongate dune that tends to form immediately inland from ocean or lake beach where vegetation traps windblown sand.

Forest decline: Increased mortality and reduced growth of trees due to factors such as tree age and environmental stress.

Foundation species: A lower trophic level species whose abundance determines much of the upper trophic level structure of the ecosystem.

Founder effect: Influence of the small group of individuals forming a new wild or captive population on the genetic structure of the resulting population.

Functional response: A change in the effort exerted by a predator in hunting a particular prey species, in response to changed prey abundance.

Future use value: See *ecological resources.*

Fynbos: The evergreen shrubland ecosystem type characteristic of the Cape Region of South Africa.

G

Gaia Hypothesis: The suggestion that the global environment is strongly self-regulating in a fashion that maintains conditions favorable for life.

Game cropping: The commercial harvesting of wildlife from unconfined populations.

Game farming: A ranching operation that mixes livestock with selected species of wildlife.

Gap analysis: Determination of which species are inadequately protected by existing or planned sets of preserves.

Gene: Unit of the genetic molecule, DNA (deoxyribose nucleic acid), that contains the coded instructions for assembly of particular proteins.

Gene flow: Spread of alleles through populations by dispersal and interbreeding of individuals.

Gene pool: The genetic makeup of a population, in terms of number of alleles per gene and the frequency of heterozygous allele combinations.

Genetically effective population size: The size of an ideal breeding population (1:1 sex ratio, random mating, randomly varying number of offspring per pair) to which an actual population is equivalent.

Genetic bottleneck: Greatly reduced genetic variability in a population due to its past reduction to a very small size at which many alleles were lost by genetic drift.

Genetic drift: The random fluctuation in the frequency of an allele due to accidents that affect the survival and reproduction of individuals.

Genetic swamping: Extinction or loss of identity of a species through extensive interbreeding with one or more related forms.

Geographic information systems: The computerized recording of data for a region, using geographic coordinates as the primary indexing system.

Global Environmental Monitoring System (GEMS): UNEP program of collecting and maintaining a global database on climate, natural resources, biotic diversity, and environmental quality indicators.

Global Plan of Action for the Conservation, Management, and Utilization of Marine Mammals: A set of guidelines prepared by United Nations agencies and international conservation groups for protection and sustained use of marine mammals.

Global Strategy for Conserving Biological Resources: A companion plan to the 1991 revision of the *World Conservation Strategy* that focusses in detail on protection of biodiversity.

Grazing lawns: Dense carpets of actively growing grasses that are kept in an actively growing, juvenile state by herbivore grazing.

Gyres: Large, circular systems of water circulation that occupy the major ocean basins.

H

Habitat: The conditions of the site occupied by a population or community of organisms.

Habitat reconstruction value: See *ecological resources.*

Hacking: Reintroduction of captive-reared animals to the wild at acclimation sites where food is provided regularly.

Hamburger connection: Clearing of tropical forest for ranching operations designed to export beef to developed countries for use in fast foods.

Harvestable value: See *ecological resources.*

Hawk Mountain Sanctuary: A raptor observation and research center located on Kittatinny Ridge in northeastern Pennsylvania.

Headstarting: Raising young turtles to the age of about a year, before their release in areas believed to be optimal habitat.

Herbivores: Organisms that feed exclusively on green plants.

Herbivory hypothesis: The postulate that cycles of small mammals are caused by interaction of the animals with their food plants.

Heterozygosity: A condition in which the two examples of a particular gene in an individual are of different alleles.

Highwall: A cliff left as the result of strip mining on a hillside.

Homozygous(ity): A condition in which both examples of a particular gene in an individual are of the same allele.

Hot deserts: See *desert.*

Hypolimnion: The deep cold, high-density bottom water zone in a lake.

I

Iberian Gene Bank: A cryogenic seed storage facility in Spain that preserves species of the Mediterranean Region.

Igapo forests: Forests of the periodically flooded zone bordering low-sediment, blackwater rivers in the Amazon Basin.

Inbreeding: The mating of closely related individuals.

Inbreeding depression: Poor survival or reproduction due to the pairing of deleterious alleles as a result of mating of close relatives.

Increasers: Native rangeland plants that become more abundant under grazing pressure by livestock.

Indirect use values: See *ecological resources.*

Integrated pest management: A strategy of control of pest species that combines a variety of cultural, biological, and chemical approaches.

Inter-American Tropical Tuna Commission: The international organization regulating tuna harvests in the eastern tropical Pacific Ocean.

Interglacial: The period(s) between major episodes of continental glaciation.

International Baltic Sea Fisheries Commission: The international organization regulating fishing harvests in the Baltic Sea.

International Biological Program (IBP): Program of basic research in ecosystem dynamics and evolution in the 1960s and early 1970s.

International Board for Plant Genetic Resources: International coordinating committee for organizations engaged in preservation of plant biodiversity.

International Commission for the Conservation of Antarctic Living Marine Resources: The international organization regulating fisheries harvests in the Antarctic Marine Ecosystem.

International Convention for the Prevention of Pollution from Ships (MARPOL Convention): A 1973 international agreement restricting discharge of polluting wastes by ships, and amended specifically in 1978 to prohibit all dumping of plastics.

International Council for Bird Preservation: Coordinating organization for global conservation efforts for birds.

International Geosphere-Biosphere Program: An extended collaborative effort among scientific organizations in different countries to study the interaction of biological and geological processes at the global level.

International Pacific Halibut Commission: The international organization regulating halibut harvests in the Pacific Ocean off Canada and the United States.

International Species Inventory System: A listing of animals held in zoos and aquaria throughout the world.

International Union for the Conservation of Nature and Natural Resources: See *World Conservation Union.*

International Whaling Commission: International organization formed to regulate the harvest of whales by member nations.

Intertropical convergence: The belt of most intense solar heating of the earth's surface, into which northern and southern trade winds flow.

Invaders: Non-native plant species that tend to colonize rangelands that are grazed by livestock.

Island biogeography: The study of processes of colonization, evolution, and extinction in insular biotas.

K

Kelp forests: Shallow marine communities dominated by giant brown algae, or kelps.

Keystone exotics: Foreign species capable of restructuring almost completely the ecosystems they invade.

Keystone mutualists: Tropical forest plant species that are essential to the survival of many species of frugivorous birds and mammals.

Keystone predator: An animal that can cause an almost total restructuring of the ecosystem by predation influences.

Kipukas: Islands of native vegetation that exist in Hawaii where extensive lava flows have isolated them.

Krill: Large marine zooplankton, primarily the small shrimp, *Euphausia superba,* that abound in cold upwelling waters.

L

Lampricide: A pesticide used to kill lamprey larvae during their early life in streams.

Landscape ecology: A branch of ecology that concentrates on how the interactions of the different types of ecosystems in a certain area influence their own structure and dynamics and that of the total area.

La Niña: The stage in the long-term oscillation of water in tropical ocean basins when warm surface water becomes concentrated in the western part of the basin and upwelling dominates in the eastern.

Large Marine Ecosystems: Ocean areas with more or less natural boundaries formed by surface currents and submarine topography.

Late successional species: Species that are adapted to stages of biotic succession approaching the climax.

Lianas: Massive vines that expose their foliage in the tropical forest canopy.

Load-On-Top (LOT): A procedure used by oil tanker ships to concentrate and retain oil residues when ballast water is discharged, so that the new oil cargo incorporates these residues.

Logistic equation: Mathematical relationship describing the S-shaped curve of growth of a population toward the limit set by resources.

London Dumping Convention: A 1972 international agreement to limit disposal of hazardous materials in lakes and the ocean by signatory nations.

Long-distance migrants: See *migratory birds.*

M

Macroclimate: The general regional climate, or the climatic conditions prevailing outside an ecosystem.

Mainland Beach Detachment Hypothesis: The theory that barrier islands arose by the isolation of coastal beach and dune systems from the mainland by rising post-glacial seas.

Man and the Biosphere Program: UNESCO program, begun in the early 1970s, emphasizing applied study of sustainable management of natural ecosystems for human welfare.

Marine Mammal Protection Act: The 1972 United States federal law prohibiting the harvest or killing of marine mammals except under specified conditions.

Maximum sustained revenue: The greatest net profit that can be obtained on a continuing basis by harvesting a population.

Maximum sustained yield: The greatest harvest that can be taken from a population indefinitely.

Mean heterozygosity: The average percent of the genes that are represented by different alleles in individuals of a population.

Megafauna: Large-bodied animals, either living or extinct.

Metapopulation: An overall regional population comprising many local populations in isolated patches of habitat.

M-44 device: A metal tube that uses a shotgun shell to shoot cyanide into the mouth of an animal, such as a coyote, that pulls on the baited end.

Microclimate: The climatic conditions prevailing within an ecosystem.

Mid-grass prairie: See *prairie.*

Migratory birds: Birds that move seasonally between breeding and non-breeding ranges in different areas. **Weather migrants** move varying distances in direct response to severe weather conditions, **short-distance migrants** show more regularly timed movements between ranges up to a few hundred kilometers apart, and **long-distance migrants** move between ranges separated by thousands of kilometers.

Migratory Bird Treaty Act: The 1918 agreement between the U.S. and Canada defining protection and management policy for migratory birds.

Minimum Critical Size of Ecosystems Project: The experimental study initiated in the Amazon Basin to determine how fast species are lost from newly created tropical forest fragments of different sizes.

Minimum viable population: The smallest number of breeding individuals that has a specified probability of surviving for a certain time, without losing its evolutionary adaptability.

Mobile links: Animals of tropical forest that carry pollen from individual to individual, or seeds from mature tree to potential establishment sites.

Moist dystrophic savannas: See *tropical savannas.*

Moist eutrophic savannas: See *tropical savannas.*

Monomorphism: A condition in which a gene has only a single allele.

Multiple-use modules: Conservation units with a fully protected core area surrounded by natural or semi-natural zones used in progressively more intense fashion.

Mutation: An accident in replication of the chemical structure of a gene during the process of cell division.

Mutualism: The interaction of two or more species in ways beneficial to both or all.

Mycorrhizae: Symbiotic associations between filamentous fungi and the fine roots of higher plants.

N

National effluent standards: Limits set by the United States Environmental Protection Agency on the quantities of pollutants that can be discharged into surface waters.

National Environmental Policy Act: The 1970 federal law that established the requirement for environmental impact assessment for projects supported by federal funds.

National Seed Storage Laboratory: A facility in Colorado that maintains cold-storage seed collections of agricultural plants.

National Wildlife Refuge System: A network of protected wildlife habitats administered by the U.S. Fish and Wildlife Service.

Nearctic-Neotropical Migration System: The seasonal movement pattern of birds between breeding areas in North America and wintering areas in the New World tropics or temperate South America.

Non-consumptive use: See *ecological resources.*

Non-use values: See *ecological resources.*

North American Breeding Bird Survey (BBS): A system of standardized counts of birds along 50-kilometer routes throughout the United States and Canada during the height of the breeding season.

North American Waterfowl Management Plan: A plan for cooperation by federal wildlife agencies of the U.S. and Canada to restore and maintain populations of migratory ducks and geese.

Northeast Atlantic Fishery Commission: The international organization regulating fishery harvests in northern European ocean waters.

Numerical response: A change in the density of predators, due to reproduction and mortality, resulting from change in prey abundance.

Nutrient capital: The quantity of an essential element or compound that is in active circulation within an ecosystem at a given time.

Nutrient cycling: The movement of an essential element or compound among living and non-living components of an ecosystem.

Nutrient recovery hypothesis: The postulate that lemming feeding on tundra plants causes a tie-up of nutrients essential to plant productivity and creates the 3 to 4 year cycle of abundance of lemmings and their predators.

O

Oceanic islands: Islands that have never been connected to a continent.

Off-road vehicles: All motorized vehicles designed for operation on roadless terrain.

Off-site values: See *ecological resources.*

Old-growth forests: Forests that have been undisturbed for centuries, and contain trees of large size and great age.

Oligotrophic: An ecosystem with low concentrations of nutrients and a low rate of primary productivity.

On-site economic impact: See *ecological resources.*

On-site use value: See *ecological resources.*

Open ocean: The region of the seas beyond the edge of the continental shelf.

Optimum sustained yield: The yield that provides the best compromise of all societal values.

Organophosphates: A chemical group of pesticides that have very high toxicity, but a short environmental life.

Outbreeding depression: Reduced fitness of offspring produced by hybridization of distantly related individuals.

P

Palearctic-African Migration System: The seasonal movement pattern of birds between breeding areas in Europe or western Asia and wintering areas in sub-Saharan Africa.

Palearctic-Pacific Migration System: The seasonal movement pattern of birds between breeding areas in eastern Asia and wintering areas in southeast Asia, Australia, and Pacific island regions.

Palouse prairie: See *prairie.*

Pass: An inlet between barrier islands, which connects the open ocean and the sound on the inland side of the islands.

Pastoralism: A subsistence food production system involving the herding of animals such as cattle, sheep, goats, and camels.

Penetrometer: Device for measuring the resistance of soil to entry by a solid object.

Permafrost: A permanently frozen subsoil zone that occurs extensively in the arctic tundra.

Pesticide resistance: The genetically-based ability of a pest to avoid, tolerate, or inactivate the pesticide chemical.

Phenoxy herbicides: Plant-killing pesticides that are similar in structure to plant growth hormones.

Photosynthesis: The process by which some producer organisms use light energy for manufacturing organic molecules.

pH scale: A logarithmic index, ranging from 0 to 14, of concentration of acid hydrogen ions, in which acidity increases as the scale value decreases.

Physical and chemical weathering: The changes in non-living conditions of a habitat due to chemical reactions, fluctuations in temperature, and other abiotic processes.

Phytoplankton: Tiny single-celled or colonial green plants that live free-floating in the waters of aquatic ecosystems.

Pioneer community: A group of species that colonize a new or disturbed habitat and initiate biotic succession.

Plank buttresses: Flanged extensions of the trunk base that connect with lateral roots of tropical rain forest trees.

Playa: The bed of a shallow, temporary desert lake.

Polychlorinated biphenyls: A class of petroleum-derivative industrial chemicals related to the chlorinated hydrocarbon pesticides.

Polyethylene spherules: Small pellets of polyethylene that are the raw material for manufacture of molded plastic objects.

Polymorphism: The percent of genes in a population that have more than one allele.

Polynuclear aromatic hydrocarbons: Organic molecules consisting of two or more fused benzene rings.

Polystyrene spherules: Small pellets that are the raw materials for manufacture of polystyrene objects.

Population: The individuals of a species that occur together and interact with each other in a defined area of habitat.

Population vulnerability analysis: Determination of the probability that a certain number of breeding individuals can survive for a specified time without losing their evolutionary adaptability.

Potholes: Small prairie ponds and marshes of glacial origin.

Prairie: A grassland vegetation type of temperate regions, dominated by perennial bunch grasses. In North America **tall-grass prairie** characterized the eastern and southern Great Plains, **mid-grass prairie** the central plains, **short-grass prairie** the western plains, and **palouse prairie** the Columbia Plateau.

Predation hypothesis: The postulate that cycles of small mammals are caused by their interaction with predators.

Predator selectivity: The extent to which a predator concentrates on substandard prey.

Primary forest: Mature tropical stands that are free from signs of disturbance and exhibit high biomass and diversity.

Primary succession: See *biotic succession.*

Prior appropriation doctrine: Water rights based on the principle of first use by diversion of surface waters for a beneficial purpose.

Producers: Organisms that store energy in new organic matter through the processes of photosynthesis or chemosynthesis.

Propagule: Any form of reproductive body that serves as a dispersal unit for a plant or animal.

Public trust doctrine: The concept that the public has an overriding right to natural surface waters that might be destroyed by excessive diversion.

R

Rain shadow: A region of low rainfall on the lee side of a mountain range, caused by warming and decrease in relative humidity of air after it passes over the mountains.

Rapid resurgence: The quick recovery of the population of a target pest after a pesticide treatment.

Recovery plans: Procedures formally adopted to increase the abundance and distribution of endangered species so that they can be reclassified as non-endangered.

Recreational value: See *ecological resources.*

Recycling: The removal of eggs or young from a pair of animals to stimulate a new breeding cycle, leading to a replacement clutch or litter.

Red Data Books: Worldwide listings prepared by the World Conservation Union of endangered species, their ecology, and their conservation status.

Regional biota: The plant and animal species that occur within a geographical area, and are available to colonize its ecosystems.

Reserved water rights: The concept that preservation of natural surface waters is implicit when areas are set aside as parks, protected forests, and wildlife areas.

Resilience: The ability of an ecosystem to restore its normal structure after some disturbance.

Rinderpest: A highly virulent virus disease of cattle and related wild ungulates.

Riparian: Associated with streams or their floodplains.

Riparian doctrine: Water rights based on the principle of preservation of undiminished and undefiled surface waters.

Roe stripping: A fishing practice involving the removal of egg masses from female fish and discarding the rest of the body.

Royal Botanic Gardens: An institution at Kew, England, that maintains a large living collection of plants and a large cryogenic seed collection.

S

Salinization: The accumulation of salts in a soil or ecosystem at large.

Scientific research value: See *ecological resources.*

Seagrass beds: Submerged estuarine or marine communities dominated by types of eelgrass or seagrass.

Secchi disk: A black-and-white metal disk that is lowered into a lake to measure water transparency by the depth at which it can be seen.

Secondary forest: Forest stands recovering from disturbances such as fire, hurricane impacts, or cutting by humans.

Secondary outbreak: Population explosion of a species that was not a pest due to killing of natural biological control species by a pesticide treatment.

Secondary succession: See *biotic succession.*

Seed-tree cutting: Logging of all but a scattering of mature trees that are left to provide the seed for forest regeneration.

Selective harvesting: Harvesting of individual mature trees at frequent intervals.

Shark finning: A fishing practice consisting of the removal of fins and the discard of the remaining fish.

Shelterwood cutting: Removal of mature trees by logging over a period of years so that a partial canopy favoring growth of new trees is maintained.

Shifting cultivation: Farming systems in which a plot of tropical forest is cleared and farmed for two to several seasons, and then abandoned to succession.

Short-distance migrants: See *migratory birds.*

Short-grass prairie: See *prairie.*

Shrub steppe: The prairie vegetation characteristic of the northern Great Basin and Columbia Plateau in North America.

SLOSS argument: The controversy over whether a single(S) large(L) preserve or(O) several(S) small(S) preserves of equal total area is most effective in long-term preservation of biotic diversity.

Soil Conservation Service: A federal agency established in 1935 to evaluate land use potential and promote sound land use practices.

Sound: A estuarine water area lying between a barrier island and the mainland.

Speciation: The separation of one evolutionary lineage into two or more daughter lineages.

Species: Groups of actually or potentially interbreeding populations that are reproductively isolated from other such groups.

Species-area curve: A graph of the numbers of species in relation to the sizes of the areas censused constructed to determine the rate at which species number changes with change in area of habitat.

Species Survival Plan: Outline for coordination of captive management of endangered animals developed by the American Association of Zoological parks and aquariums.

Spermaceti: A waxy material from the oil deposit in the head of the sperm whale, formerly used in the making of fine candles.

Stability: The ability of an ecosystem to resist change in structure and dynamics in the face of disrupting influences.

Staging areas: Locations at which birds gather prior to the first major leg of migration to store body fat for migratory flights.

Stochastic variability: Chance events that lead to variation in some quantitative feature of an ecosystem or its components. **Demographic stochasticity** refers to random variation in mortality and survivorship among individuals in a population, **environmental stochasticity** to random variation in habitat conditions.

Stopover areas: Locations where birds stop to replenish their fat stores partway through migration.

Strip mining: Gaining access to a coal or other mineral deposit by removal of overlying soil and rock. **Area strip mining** is removal of the soil and rock in a long trench across level terrain and dumping it into the adjacent trench which has just been mined. **Contour strip mining** is removal above an outcrop of coal or ore on a hillside, leaving a cliff or highwall.

Subspecies: A geographically defined set of local populations of a species that differ from other such populations in average characteristics.

Sustainable Biosphere Initiative: A framework for focussing ecological research on questions of sustainable use of the global environment, prepared by the Ecological Society of America in 1991.

Sustained yield: Any level of harvest that can be taken from a population indefinitely.

Symbiotic: The living of two or more species in intimate association with each other.

T

Tall-grass prairie: See *prairie*.

Tennessee Valley Authority: A federal agency created in 1933 to effect flood control, produce hydroelectric power, promote soil conservation, and encourage development in the Tennessee Valley.

Territoriality: Defense of an area by one or more individuals against others of the same, related, or ecologically similar species.

The Nature Conservancy (TNC): Private conservation group whose efforts center on identifying and acquiring sites of major value for preservation of biodiversity.

Theory of evolution by natural selection: The thesis, proposed by Charles Darwin and Alfred Wallace, that species have changed through time as competition among individuals with inheritable differences resulted in some leaving more offspring than others.

Thermal pollution: Release of heated water from the cooling systems of power plants and certain industries into aquatic ecosystems.

Thermocline: A relatively thin zone in a lake where temperature and density change rapidly with depth.

Threatened species: Species, subspecies, or populations likely to become endangered within the foreseeable future.

Threshold of protection: The average capacity, based on availability of food and cover, of a habitat to carry animals through an unfavorable season.

Top carnivores: See *carnivores*.

Transcendentalism: A philosophy of rejection of material goals and a seeking of harmony and beauty through the contemplation of nature.

Translocation: The release of individuals into the wild to create or enlarge a wild population.

Travel-cost method: A method to determine the consumer's surplus by analyzing the distance people travel to use a resource.

Tributyltin: An organic tin-containing compound that is an ingredient of anti-fouling paints for marine vessels.

Trophic levels: Groups of organisms the same number of feeding steps away from the energy input to an ecosystem.

Tropical deciduous forests: Forests in the tropics that lose most or all of their leaves during the dry season.

Tropical montane forests: Forests in the tropics that occur at intermediate elevations in mountains.

Tropical rain forests: Forests in the tropics that receive abundant rain in every month.

Tropical savannas: Ecosystems with a continuous stratum of perennial grasses and forbs, typically with a scattering of shrubs or trees. **Moist eutrophic savannas** have high rainfall and fertile soils, **dry eutrophic savannas** low rainfall and fertile soils, **moist dystrophic savannas** high rainfall and infertile soils, and **dry dystrophic savannas** low rainfall and infertile soils.

Tropical woodlands: Ecosystems with a well developed ground layer of grasses, but with trees that almost form a complete canopy.

Tundra: An ecosystem type, dominated by herbaceous plants and low shrubs, typical of cold regions. **Arctic tundra** occurs at high latitudes and **alpine tundra** at high elevations in mountains.

Turnover of species: The amount of change in the composition of an island community due to colonizations and extinctions during a certain time period.

Turtle excluder device: A trap-door arrangement that allows turtles and other large animals to escape trawl nets of fishing boats.

U

Undiscovered value: See *ecological resources*.

United Nations Conference on Environment and Development: The 1992 United Nations conference in Brazil to draft an environmentally sound plan for development into the 21st century.

United Nations Development Program (UNDP): United Nations agency that assists Third World nations in planning and administration of developmental efforts.

United Nations Educational, Scientific, and Cultural Organization (UNESCO): United Nations agency, created in 1945, that has sponsored basic programs of research and education involving interaction of humans with global ecosystems.

United Nations Environment Program (UNEP): United Nations agency, founded in 1972, that has focussed on applied problems of human environments.

United States Bureau of Reclamation: The federal agency established in 1902 to provide sound management of federal land and water resources.

United States Environmental Protection Agency: The federal organization established in 1970 to regulate the discharge of toxic materials into air, water, and soil.

United States Forest Service: The federal agency formed in 1905 to administer federal forest lands.

United States National Parks and Monuments: The system of federal parks and national monuments initiated in the late 1800s and placed under the United States Park Service in 1916.

United States National Wildlife Refuge System: The federal network of areas set aside for the protection of wildlife resources.

Upwellings: Ocean areas, usually near the edges of continents, where deep, cold, nutrient-rich water rises to the surface.

V

Valley grassland: The original prairie vegetation of the central valley and southern coastal zone of California.

Varzea forest: Forest of the periodically flooded zone bordering tropical whitewater rivers of the Amazon Basin that carry heavy sediment loads.

Vernal pools: Temporary ponds formed in winter and spring.

W

Weather migrants: See *migratory birds*.

Western Hemisphere Shorebird Reserve Network: A network of protected breeding, staging, stopover, and wintering areas for shorebirds in North and South America.

Wilderness Act: The 1964 federal law authorizing a formal system of wilderness areas in national forests, parks, and wildlife refuges.

Woodlands: Wooded ecosystems in which trees do not form a continuous canopy, and grasses and herbs still form a well developed ground layer.

World Bank: United Nations agency that funds developmental efforts in Third World nations.

World Conservation Monitoring Center (WCMC): Organization supported by the World Conservation Union to compile and publish information on the status of endangered organisms and on global conservation efforts.

World Conservation Strategy: A framework for integrating conservation into national development efforts, prepared in 1980, and revised in 1991, with support from United Nations agencies and the World Conservation Union.

World Conservation Union: The private organization, also known as the **International Union for the Conservation of Nature and Natural Resources (IUCN),** that coordinates many global conservation efforts on behalf of other conservation groups, national governments, and international agencies.

World Heritage Fund: See *World Heritage Program*.

World Heritage in Danger List: See *World Heritage Program*.

World Heritage List: See *World Heritage Program*.

World Heritage Program: A UNESCO program established in 1972 by the **Convention Concerning the Protection of the World Cultural and Natural Heritage.** It creates a **World Heritage List** of areas or sites deemed to be of major significance for all of humankind, provides for the placement of sites on a **World Heritage in Danger List** if they become threatened, and maintains the **World Heritage Fund** for assistance in protection of listed sites.

World Resources Institute: Private organization that collaborates with other conservation groups to analyze global environmental problems and publish information on the state of the environment.

World Wide Fund for Nature: See *World Wildlife Fund*.

World Wildlife Fund: Private conservation organization, known internationally as the **World Wide Fund for Nature,** that supports research designed to promote the survival and recovery of endangered plants and animals.

Z

Zooplankton: Tiny, weakly swimming animals that live in the open waters of aquatic ecosystems.

Zooxanthellae: Single-celled symbiotic algae that live within the cells of coral animals.

Figure 2.3 From P. S. Martin, "The Discovery of America" in *Science,* 179:969–974, 1973. Copyright © 1973 by the American Association for the Advancement of Science, Washington, DC. Reprinted by permission.

Figure 2.8 From S. L. Pimm, et al., "On the Risk of Extinction" in *American Naturalist,* 132:757–785, 1988. Copyright © 1988 University of Chicago Press, Chicago, IL. Reprinted by permission.

Figure 3.11 From J. T. Curtis, "The Modification of Mid-Latitude Grasslands and Forests by Man" in W. L. Thomas, *Man's Role in Changing the Face of the Earth.* Copyright © 1956 The University of Chicago Press, Chicago, IL. Reprinted by permission.

Figure 5.6 From H. E. Dregne, *Desertification of Arid Lands.* Copyright © 1983 Harwood Academic Publishers, Montreaux, Switzerland. Reprinted by permission.

Figure 6.9 From George W. Cox and M. D. Atkins, *Agricultural Ecology.* Copyright © 1979 W. H. Freeman and Company Publishers, New York, NY. Reprinted by permission of the authors. Data from Ruthenberg, *Farming Systems in the Tropics,* Oxford University Press, Oxford, England, 1977; Nye and Greenland 1961; and B. Andreae 1972, *Landwirtschaftliche Betreiebsformen in den Tropen,* Parey, Hamburg, and Berlin.

Figure 6.10 From S. W. Barrett, "Conservation in Amazonia" in *Biological Conservation,* 18:209–235, 1980. Copyright © 1980 Elsevier Applied Science Publishers Ltd., Barking, Essex, England. Reprinted by permission.

Figure 7.2 From S. Niewolt, *Tropical Climatology.* Copyright © 1977 John Wiley & Sons, London, England. Reprinted by permission of John Wiley & Sons, Ltd.

Figure 7.4 From A. R. E. Sinclair, "Dynamics of the Serengeti Ecosystem" in A. R. E. Sinclair and M. Norton-Griffiths, Editors, *Serengeti: Dynamics of an Ecosystem.* Copyright © 1979 University of Chicago Press, Chicago, IL. Reprinted by permission.

Figure 7.5 Source: A. R. E. Sinclair, "Dynamics of the Serengeti Ecosystem" in A. R. E. Sinclair and M. Norton-Griffiths, Editors, *Serengeti: Dynamics of an Ecosystem.* Copyright © 1979 University of Chicago Press, Chicago, IL.

Figure 8.10b Reprinted with permission from *Decline of the Sea Turtles: Causes and Prevention,* c. 1990 by the National Academy of Sciences. Published by National Academy Press, Washington, DC.

Figure 10.5 African Elephant & Rhino Specialist Group. Reprinted by permission.

Figure 11.1 From J. P. Myers, et al., "Conservation Strategy for Migratory Species" in *American Scientist,* 75:19–26, 1987. Copyright © 1987 American Scientist, Research Triangle Park, NC. Reprinted by permission of *American Scientist,* journal of Sigma Xi, The Scientific Research Society.

Figure 11.3 From F. Moore and P. Kerlinger, "Stopover and Fat Deposition by North American Wood-Warblers (Parulinae) Following Spring Migration Over the Gulf of Mexico" in *Oecologia,* 74:47–54, 1987. Copyright © 1987 Springer-Verlag, Heidelberg, Germany. Reprinted by permission.

Figure 11.9 J. J. Brett, Hawk Mountain Sanctuary Association, copyright 1986, 1991. Reprinted by permission.

Figure 12.1 Modified from T. K. Fuller, "Population Dynamic of Wolves in North-Central Minnesota" in *Wildlife Monographs,* 105:1–41, 1989, with permission of The Wildlife Society, Inc., Bethesda, MD.

Figure 13.8 Reprinted with permission from *Pesticide Resistance: Strategies and Tactics for Management,* 1986. Published by National Academy Press, Washington, DC.

Figure 15.5 From P. Beaumont, "Man's Impact on River Systems: A World-Wide View" in *Area,* 10:38–41, 1978. Copyright © 1978 Institute of British Geographers, London, England. Reprinted by permission.

Figure 16.7 Based on a map in W. Roy Siegfried, 1985, "Krill: The Last Unexploited Food Resource" in *Optima* 33(2):67–79. Reproduced courtesy of *Optima,* published by Anglo American Corporation and De Beers.

Figure 17.4 From F. H. Nichols, et al., "The Modification of an Estuary" in *Science,* 231:567–573, 1986. Copyright © 1986 American Association for the Advancement of Science, Washington, DC. Reprinted by permission of the publisher and authors.

Figure 17.6 "From *Environment,* 29:6-11, 30–33, 1987. Reprinted with permission of the Helen Dwight Reid Educational Foundation. Published by Heldref Publications, 1319 Eighteenth St., N.W., Washington, D.C. 20036-1802. Copyright © 1987."

Figure 17.8 From "Tropic Structure and Productivity of a Windward Coral Reef Community on Eniwetok Atoll" by H. T. Odum and E. P. Odum, in *Ecological Monographs,* 1957, *25,* 291–320. Copyright © 1957 Ecological Society of America, Tempe, AZ. Reprinted by permission.

Figure 19.9 From D. A. Bolze and M. B. Lee, "Offshore Oil and Gas Development Implications for Wildlife in Alaska" in *Marine Policy,* 13:231–248, 1989. Copyright © 1989 Butterworth Scientific Ltd., Surrey, England. Reprinted by permission of the publisher and authors.

Figure 20.1 From G. J. MacDonald, "Climate Change and Acid Rain," © The MITRE Corporation, 1986. Reprinted by permission.

Figure 21.3 a & b Modified from L. K. Boerma and J. A. Gulland, 1973, *J. Fish. Res. Bd. Canada,* 30:2226–2235, Department of Fisheries and Oceans, Ottawa, Ontario, Canada. Used by permission.

Figure 21.9 From D. R. McCullough, *The George Reserve Deer Herd: Population Ecology of a K-Selected Species.* Copyright © 1979 University of Michigan Press, Ann Arbor, MI. Reprinted by permission.

Figure 22.8 From D. S. Brookshire, B. Ives, and W. Schulze, "The Valuation of Aesthetic Preferences," *Journal of Environmental Economics and Management,* Vol. 3, No. 4, 1976, pp. 325–346. Copyright © 1976 Academic Press, Orlando, FL. Reprinted by permission of the publisher and authors.

Figure 23.4 From M. D. Fox and B. J. Fox, "The Susceptibility of Natural Communities to Invasion" R. H. Groves and J. J. Burdon, (editors), *Ecology of Biological Invasions.* Copyright © 1986 Cambridge University Press, New York, NY. Reprinted by permission.

Figure 24.9 a & b From C. Santiapillai and W. S. Ramono, "Tiger Numbers and Habitant Evaluation in Indonesia" in R. L. Tilson and U. S. Seal, (editors), *Tigers of the World: The Biology, Biopolitics, Management, and Conservation of an Endangered Species.* Copyright © 1987 Noyes Publications, Park Ridge, NJ. Reprinted by permission.

Figure 25.4 From M. Shaffer, "Minimum Viable Populations: Coping with Uncertainty" in M. E. Soule, (Ed.), *Viable Populations for Conservation.* Copyright © 1987 Cambridge University Press, New York, NY. Reprinted by permission.

Figure 25.5 From G. E. Belovsky, "Extinction Models and Mammalian Persistence" in M. E. Soule, (Ed.), *Viable Populations for Conservation.* Copyright © 1987 Cambridge University Press, New York, NY. Reprinted by permission.

Figure 26.7 From J. M. Diamond, "The Island Dilemma: Lessions of Modern Biogeography Studies for the Design of Nature Preserves" in *Biological Conservation,* 7:129–146, 1975. Copyright © 1975 Elsevier Applied Science Publishers Ltd., Barking, Essex, England. Reprinted by permission.

Figure 26.8 From H. Salwasser, et al., "The Role of Interagency Cooperation in Managing for Viable Populations" in M. E. Soule, (Ed.), *Viable Populations for Conservation.* Copyright © 1987 Cambridge University Press, New York, NY. Reprinted by permission.

Figure 26.10 From J. M. Scott, "Species Richness" in *BioScience,* 37:782–788, 1987. Copyright © 1987 American Institute of Biological Sciences, Washington, DC. Reprinted by permission.

Figure 27.8 C. D. Keeling, R. B. Bacastow, A. F. Carter, S. C. Piper, T. P. Whorf, M. Heimann, W. G. Mook, and H. Roeloffzen, "A Three Dimensional Model of Atmospheric CO_2 Transport Based on Observed Winds: Observational Data and Preliminary Analysis," Appendix A, in *Aspects of Climate Variability in the Pacific and the Western Americas,* Geophysical Monograph, American Geophysical Union, vol. 55, 1989 (Nov).

Figure 27.9 Adapted from *THE AGES OF GAIA, A Biography of Our Living Earth,* by James Lovelock, by permission of W. W. Norton & Company, Inc. Copyright © 1988 by The Commonwealth Fund Book Program of Memorial Sloan-Kettering Cancer Center.

Figure 28.4 "Figure from *Nature and Resources,* Vol. XXII. © UNESCO 1986. Reproduced by permission of UNESCO."

Ex situ care of endangered species, 293
Extinction curves and colonization of new species, 301
Extinction of species, 3, 13–22
　defined, 13
　deforestation of tropical forests and, 66–67
　early humans and, 14–16
　factors in vulnerability of, 20–21
　global rates of, 21
　hunting and, 14–16, 94
　as normal process, 14
　recent, 16–19
Extractive reserves, 67
Exxon Valdez oil spill, 220, 221–23(figs.), 229

f

Facilitation process, 27(fig.)
Famine, 50
Faunal relaxation, 301, 302
Federal Environmental Pesticide Act of 1972, 144
Federal Insecticide, Fungicide, and Rodenticide Act (FIFRA), 144
Feral populations, 52–53, 271–72
Finches, Galápagos Island, 91
Fire, ecosystem role of, 10, 11(fig.)
　in chaparral shrublands, 28–29
　in creating feeding habitat for whooping cranes, 295(fig.)
　in grasslands, 42
　predation and, 130
　in redwood and sequoia groves, 31
　in tall-grass prairie ecosystem, 38, 44(fig.)
　in tundra ecosystem, 43–44
　in Yellowstone National Park, 32, 33(fig.), 34(fig.)
Fire Island National Seashore, New York, 83(fig.)
Firetree (*Myrica faya*), 97
Fish
　anadromous and catadromous species of, 162
　coral reef, 198, 199(fig.)
　effect of acid deposition on, 236(fig.), 237(table)
　introduction of exotic, 152, 154, 157–58
　types of, found in various oceanic zones, 175, 178
Fisheries
　drift gill net, 185–86
　estuarine, 189–90
　harvesting influences in marine, 182–84
　harvest trends in marine, 178, 179(fig.)
　managing offshore marine, 186
　net primary productivity of marine ecosystems and potential sustainable yields of, 175(table)
　overexploitation of, 152, 154, 164, 178. 182
　Peruvian anchoveta, 179, 180, 183
　shellfish, 189–90, 195
Fishery Conservation and Management Act, 186
Fishing industry
　discarded catch of, 184–85
　effects of, on sea turtles, 85–86
　effects of pesticides on, 142
　Peruvian anchoveta, 179, 180, 183
　tuna, effects of, on cetaceans, 210, 211(fig.)
Fixed quota harvests, 248, 249, 250(fig.)
Florida Keys, 84, 190(fig.)
Florida oilspill, 222
Florida panther, 305, 306(fig.)
Flyways of migratory waterfowl, 118, 119(fig.), 121(fig.)
Food and Agricultural Organization (FAO), 324, 325
Food chains
　extinction vulnerability and position on, 21
　ocean, 175, 178, 181–82
　polychlorinated biphenyls as pollutant in, 224, 225(fig.)
Food supply-predation hypothesis for vertebrate population cycles, 41
Foredune ridge, 81
Forest decline, 238
Forests. *See also* Temperate forests and shrublands; Tropical moist forests; Tropical savannas and woodlands
　clearcutting of, 256
　deforestation of (*see* Deforestation)

fragmentation of, 34, 35(fig.)
igapo, 59
kelp, 195(fig.), 196
management of, for sustained yield, 256
net primary productivity of, 61(table), 63
old-growth, 31–32
pollution, acid deposition, and decline of, 238–39
primary, 60, 61
secondary, 60
varzea, 59, 171
Forest Service, United States, 316
Fossil fuels
　acid deposition and combustion of, 232–33
　atmospheric carbon dioxide levels and, 321(fig.)
Foundation species, 182
Founder effect, genetic, 283
Fragmentation of ecosystems
　migratory bird declines and, 117
　species extinction caused by, 18
　in temperate forests and ecosystems, 34, 35(fig.)
　in tropical savannas and woodlands, 78
France, oil tanker spills off coast of, 222, 223, 224
Frontier, closing of, in United States, 316–17
Functional response of predators to prey density, 127, 128(fig.)
Future use values of ecological resources, 261(fig.), 263
　determining, 265(table)

g

Gaia Hypothesis, 321–22
Galápagos Islands, 90, 92(fig.), 95–96
　exotic species on, 18, 270
　low species diversity and high endemicity on, 91, 95
　tourism on, 95
Game cropping, 256–57, 260
Game ecology, A. Leopold's work on, 317–18
Game farming, 256–57
Gap analysis, 307
Gene banks, 285–86
Gene flow, 279
Gene pool, 279, 280(fig.)
Genes
　banks of, 285–86
　changes of, in small populations, 279–83
　conserving variability in, 284
　in populations, 279, 280(fig.)
　problems of, in captive populations, 283
　in small, wild populations, 284–85
Genetically effective population size, 284
Genetic bottleneck, 281–82, 284
Genetic drift, 279–80, 281(fig.)
Genetic swamping
　impact of exotic species and potential for, 274
　species extinction and, 13, 14(fig.), 18
Genetic variability in wild species
　conservation of, 262
　determining minimum population required for, 293
Geographic information systems (GIS), 307, 308(fig.), 333
Georges Bank fishery, 182
Giant panda bear, 109
Giraffe, 76
Glacier National Park, 135
　grizzly bears in, 105, 106
Glen Canyon National Recreation Area
　determining consumer's surplus for, 264(fig.)
　determining future use value, 265(table)
Global climatic warming, 50, 114, 320, 321(fig.)
　lake ecosystems and, 159
Global conservation programs. *See* Conservation programs, global
Global environmental change, modern era of, 320–22
Global Environmental Monitoring System (GEMS), 325
Global Plan of Action for the Conservation, Management, and Utilization of Marine Mammals, 217

Global Strategy for Conserving Biological Resources, 333
Glossary, 335–43
Goats, plant species destruction caused by, 18, 96, 98
Gorillas
　illegal hunting of, 108–9
　ranges of three subspecies, in Africa, 109(fig.)
Grand Canyon National Park, 330, 331(fig.)
　problem of feral burros in, 52, 53(table)
Grant's gazelle, 76
Grasses
　coastal, 81(fig.), 83, 88
　estuarine, 189, 191
　exotic, 97, 273
Grasslands, 37–41
　desert and valley, 38
　distribution of, 38(fig.), 39(fig.)
　human impact on, 41(table), 42
　population cycles in, 39–41
　pothole wetlands on, 150, 151(fig.)
　restoration of, 44–45
Grasslands National Park, Saskatchewan, Canada, 44
Gray wolf
　case study of predation by, on Isle Royale, Minnesota, 131–33
　reintroduction of, 135
Grazing. *See* Livestock grazing; Overgrazing
Grazing lawns, 74, 75(fig.)
Great auk, 17
Great Barrier Reef, Australia, 201
Great Basin Desert, 49
Great Depression, 317
Great Lakes, case study of, 154–55
Great Lakes Fishery Commission, 160
Great Plains, North American, 38–39, 42, 150, 317–19
Green sea turtle, 85
Grizzly bears, 104, 105–6
　characteristics of attacks on humans by, 106(table)
　populations of, and recovery targets for, 105(table)
Guam, 17–18
Gulls, effects of pesticides on, 141–42

h

Habitat, 4
　carrying capacity of, 246–47
　critical, 289
　fragmentation of, in tropical savannas and woodlands, 78
　migratory birds and destruction of, 116–18
　species extinctions caused by disruption of, 18
　suitability of, for exotic species, 269
　vulnerability to extinction linked to, 21
Habitat islands and island biogeography, 302–3
Habitat reconstruction value of ecological resources, 263
Hacking procedure in breeding of birds, 298
Halibut fish industry, 185
Halocarpus bidwillii, 279–80, 281(fig.)
Hamburger connection to tropical deforestation, 64
Haplochromines (fish), 157–58
Harbor seal, 212
Harp seal, 212, 214
Harvestable value of ecological resources, 260, 261(fig.)
Harvesting strategies. *See* Sustained yield exploitation
Hawaiian Islands, 90, 92(fig.), 97–98, 274, 298
　designing natural preserves in, 307
　extinction of native species in, 16(table)
　low species diversity and high endemicity on, 91
　managing exotic species in, 275
Hawaiian monk seal, 212, 213(fig.)
Hawk Mountain Sanctuary, Pennsylvania, 121(fig.), 122(fig.), 140, 141
Hawks
　hunting of, 121–22
　selectivity for prey of, 127
　shell-thinning and reproductive failure in, 140–41

Load-On-Top (LOT) procedure for oil tankers, 220
Loggerhead turtles, 85, 86(fig.)
Logistic equation, 247, 248(fig.)
Logistic harvesting models, 247–52
London Dumping Convention, 229–30
Long-distance migrants, bird species as, 114
Long Island Sound, 194
Lovelock, James, 321–22
Lynx, 40–41, 135

m

M-44 device, coyote control using, 134
Macroclimate, 6
Madagascar rosy periwinkle (*Catharanthus roseus*), 260
Mainland beach detachment hypothesis for barrier
    island formation, 82
Malaria, use of DDT to control, 142
Malaysia, 109
Man and the Biosphere Program (MAB), 324, 325
Manatees, 205, 214, 215(fig.)
Mangrove swamps, 190(fig.), 191
Mara National Park, Kenya, 74–76
Marbled notothenia (fish), 178–79
Marine ecosystems. *See* Coastal marine ecosystems;
    Oceanic ecosystems
Marine Mammal Protection Act of 1972, 211,
    214–16
Marine mammals, 205–16
    cetaceans, 205, 210–12
    conflicts between fisheries and, 214, 216
    die-offs of, due to pollution, 228–29
    effect of oil spills on, 223–24
    four orders of, 205, 206(table)
    other types of, 214, 215(fig.)
    pinnipeds, 205, 212–14
    whales, 205, 206–10
MARPOL Convention, 230
Marsh, George Perkins, 316
Marshes
    freshwater, 149, 150, 158–59
    saltwater, 189(fig.), 190–91, 222
Masai people, Kenya, 77
Maximum sustained revenue (MSR), 246, 250(fig.)
Maximum sustained yield (MSY), 246, 250(fig.)
Meadow voles, 40
Mean heterozygosity, 279
Medicinal plants, 260
Megafauna. *See* Large animals
Mekong River, 162, 164, 171
Mercury, 225
Meru National Park, Kenya, 111
Metapopulation, 117
Microtine rodents, 40
Mid-grass prairie, 38
Migration as adaptation to tropical savannas and
    woodlands, 71–72
Migratory birds, 113–25
    on breeding grounds, 117
    coastal environments and, 83–84
    conservation of, 122–23
    effect of wetlands diversion and drainage on,
        150–52
    hunting of, 118–22
    long-term trends of, 116–17
    migration hazards, 122
    migration patterns of, 114, 118, 119(fig.),
        121(fig.)
    on non-breeding grounds, 117–18
    staging and stopover areas for, 114–16
Migratory Bird Treaty Act of 1918, 122
Migratory species
    birds (*see* Migratory birds)
    extinction vulnerabilities of, 21
    on tropical savannas and woodlands, 72, 73(fig.),
        76
Mimosa pigra, introduction of, 272, 273(fig.)
Minimum Critical Size of Ecosystems Project, 304
Minimum viable population (MVP), determining,
    291–93
Mining
    effects of, on estuaries, 192
    effects of, on rivers and streams, 165–66,
        167(table)

Miombo woodlands, 73
Mississippi flyway, 118, 119(fig.)
Mkomazi Game Reserve, Tanzania, 78, 303
Mobile links, 62
Moist dystrophic savannas, 73
Moist eutrophic savannas, 73
Mojave Desert, 49
    effect of off-road vehicles in, 54(table), 55(table)
    effect of overgrazing in, 51, 52(fig.)
Mono Lake, California, 150(fig.), 151(fig.)
Monomorphic alleles, 279
Montane forest, tropical, 59
Montrose Chemical Corporation, 141–42
Moose, predation of, by gray wolves, 132, 133(table)
Mopane tree (*Colospermum mopane*), 73
Mountain lions, North American, 104, 127,
    130(fig.). *See also* Lions, African
Mourning dove (*Zenaida macroura*), 30
Muir, John, 316(fig.)
Mule deer (*Odocoileus hemionum*)
    agricultural damage caused by, 109
    ecosystem factors and predation of, 129–30
    as example of early successional species, 28,
        29(fig.)
Multiple-use modules for nature preserves, 305–6
Mussels, 165(fig.), 263
Mutation, genetic, 279
Mycorrhizae, 63

n

Nairobi National Park, Kenya, 77, 78
Nakuru National Park, Kenya, 79
Namib Desert, 49(fig.)
National effluent standards on water quality, 173
National Elk Refuge, U.S., 112
National Environmental Policy Act of 1970, 320
National Estuarine Research Reserve Program, 201
National forests, U.S., 316
    recreational value of, 262(fig.)
National Marine Sanctuary, 201
National parks and monuments
    African, 9–11, 56, 74–76, 77, 78, 104, 111
    mammal groups in North American, 291(table)
    United States, 317 (*see also* names of individual
        parks)
National Seed Storage Laboratory (NSSL), 286
National Wildlife Refuge System, United States,
    120(fig.), 121(fig.), 122(fig.)
Natural preserves, 300–309
    African wildlife (*see* Wildlife parks, East African)
    ecological strategies for design of, 307–8
    edge effects and surroundings of, 306–7
    evaluating adequacy of, 303–4
    habitat islands, ecosystem fragmentation, and,
        302–3
    international distribution of, 324(table)
    island biogeographic theory and design of, 300,
        301–2
    landscape ecology and design of, 300
    reintroduction of predators to, 135
    shortcomings of island biogeographic theory,
        307
    size, shape, and spacing of, 304
    SLOSS controversy on, 304–5
    spacing and interconnection of units of, 305–6
    U.S. marine, coral reef, and estuarine, 201
    U.S. refuge system, 120–22
Natural selection, Darwin's theory of evolution by,
    314–15
Nature, western concepts and attitudes about, 314
Nature Conservancy (TNC), 44, 326
Nearctic-Neotropical Migration System, 114
Nest parasitism, migratory bird declines and, 18(fig.),
    117(fig.)
Net primary productivity (NPP)
    of estuaries, 189
    of forests, 61(table), 63
    of marine environments, 174, 175(table)
New Zealand
    genetic swamping of species in, 18
    genetic variability in trees of, 279–80, 281(fig.)
    introduced exotic species in, 99, 273

Ngorongoro Crater, Tanzania, 332(fig.)
Nihoa millerbird, 298
Nile perch, 157, 158(fig.), 273
Nile River, 163
Niobrara Valley Preserve, Nebraska, 44
Nipomo Dunes, California, 83
Nitrogen emissions in North America, 232(fig.)
Non-consumptive use values, 259
Non-use values, 259
North America
    acid precipitation in regions of, 234(fig.)
    coastal ecosystems, 81–83, 84, 87, 88
    desertification in, 50, 51(fig.)
    deserts of, 48(fig.), 49
    exotic species introduced into, 269
    extinct large animals of, 14, 15(fig.)
    grassland regions of, 38–39, 41–42
    Great Lakes of, 154–55
    migratory birds of, 114–22
    parks and preserves in, 324
    problems of large animals in, 104–6, 109,
        111–12
    recent extinctions in, 17–18
    sulfur and nitrogen emissions in, 232(fig.)
    temperate forest/shrubland fragmentation in, 34,
        35(fig.)
North American Breeding Bird Survey (BBS),
    116–17
North American Waterfowl Management Plan, 122,
    123(fig.)
Northeast Atlantic Fishery Commission, 186
Northern fur seal, 212
North Sea, 182, 186
Numerical response of predators to prey density, 127,
    128(fig.)
Nutria (*Myocastor coypus*), 272
Nutrient capital, 8
Nutrient cycling in ecosystems, 7(fig.), 8, 319
    in tropical moist forests, 63–64
Nutrient recovery hypothesis for vertebrate
    population cycles, 40

o

Ocean currents, deserts and, 48
Oceanic ecosystems, 174–87. *See also* Coastal marine
    ecosystems
    discarded catches of commercial fishing in,
        184–85
    drift gill net fisheries in, 185–86
    food chain in, 181–82
    harvesting influences in complex, 182–84
    managing offshore fisheries in, 186
    Peruvian anchoveta industry, 179, 180, 183,
        248–49
    productivity of oceanic regions in, 174,
        175(table)
    trends in marine fishery harvests in, 178,
        179(fig.)
    zones of, 175, 178
Oceanic islands, 90, 92(fig.)
    exotic species in, 18, 270
Off-road vehicles (ORVs), impacts of, on deserts,
    54–55
Off-site use value of ecological resources, 262–63
'Ohi'a tree (*Metrosideros polymorpha*), 97
Oil
    acid deposition and combustion of, 232–33
    Alaska offshore basins for gas and, 229(fig.)
    Amoco Cadiz grounding and marine pollution by,
        224
    controlling pollution caused by, 229–30
    effects of, on marine ecosystems, 221–23
    effects of, on seabirds and mammals, 223–24
    effects of exploring for, on tundra ecosystems, 43
    marine pollution by, 220–21
    quantities of, entering marine environments,
        220(table)
Oil tanker accidents, 220, 221(fig.), 222(fig.),
    223(fig.), 224(table), 229
Okapi, genetic variability of populations of, 283
Okavango River and delta, Botswana, 77, 160, 162,
    163(fig.)

Old-growth forests as ultra-climax ecosystem, 31–32
Oligotrophic lakes, 152, 153(fig.)
    ultraoligotrophism, 156–57
Olmstead, Frederick Law, 315
Onchocerciasis (river blindness), 142–43
On-site economic impact of ecological resources, 262
On-site use value of ecological resources, 262
Open ocean region, 175
    oil discharged into, 221
Opossum shrimp (Mysis relicta), 152, 156
Optimum sustained yield, 246
Organophosphates, 142–43
Ospreys, 140(fig.)
Outbreeding depression, 283, 285
Outer Banks, U.S. Atlantic coastline, 87
Overexploitation
    of lake fisheries, 152, 154
    of ocean fisheries, 178
    of river fisheries, 164
Overgrazing, 9
    deserts and effects of, 50, 51, 52, 53(fig.)
    grasslands and effects of, 42
    savannas and effects of, 77
Ozone layer, atmospheric, 320

p

Pacific flyway, 118, 119(fig.)
Pacific Northwest old-growth forests, 31–32
Pacific sardine fishery, 182
Pacific whiting fish (Merluccius productus), 255
Padre Island National Seashore, 85(fig.), 86
Palearctic-African Migration System, 114
Palearctic-Pacific Migration System, 114
Palouse prairie, 38, 41
Panama, 91, 152
Pantanal wetland complex, 162
Paperbark tree (Melaleuca quinquenervia), 273
Paraguay River, 162
Parasites, success of exotic species and effects of, 270–71
Parks. See also National parks and monuments;
    Wildlife parks, East African
    distribution of international, 324(table)
    origin of idea for, 315
Pass, coastal, 82
Passenger pigeon (Ectopistes migratorius), 19(fig.), 30
Pastoralism, 72, 77
Peacock bass fish (Cichla ocelaris), 152, 273
Pea Island National Wildlife Refuge, 87
Pelican Island, Florida, 317(fig.)
Peppertree (Schinus terebinthefolius), 273
Pere David's deer, 282
Peregrine falcon, 140(fig.), 141, 298
Permafrost, 39
Persian Gulf War, 220
Peru
    anchoveta fishery in, 179, 180, 183, 248–49
    sustained-yield timber harvesting in, 68
Pesticides, 137–46
    alternatives to, 144–45
    carbamates as type of, 142
    R. Carson's book on, 319–20
    chlorinated hydrocarbons, characteristics and
        effects of, 137, 138–42
    control of exotic species using, 275
    herbicides as, 143–44
    mammalian, 143
    organophosphates as type of, 142–43
    resistance to, 144(fig.)
    use regulation of, 144
Petroleum. See Oil
Phenoxy herbicides, 143
Philippine Islands, 198
Photochemical smog, 238
Photosynthesis, 6
pH scale, 233(fig.)
Physical and chemical weathering in biomes, 27
Phytoplankton in ecosystems, 5
    eutrophication and, 156
Pigs, damage to vegetation caused by, 98
Pike killifish (Belonesox belizanus), 273
Pinchot, Gifford, 316
Pinnipeds, 205, 206(table), 212–14

Pioneer community, 26
Piping plover, 84(fig.)
Plank buttresses, 63
Plants. See also Endangered and threatened species;
    Endemic species; Exotic plants and animals;
    Trees
    of coastal ecosystems, 83
    damage of desert, by off-road vehicles, 54
    decreasers and increasers, 42
    extinction of endemic, 18 (see also Extinction of
        species)
    in grasslands and tundras, 39
    harvesting wild (see Sustained yield exploitation)
    on islands, 93(table), 94
    of marine estuaries, 189, 190–91
    in tropical savannas and woodlands, 72
Plastics as pollutants, 227–28, 230
Platte River, Nebraska, 167
Pleistocene period and species extinctions, 14,
    15(fig.)
Poaching of wildlife, 78–79, 107–9, 110–11
Point Pelee National Park, Ontario, Canada, 262
Polar bears, 104, 205
Pollutants in aquatic ecosystems, 9, 219–30
    acid deposition as (see Acid deposition)
    controlling chemical, 229–30
    heavy metals and cyanide, 225
    interactions of, 228–29
    lakes, 152–56
    marine oil as, 220–24, 229(fig.)
    ocean, 198
    pesticides, 141–42
    plastics, 227–28, 230
    radioisotopes, 226–27
    rivers and streams, 164–65
    selenium, 226
    synthetic organic, 224–25
Pollution, environmental, in post-World War II era,
    319–20
Polychlorinated Biphenyls (PCBs) as pollutant, 165,
    224–25
Polyethylene spherules as pollutants, 227
Polymorphic alleles, 279
Polymorphism, 279, 281–82
Polynuclear aromatic hydrocarbons (PAHs) as
    pollutants, 225
Polystyrene spherules as pollutants, 227
Ponds. See Lakes, ponds, and marshes
Population, 4
    age-structured, 251, 252(fig.)
    carrying capacity and regulation of, 246–47
    conserving genetics of small wild, 284–85
    genes in, 279
    genetically effective size of, 284
    genetic changes in small, 279–83
    genetic problems in captive, 283
    growth of human, in post-World War II era,
        319–20
    harvest management models based on, 247–55
    metapopulation, 117
    size of, and extinction threats, 19(fig.), 20(fig.)
    translocation of, 298
Population Bomb, The (Ehrlich), 320
Population demography models, 291–92
Population vulnerability analysis (PVA), 291–93
Potholes, prairie, 150, 151(fig.)
Prairie. See Grasslands
Prairie dogs, 42, 43(fig.), 293
Precipitation
    acid, 232
    effect of seasonal, on tropical savannas and
        woodlands, 71, 72(fig.)
Predation hypothesis for vertebrate population cycles,
    41
Predators and predation, 5, 126–36
    ecology and behavior of, 127–29
    ecosystem context for, 129–30
    gray wolf as, 131–33
    introduced species of, 94
    keystone, 196
    livestock depredations and, 133–34
    migratory bird decline caused by, 117
    reintroduction of predators to parks and
        preserves, 135
    success of exotic species and effects of, 270–71

Preserves. See Natural preserves
Prezwalski's horse, 55
Primary forest, 60, 61
Primary succession, 26
Prior appropriate doctrine, 172
Private conservation organizations, 325–26
Producers in ecosystems, 4
Productivity of ecosystems
    estuarine, 189, 190
    of forests, 61(table), 63
    in oceanic ecosystems, 174, 175(table), 178
Public trust doctrine, 172

r

Rabbit, introduction of, into Australia, 270(fig.)
Radioisotopes as pollutants, 226–27
Rainfall
    acidic, 231–42
    effect of seasonal, on tropical savannas and
        woodlands, 71, 72(fig.)
Rain forest, 59. See also Tropical moist forests
Rain shadow, 48, 49
Ramsar Convention, 160
Ranching. See also Livestock; Livestock grazing
    game cropping vs., 257
    tropical deforestation and, 64
Rapid resurgence of insect pests, 144
Raptors
    captive breeding of, 293, 298
    hawk hunting, 121–22
    migration of, 121(fig.), 122
    peregrine falcon, 140(fig.), 141, 298
    selectivity of prey by, 127
    shell-thinning and reproductive failure in,
        140–41
Recovery plans for endangered species, 289–90
Recreational value of ecological resources, 261(fig.),
    262–63
Recycling in bird breeding, 293
Red-cockaded woodpecker (Picoides borealis), 30, 293
Red Data Books, 291, 325–26
Red fox, 272
Red wolf, 18, 135(fig.), 269, 298
Redwood trees, 31
Reefs. See Coral reefs
Regional biota, 6, 7
Reintroduction of predators to parks and preserves,
    135
Religion, conservation attitudes and, 314
Reproduction, pesticides and failure of avian, 140–41
Reserved water rights, 172
Resilience of ecosystems, 8
Rhinoceros, 78, 110–11
    poaching of, 108, 110–11
    wild populations of, 110(table)
Rift Valley lakes of Africa, 157–58
Rinderpest, 76
Rio Beni Biosphere Reserve, 326, 327(fig.)
Riparian doctrine, 172
Rivers and streams, 162–73
    Colorado River case study, 172
    Columbia River case study, 168–70
    conservation management of, 172–73
    estuaries and inflow of, 190, 192
    fish species of, 162
    human impacts on ecosystems of, 163–71
    major tropical, 162, 171
    riparian ecosystems of, 162
Rocky Mountain National Park, lakes in, 5(fig.)
Roe stripping, 184
Roosevelt, Theodore, 316(fig.)
Royal Botanic Gardens, England, 286
Ruffed grouse (Bonasa umbellus), 30
Rwanda, gorilla preserve in, 109

s

Sacramento River, California, 192–93
Saddleback bird, 298
Sahel region, desertification and famine in, 50, 56
Salmonid fish, 152
    in Columbia River, 168–70
    in Great Lakes, 154–55

Salt cedar (*Tamarix* spp.), 274
Sand dunes
    coastal, 80, 81–82, 83
    desert, 49(fig.), 55
San Francisco Bay estuary, case study of, 192–93
San Joaquin River, California, 192–93
San Joaquin Valley, California, 226
Sargasso Sea, 84, 227
Savanna. *See* Tropical savannas and woodlands
Scientific research value of ecological resources, 261(fig.), 264
Scientific Revolution, 314
Scimitar-horned oryx, 55–56
Seabirds, 80, 83–84
    conservation of, 217
    effect of oil pollution on, 223, 224(table)
    effect of plastic pollution on, 228
    nesting areas for, 88–89
    orders of, 205
Seagrass beds in estuaries, 191
Sea lamprey, 154–55, 273
Sea lion, 205, 212, 216
Seals, 182, 183(fig.), 205, 212, 213(fig.), 216(fig.)
Sea otters, 196, 205
    effect of oil pollution on, 223(fig.)
Sears, Paul B., 318–19
Sea turtles, 84–86
    conservation strategies to protect, 86
    ecological characteristics of, 85(table)
    effects of plastic pollution on, 228(fig.)
    nesting habits of, 84, 85(fig.)
Secchi disk, 156
Secondary forest, 60
Secondary outbreaks of insect pests, 144
Secondary succession, 26
Seed-tree cutting, 256
Selective harvesting of trees, 256
Selectivity of predators, 127, 128(fig.), 129(table)
Selenium, 226
Sequoia trees, 31
Serengeti-Mara ecosystem, 74–76, 263
Serengeti National Park, Tanzania, 74–76
Sewage
    estuarine eutrophication due to, 194
    lake eutrophication due to, 152, 156
    regulations on marine dumping of, 230
    river and stream pollution due to, 164
Shallow marine waters, 195–96
Shark finning, 184
Shaumari Wildlife Reserve, Jordan, 56
Sheep, predation of domesticated, 133, 134(tables)
Shellfish fisheries, 189–90, 195
Shell-thinning in birds, 140–41, 237
Shelterwood cutting, 256
Shifting cultivation, 64, 65(fig.)
    decline in productivity of crops in, 65(table)
Shorebirds
    migration patterns of, 114, 115(fig.)
    nesting areas for, 88–89
    staging and stopover areas for, 114,
Short-distance migrants, bird species as, 114
Short-grass prairie, 38
Shrimp fisheries
    by-catch from marine, 184(fig.), 185
    estuarine, 189–90
Shrublands. *See* Temperate forests and shrublands
Shrub steppe region, 38
Sierra Club, 316
Sika deer (*Cervus nippon*), 274
Silent Spring (Carson), 138, 319–20
Siltation of rivers and streams, 164–65
Single Large Or Several Small preserves (SLOSS), 304–5
Sirenia (order), 205, 206(table)
SLOSS argument, 304–5
Snowshoe hare, 40–41
Snowy egrets, 141
Soil
    desert, 49, 50, 54
    salinized, 50
    selenium contamination of, 226
    tropical forest, 62–63
    tropical savanna and woodland, 72–73
Soil Conservation Service, 319

Soil erosion, 9
    off-road vehicles as cause of, 54
Sonoran Desert, 48, 49(fig.)
Sound, coastal, 82
South Africa
    exotic trees introduced into, 270
    national parks of, 78
South America
    bird migration paths of, 114, 115(fig.)
    exotic species introduced into, 269
    extinct large animals of, 14, 15(fig.),16
Speciation, extinction and, 13, 14(fig.)
Species-area curves, 302(fig.)
Species Survival Plan, 293
Spermaceti, 207
Sperm whale, 206, 207(fig.)
Spotted owl, 21, 31, 32(fig.)
Spruce grouse (*Canachites canadensis*), 30
Sri Lanka, 107
Stability of ecosystems, 8
Staging areas for migratory birds, 114–16
Steller's sea cow, 17, 214
Steller's sea lion, 212
Stochastic variability, 19
Stopover areas for migratory birds, 114–16, 150
Streams. *See* Rivers and streams
Strip mining, 165–66
Subtidal environments, 195
Succession. *See* Biotic succession
Suderbans Preserve, India, 104
Sulfur emissions in North America, 232(fig.), 233
Sustainable Biosphere Initiative, 333
Sustained yield, defined, 246
Sustained yield exploitation, 245–57
    carrying capacity and population regulation, 246–47
    dynamic pool harvesting models, 253–54
    early harvest models, 246
    forest management for, 256
    game cropping and wildlife conservation for, 256–57
    harvesting terminology for, 246
    harvest management in practice, 254–55
    logistic harvesting models, 247–52
    shortcomings of population models, 255
Synthetic organic chemicals as pollutants, 223–25

t

Tall-grass prairie, 38, 39(fig.), 41
    fire in, 38, 44(fig.)
    restoration of, 44, 45(fig.)
Tallgrass Prairie Preserve, Oklahoma, 44, 45(fig.)
Tansley, A. G., 317
Tanzania, 74, 77, 78, 332(fig.)
Television and radio towers, hazards of, to migratory birds, 122
Temperate forests and shrublands, 25–36
    biomass and net productivity of, 61(table)
    biotic succession in, 25, 26–27
    cyclical patterns of succession in, 32–34
    distribution of, 26(fig.)
    early and late successional species in, 28–30
    effects of acid deposition on, 238–39
    fragmentation of, 34, 35(fig.)
    maintaining successional stages in, 30–31
    old-growth forests as ultra-climax, 31–32
Tennessee Valley Authority, 319
Termites in dystrophic savannas, 73(fig.)
Terrestrial ecosystems, 5, 23
    coastal ecosystems, 80–89
    deserts, 47–57
    grasslands and tundra, 37–46
    islands, 90–100
    large animal problems in, 103–12
    migratory bird problems in, 113–25
    pesticides in, 137–46
    predators and predator management in, 126–36
    temperate forests and shrublands, 25–36
    tropical moist forests, 58–69
    tropical savannas and woodlands, 70–79
Territoriality of predators, 127
Thalassia testudinum, 191
Thermal pollution in rivers and streams, 165

Thermocline of lakes, 152, 153(fig.)
Thick-billed murres, 88(fig.)
Thompson's gazelle, 74
Thoreau, Henry D., 315(fig.)
Threshold of protection concept (T of P), 246, 247(fig.)
Tiger, 104, 285(fig.)
Timber industry
    damage caused by bears to, 109
    in old-growth forests, 31
    sustained yield in tropical forests, 68
Toothed whales, 206, 207(fig.)
Top carnivores in ecosystems, 5
Topi ungulate, 75(fig.)
Torrey Canyon oil spill, 222, 223, 224(table)
Tortoises of Galàpagos Islands, 95
Tourism, 262(fig.)
    in African wildlife preserves, 79, 104, 257
    ecotourism, 262
    on Galàpagos Islands, 95
Tragedy of the commons, 260
    use of off-road vehicles in deserts as, 55
Transcendentalism, philosophy of, 315
Translocation of endangered and threatened populations, 298
Travel cost method of determining consumer's surplus values, 264
Trees
    of Hawaiian Islands, 97
    introduced exotic, into South Africa, 270
    keystone mutualists, in tropical forests, 62
    mycorrhizal roots of, 63
    plank buttresses of, 63
    selective harvesting of, 256
    tropical savanna, 76
Tributyltin (TBT), 225
Trinidad, 90, 303
Triton snail (*Charonia tritonis*), 199
Tropical moist forests
    characteristics of, 61–64
    conservation of diversity in, 67
    deforestation of, 58, 64–67, 118
    distribution of, 59, 60(fig.)
    impact of deforestation of, 66–67
    migratory birds in, 117–18
    original vs. present-day, 60(table)
    resource management efforts for, 67–68
Tropical rivers and streams, 162, 171
Tropical savannas and woodlands, 70–79
    characteristics of, 71–72
    distribution of, 70, 71(fig.)
    ecosystem of, 10–11
    fragmentation of wildlife habitat in, 78
    human populations and preserves for, 77
    restoring ecological balance in, 78–79
    Serengeti-Mari, 74–76
    types of, 72–73
Trumpeter swans, 118, 120
Tsetse fly, 79
Tuna, 186, 211, 212(fig.)
Tundra, 37, 39–45
    distribution of, 38(fig.), 39
    human impacts on, 41(table), 42–44
    population cycles in, 39–41
    restoration of, 44–45
Tundra swan, 193(fig.)
Turnover of species, 302
Turtle extruder devices (TEDs), 86(fig.)
Turtles. *See* Sea turtles

u

Uganda, wildlife poaching in, 107, 108–9
Ultraoligotrophic lakes, 156–57
Umbolozi National Park, South Africa, 78
Undiscovered value of ecological resources, 260, 261(fig.), 262
United Nations Conference on Environment and Development of 1992, 333
United Nations Convention on the Law of the Sea, 186
United Nations Development Program, 326
United Nations Educational, Scientific, and Cultural Organization (UNESCO), 324